Secure Roaming
In 802.11 Networks

Secure Roaming In 802.11 Networks

By

Paul Goransson and
Raymond Greenlaw

AMSTERDAM • BOSTON • HEIDELBERG • LONDON
NEW YORK • OXFORD • PARIS • SAN DIEGO
SAN FRANCISCO • SINGAPORE • SYDNEY • TOKYO

Newnes is an imprint of Elsevier

Newnes is an imprint of Elsevier
30 Corporate Drive, Suite 400, Burlington, MA 01803, USA
Linacre House, Jordan Hill, Oxford OX2 8DP, UK

Copyright © 2007, Elsevier Inc. All rights reserved.

No part of this publication may be reproduced, stored in a retrieval system, or transmitted in any form or by any means, electronic, mechanical, photocopying, recording, or otherwise, without the prior written permission of the publisher.

Permissions may be sought directly from Elsevier's Science & Technology Rights Department in Oxford, UK: phone: (+44) 1865 843830, fax: (+44) 1865 853333, E-mail: permissions@elsevier.com. You may also complete your request online via the Elsevier homepage (http://elsevier.com), by selecting "Support & Contact" then "Copyright and Permission" and then "Obtaining Permissions."

 Recognizing the importance of preserving what has been written, Elsevier prints its books on acid-free paper whenever possible.

Library of Congress Cataloging-in-Publication Data
Goransson, Paul.
  Secure Roaming in 802.11 Networks/Paul Goransson, Raymond Greenlaw.
    p. cm.
  Includes bibliographical references and index.
  ISBN 978-0-7506-8211-4 (pbk. : alk. paper) 1. Wireless LANs–Security measures.
2. IEEE 802.11 (Standard) I. Greenlaw, Raymond. II. Title.
  TK5105.78.G67 2007
  621.384–dc22

                                                                              2007009361

British Library Cataloguing-in-Publication Data
A catalogue record for this book is available from the British Library.

ISBN: 978-0-7506-8211-4

For information on all Newnes publications
visit our Web site at www.books.elsevier.com

07 08 09 10   10 9 8 7 6 5 4 3 2 1

Printed in the United States of America

**Working together to grow
libraries in developing countries**

www.elsevier.com | www.bookaid.org | www.sabre.org

ELSEVIER   BOOK AID International   Sabre Foundation

This book is dedicated—

To our families

Contents

Preface xv
Acknowledgments xix
Authors' Note xxi
About the Authors xxiii

Chapter 1: Introduction 1
   1.1   Introduction 1
   1.2   Basic Networking Terminology and Conventions 3
   1.3   Setting the Scene 5
        1.3.1   Introduction 5
        1.3.2   Precellular Wireless Networks and the Birth of the Cellular Concept . 6
        1.3.3   802.11 Arrives on the Scene 8
   1.4   Different Notions of Roaming 8
   1.5   Big Cells, Little Cells 12
        1.5.1   Introduction 12
        1.5.2   RF Technology and Transmit Power 13
        1.5.3   Number of Active Users in a Cell 13
        1.5.4   Overview of Quality of Service Requirements 14
        1.5.5   Complexity of Network Design and Implementation 16
        1.5.6   Frequency and Speed of Roaming 16
        1.5.7   Impact on 802.11 Cell Size 16
   1.6   Authentication, Authorization, Accounting, and Roaming 17
   1.7   How Fast Do We Roam on the Range? 19
        1.7.1   Speed of Travel 20
        1.7.2   Size of Overlapping Cell Coverage Area 20
        1.7.3   Application's Tolerance for Disruption in User Data Flow 20
        1.7.4   Complexity of Accomplishing the Roam 21
   1.8   Taxonomy for Roaming 21
        1.8.1   Network Cell Models 21
        1.8.2   A Handoff Model 23
        1.8.3   Hard Handoffs 26
        1.8.4   Soft Handoffs 26
        1.8.5   Comparison of 802.11 and Cellular Roaming 26

	1.9	Organization of the Book	27
		References	28

Chapter 2: Cellular Telephony: Wireless Roaming Pioneers — 29

	2.1	Introduction	29
	2.2	The Future of Computing	30
	2.3	Basic Concepts	31
	2.4	Early History of Radio Telephony	32
		2.4.1 Introduction	32
		2.4.2 Precellular Era	32
		2.4.3 Advanced Mobile Phone System	35
		2.4.4 Analog Systems in Europe and Japan	38
	2.5	The Digital Revolution	40
		2.5.1 Introduction	40
		2.5.2 Global System for Mobile Communications	41
		2.5.3 North American TDMA (IS-54)	46
		2.5.4 Japanese Systems	47
		2.5.5 Focus on CDMA and Soft Handoff	48
	2.6	Soft Versus Hard Handoffs in Various Cellular Technologies	50
	2.7	The Quest for Convergence	51
		2.7.1 Introduction	51
		2.7.2 High-Speed Circuit-Switched Data	51
		2.7.3 General Packet Radio Service	52
		2.7.4 Roaming for Data Applications in GPRS	55
		2.7.5 Enhanced Data Rates for Global Evolution	57
	2.8	Summary	60
		References	60

Chapter 3: Roaming in 802.11 WLANs: General Principles — 63

	3.1	Introduction	63
	3.2	Primer on the 802.11 Standard	64
		3.2.1 Introduction	64
		3.2.2 Beacons and Probes	64
		3.2.3 Channels in 802.11	66
		3.2.4 Basic Service Set	68
		3.2.5 Active and Passive Scanning	69
		3.2.6 Association	72
		3.2.7 Contention-Based Access	74
		3.2.8 Rate Adaptation in 802.11	76
		3.2.9 Other 802.11 Frames	77
	3.3	Introduction to 802.11 Roaming	79
		3.3.1 Extended Service Set	79
		3.3.2 Example of Multiple ESSs in Operation	80
		3.3.3 Phases of 802.11 Roaming	81

3.4	Local Roaming		86
	3.4.1	Introduction	86
	3.4.2	Scanning Tradeoffs	87
	3.4.3	Assumptions about Local Roaming and IP Subnets	88
3.5	Global Roaming		88
	3.5.1	Introduction	88
	3.5.2	Multiple Alternative SSIDs	90
3.6	Mobile IP and Its Role in 802.11 Roaming		92
	3.6.1	Introduction	92
	3.6.2	Review of the Mobile-IP Architecture	92
	3.6.3	802.11 Global Roaming with Mobile IP	95
	3.6.4	Alternatives to Mobile IP	96
3.7	Those Pesky Laws of Physics		97
	3.7.1	Picocells—A Double-Edged Sword	97
	3.7.2	Limitations on Avoiding Channel Overlap	97
	References		98

Chapter 4: Dynamics of 802.11 Task Groups 101

4.1	Introduction		101
4.2	Evolution of an IEEE Standard		102
	4.2.1	Introduction	102
	4.2.2	New Standards	103
	4.2.3	Chairs and More on Balloting	104
	4.2.4	Timeline of 802.11 Task Groups	106
4.3	Battle for Speed, Cost, and Market Dominance		106
	4.3.1	Market Dynamics	106
	4.3.2	Innovation	107
	4.3.3	Recent History	108
4.4	The 802.11 Standard's Physical Layer		108
	4.4.1	Introduction	108
	4.4.2	Deployments and the Players	109
4.5	Fast Secure Roaming Task Groups		111
	4.5.1	Introduction	111
	4.5.2	Basic Architectural Services	111
	4.5.3	Workgroup Foci	112
4.6	802.11i Security		112
	4.6.1	Introduction	112
	4.6.2	WEP's Limitations	114
	4.6.3	802.11 Cipher	114
	4.6.4	Preauthentication	116
4.7	802.11e Quality of Service		117
	4.7.1	Introduction	117
	4.7.2	Mixed Environment Terminology	118
	4.7.3	Miscellaneous Issues	120

Contents

- 4.8 802.11k Radio Resource Measurement Enhancements 120
  - 4.8.1 Introduction 120
  - 4.8.2 Station Statistics Request 122
  - 4.8.3 Additional Reports Important to Roaming 123
- 4.9 802.11r Roaming 124
  - 4.9.1 Introduction 124
  - 4.9.2 Basic Service Set Transition Pre-802.11r 125
  - 4.9.3 How 802.11r Handles the BSS Transition 125
- 4.10 Other 802.11 Subgroups 127
- 4.11 Wi-Fi Alliance Versus IEEE 802.11 128
  - 4.11.1 Introduction to the Wi-Fi Alliance 128
  - 4.11.2 Wi-Fi Alliance Certification 129
  - 4.11.3 IETF Versus IEEE 802 130
  - References 131

Chapter 5: Practical Aspects of Basic 802.11 Roaming 133
- 5.1 Introduction 133
- 5.2 The Driver and Client in an 802.11 Station 134
  - 5.2.1 Introduction 134
  - 5.2.2 Driver Functions 134
  - 5.2.3 Client Functions 139
  - 5.2.4 Scanning Considerations 141
  - 5.2.5 Background Scanning 144
  - 5.2.6 Scanning Timers 149
  - 5.2.7 Pitfalls with Scanning: Real-Life Example 151
  - 5.2.8 SyncScan: Enhancement to Background Scanning 152
- 5.3 Detailed Analyses of Real-Life Roams 153
  - 5.3.1 Introduction 153
  - 5.3.2 Tools Used in Roaming Study 153
  - 5.3.3 Wi-Fi Alliance's WPA Simple Roaming Test 154
- 5.4 Dissection of a Global Roam 156
  - 5.4.1 Test-Bed Description for Global Roam 156
  - 5.4.2 Test Results for Global Roam 156
- 5.5 Dissection of a Local Roam 159
  - 5.5.1 Test Description for Local Roam 159
  - 5.5.2 Test Results for Local Roam 159
  - 5.5.3 A Closer Look at Roaming Delay 162
- 5.6 Access-Point Placement Methodologies 164
  - 5.6.1 Introduction 164
  - 5.6.2 Access-Point Placement 165
  - 5.6.3 Site Surveys for Access-Point Placement 165
  - 5.6.4 Self-Monitoring 802.11 Networks 166
  - References 167

Chapter 6: Fundamentals of User Authentication in 802.11 169
- 6.1 Introduction 169
- 6.2 802.1X Port-Level Authentication 170
  - 6.2.1 Introduction 170
  - 6.2.2 Flexibility of 802.1X 171
  - 6.2.3 Evolution of 802.1X 172
- 6.3 The AAA Server 173
  - 6.3.1 Introduction 173
  - 6.3.2 Remarks about TACACS and DIAMETER 174
  - 6.3.3 The RADIUS Protocol 174
  - 6.3.4 The RADIUS Protocol in Use 176
- 6.4 The Extensible Authentication Protocol 177
  - 6.4.1 Introduction 177
  - 6.4.2 The EAP State Machine 178
  - 6.4.3 Prominent EAP Methods 179
- 6.5 Flexible and Strong Authentication in 802.11 184
  - 6.5.1 Introduction 184
  - 6.5.2 Basic Authentication Process in 802.11 184
  - 6.5.3 Discussion of Tunneling in EAP Methods 186
- 6.6 Other 802.11 Authentication Methodologies 187
  - 6.6.1 Introduction 187
  - 6.6.2 MAC-Based Authentication 187
  - 6.6.3 Web-Based Authentication 188
- 6.7 Network Access Control 188
  - 6.7.1 Introduction 188
  - 6.7.2 Cisco Network Admission Control 189
  - 6.7.3 Trusted Network Connect 190
  - 6.7.4 Microsoft's Network Access Protection 191
- 6.8 Summary 192
- References 192

Chapter 7: Roaming Securely in 802.11 195
- 7.1 Introduction 195
- 7.2 The 802.11 Security Staircase 196
  - 7.2.1 Introduction 196
  - 7.2.2 Evolution of Security Technologies 197
- 7.3 Preauthentication in 802.11i 198
  - 7.3.1 Introduction 198
  - 7.3.2 Steps Involved in 802.11i Preauthentication 198
- 7.4 Detailed Analysis of Real-Life Secured Roams 200
  - 7.4.1 Introduction 200
- 7.5 Dissection of a WPA-PSK Protected Roam 201
  - 7.5.1 Test Description for WPA-PSK Roam 201

		7.5.2	Test Results for WPA-PSK Roam	202
	7.6	Dissection of a WPA2 Enterprise Roam		203
		7.6.1	Introduction	203
		7.6.2	Test Description for WPA2 Enterprise Roam	204
		7.6.3	Test Results for WPA2 Enterprise Roam	206
	7.7	Dissection of an 802.11i Preauthentication		210
		7.7.1	Introduction	210
		7.7.2	Test Description for Preauthentication	210
		7.7.3	Test Results for Preauthentication	210
	7.8	Summary		218

Chapter 8: Proprietary Solutions for Roaming in 802.11 Networks ... 219

	8.1	Introduction		219
	8.2	Voice over Wireless IP Roaming		220
		8.2.1	Introduction	220
		8.2.2	Voice over Wireless IP Primer	220
		8.2.3	Voice over IP Protocols	222
		8.2.4	Voice over Wi-Fi—VoIP's Newest Child	223
		8.2.5	Spectralink Voice Priority (SVP)	224
	8.3	Opportunistic Key Caching		227
		8.3.1	Introduction	227
		8.3.2	Cisco's Centralized Key Management (CCKM)	228
	8.4	Centralized Wireless Switch Architectures		229
		8.4.1	Introduction	229
		8.4.2	MAC Processing	229
		8.4.3	LWAPP, CAPWAP, and SLAPP	230
		8.4.4	Using Tunnels to Keep Roams Local	231
	8.5	Summary		233
		References		234

Chapter 9: The 802.11 Workgroups' Solutions for Fast Secure Roaming ... 235

	9.1	Introduction		235
	9.2	Overview of the 802.11r Standard		236
	9.3	Detailed Concepts and Terminology of 802.11r		238
		9.3.1	Introduction	238
		9.3.2	Architectural Elements of 802.11r	240
		9.3.3	New Security Concepts	242
		9.3.4	Resource Reservations	245
		9.3.5	Information Elements	247
	9.4	Protocol Exchanges in 802.11r		250
		9.4.1	Introduction	250
		9.4.2	Fast BSS Transition Over the Air, No QoS, and No Security	251
		9.4.3	Fast BSS Transition Over the DS, No QoS, and No Security	252

Contents

	9.4.4	Fast BSS Transition with QoS and Security	254
9.5	The 802.11k Standard Applied to Roaming		257
	9.5.1	Introduction	257
	9.5.2	New Information Elements Defined by 802.11k	258
	9.5.3	Utility of 802.11k	261
	9.5.4	Limitations of 802.11k	264
9.6	Concluding Remarks		264
	References		265

Chapter 10: Roaming between 802.11 and Other Wireless Technologies 267

- 10.1 Introduction 267
  - 10.1.1 Vertical Versus Horizontal Roaming 268
- 10.2 Ideal Roaming Experience 270
  - 10.2.1 Introduction 270
  - 10.2.2 Enterprise Data User 270
  - 10.2.3 Voice User 271
- 10.3 IEEE 802.16: WiMAX 271
  - 10.3.1 Interactions between 802.11 and 802.16 272
- 10.4 IEEE 802.15.1: Bluetooth 273
  - 10.4.1 Introduction 273
  - 10.4.2 Bluetooth's Relationship with 802.11 273
- 10.5 Relevant Standards Bodies and Industry Organizations 274
- 10.6 Third Generation Partnership Program 276
  - 10.6.1 Introduction 276
  - 10.6.2 Vertical Roaming Issues 277
  - 10.6.3 Interworking between 3GPP and 802.11 279
  - 10.6.4 Five Levels of Interoperability 280
  - 10.6.5 Seamless Roaming between CS and PS Services 282
- 10.7 UMA: A Transitional Step for 3GPP 286
- 10.8 Third Generation Partnership Program 2 287
- 10.9 The 802.21 Standard 289
- 10.10 Summary 294
  - References 294

Chapter 11: Future Directions 297

- 11.1 Introduction 297
- 11.2 Survey of Ongoing Work Related to 802.11 298
  - 11.2.1 Introduction 298
  - 11.2.2 Mobile 802.11 298
  - 11.2.3 Security 303
  - 11.2.4 Quality of Service 305
- 11.3 A Mobility Model for Studying Wireless Communication 307
  - 11.3.1 Introduction 307

		11.3.2	The Mobility Model	307

         11.3.2   The Mobility Model 307
         11.3.3   Problem Definitions 311
  11.4   Conclusions 312
         References 313

Appendix A: Acronyms and Abbreviation *315*
Appendix B: List of Figures *327*
Appendix C: List of Tables *332*
Index *333*

Preface

Book Setting in Wireless Networking

In this book, we provide the reader with up-to-date coverage of the fundamentals on roaming securely in 802.11 networks. In addition, we place wireless technology in a historical context within the field of networking and provide basic material on networking in general and on wireless in particular. Once thought of as a fad and a fleeting branch of networking, wireless has evolved to the stage where it has become clear that wireless is here to stay, and furthermore, it is clear that 802.11 will be the dominant wireless technology for many years to come. Users have come to enjoy the mobility that wireless provides: from students across the nation sitting out on their campuses computing while communing with nature to *Chief Executive Officers (CEOs)* downloading their email while grabbing a cup of coffee, and then scurrying off to catch a flight, wireless is now part of our lifestyles.

Although wireless is a part of our daily routine, there are still many important issues that need to be discussed about securing wireless systems, and this book meets this important need by addressing these issues. Wireless is becoming the primary network for many installations, and the security challenges involved can no longer be ignored or handled with simple fixes. Malicious users are always devising new attacks to compromise systems, and hackers are implementing and unleashing such attacks with greater alacrity. Without sacrificing the mobility that should be inherent of wireless systems, it is desirable, and in some cases, critical to have a wireless system be as secure as its wired counterpart. This book describes the move toward that ultimate goal.

Who Should Read This Book?

Anyone with an interest in learning about safe roaming for 802.11 networks or having an interest in any of the following topics:

- networking
- security

- roaming
- wireless
- wireless networking terminology and basic definitions

should read this book.

Organization of the Book

We have organized the material for a self-study on topics related to secure roaming in 802.11 networks or for a one-semester college course. The subject matter is presented in an order and at a level appropriate for a working professional in any of the following fields: computer science, engineering, information systems, *information technology* (*IT*), or networking. The material is accessible to junior- or senior-level college students and graduate students in any of these fields. Motivated readers, for example, *Chief Information Officers* (*CIOs*) or IT managers, can glean a great deal of information from this book that will help them address concerns about wireless installations. Fitting seamlessly with the more technical issues described, we cover the basic concepts, terminology, and historical context of wireless, plus current and future trends in this field. We employ a rich set of examples from networking to illustrate key concepts. The liberal use of figures throughout the text will help the reader gain an intuitive feel for many of the abstract concepts that we present.

The table of contents was designed to provide a helpful overview of the topics covered in this book. Each chapter contains material of varying complexity. We have attempted to organize the material based on its usefulness, importance, and practical impact rather than merely by its complexity. The text is arranged in as "linear" a fashion as possible. We define concepts upon their first occurrence. The field of networking is littered with acronyms, and we define each of these also on its first occurrence in the text. A handy appendix defines all the acronyms used in this book, and the appendices for lists of figures and tables make it easy to locate these items throughout the text. The comprehensive multilevel index makes it easy for a reader to look up concepts. The reference sections provide details on text for further reading.

Teaching from This Book

Most of the material in this book can be effectively covered in a one-semester undergraduate or a graduate course, having a significant component on 802.11 networks. If this book is to be used as the primary text, if needed, an instructor can supplement our material with a significant group project on which student exercises and homework can be based. Many of the references can be used to supplement the material in the book as well, and exercises can be drawn from those resources as desired. It would also be easy to couple this book with Internet resources, and from this combination of materials a solid course on 802.11 networking could be offered.

Why Choose This Book?

The authors have a total of over 55 years of experience in computer science, with more than 30 years of experience in the field of telecommunications. We feel that we are highly qualified to write this book because of our experience and knowledge. We have published over 70 technical papers, written 13 books, developed and implemented a number of software systems, taught a wide range of courses at both the undergraduate and graduate levels, lectured throughout the world, co-authored a patent, and founded and run high-tech companies. As long-time mountaineering, endurance event, and training partners, we are used to working closely and successfully. For us, this book is a labor of love. This project is not about adding another bullet to our CVs or about adding a few dollars to our bank accounts. This project is about us collaborating on a book because we enjoy working together; we want to share our long-term experience and knowledge of computers and networking with others; and we want to present our families, friends, and Elsevier with a book that they can be proud of. Given the nature of our motivation and our ability to complete each other's sentences, we hope that you will find the quality of this book exceptional.

Of course, there are a number of other books available on networking, and in the reference sections we have listed a handful of the ones that we have found to be the most useful. There are however, few, if any, references about roaming securely in 802.11 networks that present the material in a clear, precise, and authoritative manner. We have been able to frame the 802.11 networking topics covered here broadly within the field of computing and include the technical details required of such a work. Our careful selection of topics has allowed us to present a treatment of wireless security issues that is rigorous and also practical and intuitive. The writing style used is particularly useful for system administrators, IT managers, and programmers. Although we delve into details in various parts of the book, we never lose sight of the big picture. The main features of this book are listed in the following:

- Easy-to-read
- Self-contained
- Precise presentations
- Clear examples
- Illuminating figures
- Informative historical remarks
- User-friendly index

Preface

- Comprehensive appendix of acronyms
- Highly accessible

The book contains a large number of figures and diagrams to explain and illustrate networking concepts that are being defined or discussed. These graphics enable the reader to proceed through the text, thereby eliminating the need for looking into other references. The figures are designed to be simple, uncluttered, and to-the-point, rather than to impress the reader.

Networking and wireless networking in particular is a large field that is rapidly expanding. We have made careful choices about what material to include in this book and what material to exclude. The text is not comprehensive, but it treats the topics covered in a precise and thorough manner. The reader does not require any special knowledge other than a basic understanding of computer concepts. Some experience in computer programming along with basic mathematical reasoning will be helpful for understanding the material presented. We have deliberately omitted some topics such as radiofrequency analysis, details of complex protocols, and certain aspects of hardware. More technical and specialized items are best understood after a thorough understanding of the material presented here. There are other good books available that address these peripheral topics.

Figures 9.2–9.15, 10.1, and 10.6–10.8 reprinted with permission from Draft IEEE Standard for Local and Metropolitan Area Networks: Media Independent Handover Services, IEEE P802.21/D01.00, Copyright IEEE 2006, Draft Amendment to Standard for Information Technology—Telecommunications and Information Exchange Between Systems—LAN/MAN Specfic Requirements; Part II: Wireless Medium Access Control (MAC) and Physical Layer (PHY) Specifications: Amendment 8: Fast BSS Transition, IEEE P802.11r/D2.1, Copyright IEEE 2006, Interim Contribution 802.11-06/0825r1 2006-05-31, Copyright IEEE 2006, and Interim Contribution 802.11-06/0566r2, Copyright IEEE 2006. The IEEE disclaims any responsibility or liability resulting from the placement and use in the described manner.

Suggestions and Corrections

Although we have tried to be as careful as possible, the book may still contain some errors, and certain topics may have been omitted that readers feel are especially relevant for inclusion. In anticipation of possible future printings, we would like to correct any mistakes and incorporate as many suggestions as possible. Please send comments via email to:

paulg@bondgarden.net

Acknowledgments

Thanks to our families and friends for their tremendous support throughout the years, especially while we were writing this book.

A special thanks to Harry Helms for his encouragement and support during this project. Thanks also to Rachel Roumeliotis for her help with this project.

A special thanks to Helen for her careful reading of the manuscript. Her comments were very helpful to us.

A special thanks to Jon LaRosa, Jim Burns, and Alec Rooney of Meetinghouse, Ellis Wong of Aylus Networks, and John Vollbrecht of the University of Michigan for their careful readings of draft chapters of the manuscript. Their technical expertise helped identify a number of bugs and greatly helped improve the manuscript. We are very grateful for their contributions to this work.

Heartfelt thanks to Ben Page for his help with the figures. Ben worked tirelessly for the layout, modification, and correction of figures.

Part of Ray's work on this project was supported by a Senior Fulbright Lecturing/Research Fellowship to Chiang Mai University in Thailand. The support of the United States and the Thailand Fulbright Associations is greatly appreciated.

A special thanks to Sanapawat "Bobby" Kantabutra for allowing us to include some coauthored material from the manuscript *A Mobility Model for Studying Wireless Communication* in Section 11.3, and for making Ray's stay in Thailand into a productive one.

A special thanks to Chris Williams for his help with the research described in Chapters 2 and 11. Also thanks to Chris for his help in editing the index and for securing figure and paper permissions.

Thanks to Janice Stanford for her contribution to Chapter 2.

Acknowledgments

A big thanks to Saowaluk "Yui" Rattanaudomsawat for her hard work in entering edits to the manuscript.

A big thanks to Jonnie Chandler for her assistance with editing the manuscript and for helping obtain figure permissions.

Thanks to the faculty and students at the Department of Computer Science at Chiang Mai University for making Ray feel at home in Thailand.

Thanks to Vijaya Sellaiah for her generous help in entering edits to the manuscript.

Thanks to Mirna Morrison for her assistance with the manuscript.

Thanks to Chris McCarthy for his assistance with the manuscript.

Thanks to all the employees of Meetinghouse and the faculty and staff in the school of completing at Armstrong Atlantic State University.

Thanks to Grey Goose, because without their ample supply of Vodka, we probably would not have signed this book contract.

Authors' Note

It is our hope that readers will experience the elegance and beauty of wireless networking through their study of this book. This book is our "Zen and the Art of Wireless" effort. We hope that the reader has as much fun learning this subject as we did while working together on this project.

Paul Goransson
Raymond Greenlaw

About the Authors

Dr. Paul Goransson has over 28 years of experience in the data communications field. He was the founder and President of Meetinghouse, which developed network access security software products for wireless and wired environments. Meetinghouse was acquired by Cisco Systems in 2006, where Paul currently serves as a Director of Engineering in the Wireless Networking Business Unit. He is also the owner/operator of Bondgarden Farm, a commercial beef and hay farm in southern Maine. Paul previously founded Qosnetics and QARobotics, which were later merged and subsequently acquired by Hewlett–Packard in 1999. Dr. Goransson has published technical articles in the fields of bandwidth reservation and wireless security. He received his Bachelor of Arts in Psychology from Brandeis University in 1975, his Masters of Science in Computer Engineering from Boston University in 1981, and his Ph.D. in Computer Science from the University of New Hampshire in 1994. Dr. Goransson has previously served as an adjunct Professor at the School of Computing at Armstrong Atlantic State University.

Dr. Raymond Greenlaw is the Founder and Dean of the School of Computing and Professor of Computer Science at Armstrong Atlantic State University in Savannah, Georgia. Ray is the Distinguished Professor of Computer Science at Chiang Mai University in Thailand and a Distinguished Visiting Professor at the College of Management and Technology in Kuala Lumpur, Malaysia. He is the author of 13 books in the field of computer science. His books cover complexity theory, graph theory, the Internet, networking, operating systems, parallel computing, the theory of computation, and the World Wide Web. Dr. Greenlaw has published 60 research papers and given over 160 invited lectures throughout the world. As a PI or co-PI, Ray has been awarded over $6,000,000 in grants and contracts, and his research has been supported by the following countries: Germany, Hong Kong, Iceland, Italy, Japan, Malaysia, Spain, Taiwan, Thailand, and the United States. He has won numerous international awards including three Senior Fulbright Fellowships, a Humboldt Fellowship, a Japanese Society for Promotion of Science Fellowship, two visiting Professor Fellowships from Italy, a Sasakawa Fellowship for Japanese Studies, and a Spanish Fellowship for Science and Technology. Dr. Greenlaw served as the Regional Coordinator for the State of Georgia's $100,000,000 Yamacraw Project, which was designed to make the state of Georgia a leader in the

About the Author

telecommunications field. Ray serves as a Commissioner for the Computing Accreditation Commission (CAC) of the Accreditation Board for Engineering Technology (ABET). He received a Bachelor of Arts in Mathematics from Pomona College in 1983, a Master of Science in Computer Science from the University of Washington in 1986, and a Ph.D. in Computer Science from the University of Washington in 1988.

Authors' Contact Information

Dr. Paul Goransson
Bondgarden Farm
255 Depot Road
Eliot, Maine 03903
Email address: `paulg@bondgarden.net`
World Wide Web page:
    `www.bondgarden.net`

Dr. Raymond Greenlaw
School of Computing
Armstrong Atlantic State University
11935 Abercorn Street
Savannah, Georgia 31419-1997
Email address: `greenlaw@armstrong.edu`
World Wide Web page:
    `www.cs.armstrong.edu/greenlaw`

CHAPTER 1

Introduction

1.1 Introduction

1.2 Basic Networking Terminology and Conventions

1.3 Setting the Scene

1.4 Different Notions of Roaming

1.5 Big Cells, Little Cells

1.6 Authentication, Authorization, Accounting, and Roaming

1.7 How Fast Do We Roam on the Range?

1.8 A Taxonomy for Roaming

1.9 Organization of the Book

References

1.1 Introduction

Once thought of as a fad and a fleeting branch of networking, *wireless local area networks* (*WLANs*) are firmly established in the technology field, and it is clear that 802.11 will be the dominant WLAN technology for many years to come. Users have come to enjoy the mobility that wireless provides: from students across the nation sitting out on their campuses and computing while communing with nature to *Chief Executive Officers* (*CEOs*) downloading their email while grabbing a cup of coffee and then scurrying off to catch a flight, wireless is now part of our lifestyles.

Chapter 1

Although wireless is a part of our daily routine, there are still many important issues about securing wireless systems that need to be discussed, and this book meets this important need by addressing these issues. Wireless is becoming the primary type of network for many installations, and the security challenges involved can no longer be ignored or handled with simple and ad hoc fixes. Malicious users are always devising new attacks to compromise systems, and hackers are implementing and unleashing such attacks with greater alacrity. It is desirable and in some cases critical to have a wireless system that is as secure as a wired system. This book describes the move toward that ultimate goal. The field of wireless security is laden with interesting problems and applications, and this book aims to update the reader on the issues involved.

This chapter provides the background for the fundamental notions related to roaming in wireless networks. Within the context of this book, the phrase *wireless roaming* is restricted to networks composed of possibly overlapping cells of coverage centered around the antennae that comprise the network. As shown in Figure 1.1, three antennae can be observed, where the middle cell overlaps the left and right cells. Although there is basic consensus that the phrase *wireless roaming* means terminating a connection with one antenna and initiating a connection with another, much confusion still centers around the term when one delves into the specifics.

Although cellular telephone networks are not perfectly secure, early on, the designers did incorporate security measures in the form of user authentication and encryption, and as a result, a typical user's handset is not easily modified to attack other cellular users or the network itself. *However, this is not the case for 802.11 networks.* For a number of reasons, roaming and access in 802.11 networks has historically been insecure. First, security measures were poorly addressed in the initial versions of the 802.11 standard. Second, the equipment required to initiate attacks on WLANs is readily available, and in fact, already in the hands of tens of millions of users. Each consumer who owns an 802.11-enabled laptop can mount attacks on 802.11 networks using only the hardware of the laptop and the software that can be freely downloaded from the Web. This fact motivates us to focus on issues related to *secure roaming*.

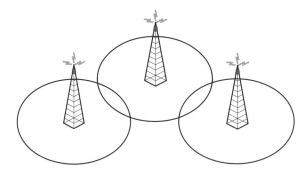

Figure 1.1: Three Overlapping Network Cells.

This book does not examine the complex and numerous *radiofrequency* (*RF*) issues involved in designing any wireless communications network. In particular, we do not present those issues related to cellular networks where different antennae of the same network operate and cooperate in close proximity to each other, with many concomitant problems relating to frequency reuse, interference, fading, and a host of other RF-specific concerns. There are previously published texts that deal with these aspects of 802.11 networks. As roaming is primarily a baseband and a protocol issue, it is reasonable to consider the treatment of the underlying RF issues only in specialized texts. Therefore, we will focus in greater depth on the protocol issues involved in 802.11 roaming, particularly in the complex field of roaming securely in 802.11 networks. In Chapter 3, we will provide a brief overview of certain RF aspects to the extent necessary to understand why the protocols and the systems that we discuss are the way they are.

All the *IP*[1]-related discussions and examples given in this book are based on *Internet Protocol Version 4* (*IPv4*). Although many of the new technologies that have been proposed, especially those in Chapter 10, presume an eventual conversion from IPv4 to *Internet Protocol Version 6* (*IPv6*), such a conversion has been slow. IPv6 is gaining some traction as evidenced by recent government announcements from the United States and Asian Pacific countries. As of this writing though, IPv4 dominates IP networks. In an attempt to make this book more relevant, we have focussed on IPv4.

Next, we look at the basic terminology that will be used throughout the text, and following that we provide a brief history of the field of wireless networking and then describe its evolution. We draw on the following two references throughout this chapter: [1] and [3].

1.2 Basic Networking Terminology and Conventions

This section defines much of the terminologies and notations used throughout the book. Our convention is to italicize a new term on its first occurrence in the text. More specialized concepts that are not defined in this section will be defined on their first occurrence in the text. Many networking terms and phrases have several and varied meanings according to different groups. Throughout the book, we try to use the most-accepted definition for terms and phrases. We also include a multitude of references where definitions used in this book and also other definitions can be found. Acronyms are also defined and emphasized on their first occurrence in the text, and the appendix on acronyms provides an alphabetized list of all the acronyms used in this book. An advanced reader may decide to skip over this section. Others may want to skim this material, and refer to specific concepts later.

---

[1] Internet Procotol.

Next, some common units of measure are defined.

A *kilobyte* is equal to 1000 bytes. The symbol K is also used to denote the value 1000. Therefore, 3K denotes 3000 bytes. The symbol *M* is used to denote 1,000,000, and *G* denotes 1,000,000,000. A *millisecond* is one-thousandth of a second, and we denote a millisecond by *ms*. *Hertz* is a measure of frequency in terms of cycles per second, and we denote Hertz by *Hz*. The expression *bits per second* is a measure of the rate of flow of information, and we denote bits per second by *bps*. *Decibels* are a measure of power and are denoted in this book using the notation *dBm*, which indicates decibels with a reference level of milliwatts, or simply *dB*, if no reference level is specified.

Next, we define some common networking terms. *Telephony* is the technology that is used in making and in providing telephone services and systems.

A *network* is a collection of two or more devices combined with a transmission medium, where the devices are capable of exchanging information using well-defined protocols. The common types of networks are as follows:

- *Local Area Network (LAN)*
- *Metropolitan Area Network (MAN)*
- *Wide Area Network (WAN)*

A *WLAN* is a LAN where the transmission medium is air. The typical maximum effective distance between a transmitter and a receiver in a WLAN is of the order of 50 m. (Note that this range is larger than ideal for seamless roaming.) The number 802.11 (read "eight-oh-two dot eleven") is the designator for the family of specifications for WLAN technology that was developed by the *Institute of Electrical and Electronics Engineers (IEEE)* and was accepted in 1997; the original technology provides 1–2-Mbps transmission rates in the 2.4-GHz range. Another important group involved in developing standards and in improving and promoting the Internet is the *Internet Engineering Task Force (IETF)*.

An *antenna* is a physical device capable of transmitting and receiving signals.

A *base station* is a physical device that processes communication data. Base stations receive and send out transmissions to coordinate communications among various devices.

An *access point* is a physical device that allows entry to a network.

A *Media Access Control address* or *MAC address* is a unique value that globally identifies a piece of networking equipment. MAC addresses are 48 bits long or equivalently 12-digit hexadecimal numbers; they are usually written as hexadecimal numbers. Each of the 12 digits can take on a value from the set $0, 1, \ldots, 9, A, B, C, D, E, F$. By convention, MAC addresses are expressed as a series of six, two-digit hexadecimal numbers that are separated either by colons or by dashes. Examples of MAC addresses are AA:07:27:19:61:BD and 12-B9-22-EE-19-54.

A *cell* is the geographic area covered by an antenna's transmission. For simplicity, cells are usually represented by circles or hexagons.

The term *multiplexing* indicates the division of a transmission medium into distinct channels in the space, time, frequency, or code dimensions, so that multiple users can simultaneously share the same medium.

Voice over Internet Protocol or *voice over IP* (*VoIP*) is a technology for transmitting and receiving speech communications using the Internet.

Next, the use of the terms *task group* and *workgroup* (or *working group*) is clarified. In all official documentation of the IETF that we have seen, the previously-mentioned terms refer only to workgroups or working groups and not to task groups. The list of current *IETF Working Groups* can be found at the *URL*[2]: www.ietf.org/html.charters/wg-dir.html

In contrast, the IEEE seems to use both these terms. For example, a workgroup seems to be at the level of 802.11 (Wireless LAN) or 802.16 (*WiMAX*[3]). These workgroups are then subdivided into task groups, which are chronologically organized and designated by letters of the alphabet, for example, 802.11g. The entire set of networking standards falls under the "Project" 802.

1.3 Setting the Scene

1.3.1 Introduction

In Figure 1.2, we show a common caricature of roaming. When talking on a mobile wireless device, in this case a cell phone, an operator of a *sports utility vehicle* (*SUV*) drives through the areas covered by two separate antennae of the same network. Notice that there is an overlapping area where communication to both antennae is possible. Most mobile phone users, other than Paul's mother, understand that they are communicating over a radio link of finite range, and when that limit is reached, their phones must roam to another cellphone tower for communication to proceed. Before we examine the details of this concept, we circumscribe the topic of roaming within the broader topic of mobile communications.

---

[2] *Uniform Resource Locator.*
[3] *Worldwide Interoperability for Microwave Access.*

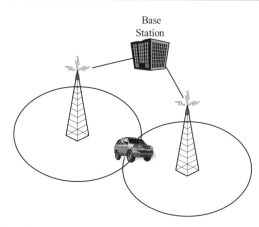

Figure 1.2: Standard Caricature of Roaming, While an SUV Driver Chats on a Cell Phone.

Not all mobile telephony involves wireless roaming. This fact was proved by Lars Magnus Ericsson—an individual whose company is intimately associated with this book's concept of wireless roaming. In 1910, Lars and his wife, Hilda Simonsson, experimented with what was probably the world's first mobile car-phone system [4]. The use of this system entailed stopping the car and reaching up with two long poles to tap into the telephone wires on the street. This mobile telephone system, although revolutionary in its own right, involved neither wireless nor roaming technology, as we define these concepts in this book. Ericsson's system was probably secure, so his car-phone receives high marks for that! Since this is not *our* concept of roaming, the next section describes how our concept of roaming evolved.

1.3.2 Precellular Wireless Networks and the Birth of the Cellular Concept

Wireless communication technology was born in the late 19th century, around 1897. Although numerous academics contributed to the scientific advances necessary for the technology to become practical, Guglielmo Marconi undoubtedly is the father of the modern commercial wireless industry. Marconi brought skills in technical innovation to bear on this industry; in addition, he had an intuition for developing markets where this nascent technology could proliferate.

The wireless industry found its first commercial success in the maritime shipping industry. The famous communication between the doomed Titanic and the Carpathia in 1912 highlighted the importance of this emerging field. World War I demonstrated the advantages that radio communication would bring on the battlefield, although it was not until the 1940s that ubiquitous use of radio technology by the United States armed forces demonstrated that this communication mechanism was a key part of any future arsenal. Back on the U.S. mainland during the prohibition years, police forces pioneered the use of radio in their battles against the recently motorized forces of organized crime.

Introduction

After World War II, the idea of commercial radio telephone service began to take root in the United States. Of course, at that time, *American Telephone and Telegraph* (*AT&T*) still enjoyed monopolistic control over all consumer communication products in the United States. However, in 1956, the U.S. Department of Justice ruled that whereas traditional telecommunications was a natural monopoly due to its public utility characteristics, mobile radio communications did not exhibit similar characteristics and, as such, should be subject to competition. This decision created an environment for the growth of the *Radio Common Carriers* (*RCCs*). By 1968, these RCCs formed nearly 500 mobile radio competitors to AT&T. Motorola was by far the best-known and the most successful of these companies. The mobile radio applications were generally highly industry specific and were devoloped for businesses such as taxicabs and construction companies, where the ability to dispatch work orders to remote workers dramatically enhanced business. As for publicly available radio telephone service, AT&T did initiate some radio telephone pilot projects in different U.S. cities during the 1950s and 1960s. These attempts certainly demonstrated that radio telephone service was technically feasible.

None of the early radio telephone pilot projects grew into widespread commercial deployments, which were attributed to several reasons. First, the early pilot programs suffered from several technical problems, which could not be fully solved for many years. Second, AT&T did not look at this arena as a mass market product, because the company's main thrust was wireline telephone service. In that area AT&T demonstrated a tremendous ability for technical innovation and price reductions, despite their comfortable monopolistic situation. To understand this point better, it should be noted that in 1968 AT&T had half of the wireless subscribers in the United States. The other half was divided among the 500 RCCs who competed with AT&T. Together, these 500 companies provided service to a national total of about 68,000 subscribers. In contrast, at that same time, AT&T had more than 100 million wireline subscribers! The radio telephone business accounted for only a minor proportion.

At best, the radio telephone service was seen as an interesting little market, so prices remained high. For example, the cost of mobile equipment on a per-user basis ranged from $1000 to $4500 in both the United States and Western Europe. The cost per uses on a monthly basis averaged about $110. The high prices slowed adoption, which in turn slowed technological innovation. A factor related to the slow growth of radio telephone service was that these early deployments were all based on a wide-area radio service model. In essence, the model assigned a user a frequency via which the user could communicate, as long as the user stayed within the range of a single, powerful antenna serving that area. This inherent inability to reuse the finite radiofrequency spectrum meant that even powerful transmitters would not enable the system to cover geographic areas that included large numbers of simultaneous users. No motivation existed to lower prices, if the technology could not scale to handle a larger customer base. In fact, the solution to this reuse problem had already been conceived by Bell Labs researchers in the 1940s' when they proposed what would later become known as *cellular communications*. However, many years would pass before a solution was actually deployed.

Ring and Young's seminal paper in 1947 addressed the problem of erosion of the RF spectrum, as more and more users vied for a finite range of frequencies [2]. The breakthrough notion involved obtaining a highly scalable degree of frequency reuse by dividing transmission and reception areas into relatively small cells. By restricting the transmission power of the transmitters in these cells, it became possible to reuse the same frequencies in not-too-distant cells. This concept was the progenitor of the cellular networks that have become such an important part of our lives since the 1980s. The evolution of cellular telephone networks through the analog (1G), digital (2G), and multimedia (3G) phases deserves a dedicated treatment, which we provide in Chapter 2.

1.3.3 802.11 Arrives on the Scene

By the early 1990s' cellular RF technology had advanced sufficiently, and the Internet explosion was far enough underway so that it seemed natural to conceive an association between the ubiquitous low-cost precepts of the Internet and low-cost, cellular-based wireless access. This association led to the establishment of the 802.11 concept. Unlike the designers of telephony-based cellular services, the IEEE decided that these cells should operate using the unlicensed RF spectrum in order to simplify deployment. If users needed to file license applications in order to deploy this new technology, it would never proliferate in the manner that the IEEE envisioned. The IEEE aimed at easy deployment of devices, without having to involve technicians with a background in RF issues. It was necessary that the technology be able to expand to Internet-scale deployment numbers and the costs of manufacturing the 802.11 equipment remained within the range of the mass market. As we now know, both these requirements have been satisfied.

1.4 Different Notions of Roaming

In this section, we examine several different notions of roaming. We begin with the usual definition of the word roaming and build on that. Webster's New World Dictionary defines roaming as follows:

Roaming: *Traveling about, especially in search of adventure: Errant, roving, wandering.*

This definition is helpful to provide an intuitive sense of roaming in the networking context. The Wi-Fi Planet Webopedia defines roaming as:

Roaming: *The extending of connectivity service of a network into a new region.*

Since the concept of roaming is of paramount importance to our work, we spend some time discussing this frequently used term. Perhaps the best way to understand roaming and to make the concept more concrete is to examine a few specific scenarios. We cite the following five examples of a mobile user of a wireless connection to illustrate how the first intuitive definition of roaming evolves to the second technical one and also to clarify the term within the content of this book:

Figure 1.3: A Person Who is Talking on a Cell Phone in a Moving Car.

1. A person who is talking on a cell phone in a moving car. Figure 1.3 depicts the situation.

2. A person who is working on an 802.11-enabled laptop and walking down a hallway. Figure 1.4 depicts the situation.

3. A person who is talking on a cell phone and reaching an area not covered by the home wireless provider. Figure 1.5 depicts the situation. In this case, several other providers are available, but the home provider is not.

4. A person who is working on an 802.11-enabled laptop and surfing the Internet, while using the wireless Internet access offered by the airline. Figure 1.6 depicts the situation.

Figure 1.4: A Person Who is Working on an 802.11-Enabled Laptop and Walking down a Hallway.

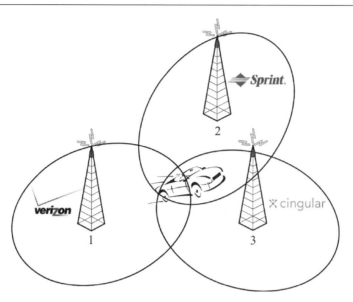

Figure 1.5: A Person Who is Talking on a Cell Phone and Reaching an Area Not Covered by the Home Wireless Provider.

Figure 1.6: A Person Who is Working on an 802.11-Enabled Laptop and Surfing the Internet, While Using the Wireless Internet Access Offered by the Airline.

Introduction

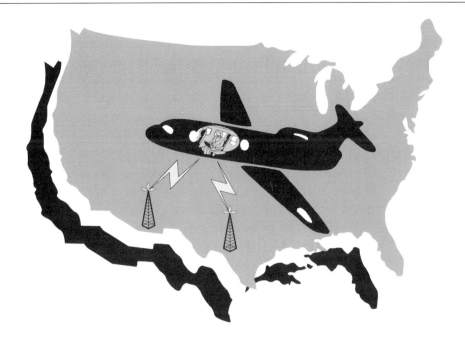

Figure 1.7: A Person Who is Working on an 802.11-Enabled Laptop and Surfing the Internet, While Using Wireless Internet Access by Making a (Currently Prohibited in Flight) Call Using a Cell Phone with the GPRS Modem in the Laptop.

5. A person who is working on an 802.11-enabled laptop and surfing the Internet, while using wireless Internet access by making a (prohibited in flight at the time of publication) call using a cell phone with the *General Packet Radio Service* (*GPRS*) modem in the laptop. Figure 1.7 depicts the situation.

In each case, we have a person in motion who is using a cell phone or laptop to tap into a wireless network. Intuitively, all five scenarios seem to be examples of wireless roaming. That is, when a wireless connection is in use, there is a definite motion taking place, which causes the user to switch antennae. Nevertheless, there are three distinct processes involved in these five examples, and only two of these processes will fit into this book's definition of wireless roaming. We consider what is actually happening by examining these examples in a more detailed manner in the following text.

Consider the first and second examples. We assume that the person is initially communicating via antenna 1 and that the movement brings the user out of range with antenna 1 and into the range of antenna 2. Furthermore, we assume that both antennae 1 and 2 are administered by the same provider and are part of the same network. The network access process of switching the radio communication from antenna 1 to antenna 2, and the back-end process of instructing the network infrastructure to receive and to relay this user's traffic from antenna 2, instead of antenna 1, is a process called *local roaming*. We further discuss this notion in Chapter 5.

We now turn our attention to the third scenario, illustrated in Figure 1.5. We assume that the driver of the car is currently communicating via antenna 1, which is administered by the user's home provider. When the driver reaches the limit of the range of antenna 1, it is assumed the only antennae within range belong to two separate providers, both of whom are different from the user's home provider. In order to begin communication via antenna 2 or antenna 3, the network must confront a new and broader problem than that described in the previous two examples. The front-end and back-end communications must switch over as described earlier, but the new issues are that this switchover must additionally include a new decision as to which of the two nonhome providers to use, and, once that decision is made and the selected provider is contacted, that provider's system must make a decision as to whether or not the user should be admitted. The latter decision may involve additional accounting issues and possibly communicating with the user's home provider. This situation is called *global roaming*. We further discuss this notion in Chapter 3.

The fourth example is illustrated in Figure 1.6. We assume that the airlines will soon provide an on board 802.11 network for business passengers, and for this example, we assume a satellite uplink for the backhaul Internet access. In this case, the users are flying through the atmosphere at 800 km per hour, while maintaining continuous Internet access, yet no roaming is taking place. Despite traveling at a high velocity, a user's laptop remains steadily connected to the same internal 802.11 access point that is located within the airplane itself. Finally, in the fifth case, shown in Figure 1.7, we assume the switch from antenna 1 to 2 does not involve a change of providers. Thus, this case is also an example of local roaming. In summary, examples one, two, and five involve local roaming, example three involves global roaming, and example four does not involve roaming at all. As this discussion indicates, the main difference between local and global roaming is whether a person is receiving service from the home provider or not.

1.5 Big Cells, Little Cells

1.5.1 Introduction

In this section, we examine a number of issues related to network cells. One of the primary differences between cellular telephony and 802.11 technology is the size of the cells. Cellular telephone network cells generally have a radius of 2–15 km, whereas 802.11 network cells typically have a radius in the tens of meters. Figure 1.8 illustrates their relative sizes. Here we review numerous factors that were initially considered by the creators and *evolvers* of cellular phone technology, and later by the IEEE, as they developed the 802.11 standard. Each of the factors that we examine impacts the design decisions that were made (and continue to be made because this process is evolving) by the cellular telephone and 802.11 groups, and several of

Introduction

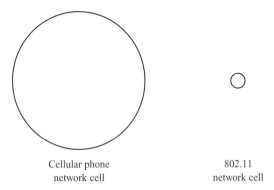

Cellular phone network cell   802.11 network cell

Figure 1.8: The Relative Sizes of a Cellular Phone Network Cell (Left) and an 802.11 Network Cell (Right).

these factors have a direct impact on the selection of the average cell size for these technologies.

1.5.2 RF Technology and Transmit Power

The most important factor in understanding cell size is the transmit power radiating from the cell antenna. One might think that if we have a sufficiently strong signal, and neglected the curvature of the earth, then the entire planet could be considered a single cell because all stations could receive the powerful signal. In fact, this notion is not as ludicrous as it appears, as long as we accept the following two small inconveniences:

- All stations in the cell receive the same signal.

- All stations can receive voice or data, but cannot transmit. This is because everyone would have to transmit on the same spectrum, which would significantly limit useable bandwidth.

Even decades old shortwave radio satisfies these requirements to a large extent. Alas, for the voice- and data cellular networks that we introduce in the early chapters of this book, these "inconveniences" are clearly unacceptable, and the individual stations need to have independent, bidirectional communications with the antenna. Thus, some type of multiplexing is required.

1.5.3 Number of Active Users in a Cell

The multiplexing mentioned in the previous section must allow multiple users in the cell to maintain independent, bidirectional communications. There are many different technologies that perform this kind of multiplexing, some of which will be introduced in Chapters 2 and 3. All these technologies have finite capabilities, so each technology will have a practical maximum number of active users that a single cell can support. Achieving a particular value for this number will constrain the cell size, perhaps even more than the RF limitations described earlier.

Consider a densely populated area. We can actively decrease the signal strength in this area because we know that the maximum number of users will be present within a short radius from the antenna, and so there is no need to transmit at a stronger power, which causes unnecessary interference with other neighboring cells. Increasing the transmit power would increase the geographic size of the cell, and, in so doing, reduce the overall user capacity of the cellular system due to a lesser degree of possible frequency reuse. In other words, as you increase transmit power, you increase cell size, and as a result, increase the number of users who need to share that spectrum. Thus, increasing the cell size merely increases the amount of multiplexing that needs to occur, which effectively limits the number of active users who can be supported in that cell.

1.5.4 Overview of Quality of Service Requirements

The term *Quality of Service* (*QoS*) refers to the delivery of guarantees of basic network service by the network to the user. This phrase most commonly refers to the parameters of *bandwidth*, *loss*, *delay*, and *jitter*. We define each of these parameters in the following text.

Bandwidth In the early analog cellular days, *bandwidth* referred to the frequency range that a user could access to send and to receive a signal. As only human voice was communicated, the bandwidth allowed for each user connection was approximately the 3000 Hz used for basic toll-quality calls from the wired telephone network. With the advent of digital cellular transmission, bandwidth referred to a certain number of bps, usually something less than the uncompressed 64-Kbps digital signal that the wired telephone network used to carry toll-quality digitized voice. When additional features such as text messaging, Web browsing, and phototransmission were added in later years to the existent feature of a cellular phone, there was a need to increase bandwidth; bandwidth needs continue to grow rapidly. For the purposes of our discussion, a *bandwidth guarantee of x* means that the network responds to a request to reserve a minimum of x bandwidth, grants at least x bandwidth, and guarantees that at least x bandwidth will be provided throughout the duration of the reservation.

Loss *Loss* is a measure of the percentage of the signal that is being sent but is not correctly received at the remote end of the connection. So, for example, if only one third of a transmission is received, we say there is a 66.7 percent loss of packets. QoS guarantees regarding loss in a digital network might guarantee no loss or some finite but nonzero loss. When a signal has built-in redundancy, such as human voice communication, a certain degree of loss is often acceptable, but when the loss reaches a certain threshold, the quality of the voice communication begins to degrade. The loss parameter is tightly coupled to bandwidth guarantees.

Delay *Delay* is the end-to-end time from when a user's signal leaves the transmitter until it is received at the remote end of the user's connection. We typically measure delay in ms. As RF signals propagate at approximately the speed of light and at such speeds the distance between any two points on the surface of the earth is negligible (Note that the speed of light is approximately one foot per nanosecond, 10^{-9} seconds.), delay is usually not a significant parameter in terrestrial analog communication. In digital communication, however, things are somewhat more complicated because a digitized signal is often buffered into packets of bits, which are then transmitted together. This buffering can introduce additional delay into a transmission. Finite amounts of end-to-end delay are unavoidable and present no problem when kept within bounds specific to the particular type of user traffic. In case of human voice transmission as an example, a common rule of thumb is that approximately 20 ms of end-to-end delay can be tolerated before adverse effects are noticed on a bidirectional communication.

Jitter *Jitter* is variance in delay and is usually measured in ms. If digitized voice data packets arrive with different amounts of delay, it is necessary to buffer the packets at the receiver end such that the packets can be played back to the receiver as a smooth stream subject to the maximum delay of any of the packets. This smoothing is accomplished by a dejittering buffer, where packets that arrive with less than the maximum delay are queued and then delivered to the user as if they had arrived with the maximum delay. Naturally, a packet that arrives after the maximum allowable delay is considered to be a lost packet; such a packet is discarded at the receiver end without being delivered to the user.

A network that expects to deliver toll-quality voice service must operate with strict QoS guarantees. As voice was always the primary service to be provided by the cellular telephone network, guarantees about loss, delay, and jitter were built into the design of all the cellular telephone technologies of the past 20 years. This situation was not the case with Internet technology, and thus, as expected, not the case with its progeny, 802.11 technology. The 1990s saw a large volume of research involving Internet QoS technology. These efforts ultimately culminated in commercial technologies that can provide QoS guarantees on the wired Internet, which are sufficient to carry voice and video traffic. Similarly, a number of extensions have been proposed to retrofit 802.11 for QoS capabilities. These extensions will be discussed in more detail in Chapter 4, when we describe the work of IEEE task group 802.11e.

Chapter 1

This book examines the challenge that stems from the fact that these QoS capabilities are being retrofitted as they were not part of the original 802.11 concept. As a result of this patching, roaming securely, while maintaining the QoS required for a voice over IP call over 802.11, seems somewhat cumbersome when compared to basic access to 802.11. In one sense saying that wireless roaming is cumbersome is unfair though because much of the complexity simply comes from the natural migration of 802.11 toward the complexity that was originally built into cellular telephone networks.

1.5.5 Complexity of Network Design and Implementation

Telephone networks have a long history of being some of the most complex systems that humans have ever designed and operated. True to this heritage, cellular phone networks require an enormous amount of custom engineering and design for each new cell deployed. As the degree of expertise involved in deployments of 802.11 ranges from a novice to the relatively sophisticated enterprise IT manager, the 802.11 standard was designed to minimize the amount of network engineering required to set up simple networks, while allowing for plant-wide installations by the more sophisticated IT manager. However, even an involved 802.11 network pales in complexity to a typical cellular telephone multicell zone. This inherent requirement for simplicity coupled with regulatory restrictions imposed by governments in terms of transmit power and other issues, forces the 802.11 cell size to remain small, thereby resulting in easier deployment.

1.5.6 Frequency and Speed of Roaming

While the smaller cell size allows the installer/designer to ignore many RF problems that a larger cell size would necessitate confronting, small cell size has the undesirable side effect of exacerbating problems related to roaming. Obviously, with all other factors mentioned in this section kept constant, a smaller cell size results in more frequent *handoffs*—a switching of communications to another antenna. We discuss handoffs in more detail in Section 1.8.2.

1.5.7 Impact on 802.11 Cell Size

Earlier we introduced five major factors that can influence the average cell size for a given cellular technology. They were as follows: signal strength, number of subscribers, QoS requirements, simplicity of deployment, and frequency and speed of roaming. Next, we review these factors as they relate to the IEEE's cell size selection for 802.11.

Clearly, the fact that 802.11 was to be based on the use of unlicensed RF spectrum implied that transmit power would be low. High-strength RF transmissions are regulated by the

Federal Communications Commission (*FCC*) in the United States and similar bodies in other countries. The strength of the RF signal could be kept small as it was always expected that independent entities would deploy 802.11 networks in relative physical proximity. In the environment of a coffee shop or an office, an effective signal range of 80 m is far less likely to conflict with a neighboring installation's deployment than a signal range of 1 km would.

A coffee shop typically provides service to dozens of simultaneous users, not to thousands. Billing and accounting were not primary concerns in the early vision of 802.11 deployments. The ability to provide QoS guarantees and frequent fast roaming were not goals in the design phase of 802.11, so the negative impact of small cell size on these two factors did not play a significant role in the selection of average cell size. Although the factors of size of the subscriber base, billing and accounting issues, and maturity of commercial deployments do not directly affect the average cell size of the cellular telephone and 802.11 paradigms, they do impact roaming, and we discuss this impact in the next section.

1.6 Authentication, Authorization, Accounting, and Roaming

This discussion relates to the domain of *Authentication, Authorization, and Accounting* (*AAA*) technology. Figure 1.9 depicts an example of an AAA server used in an 802.11 wireless network. It is useful to keep this figure in mind during the discussion in this section. For a given network that allows user access, the network has an explicit or implicit policy for each

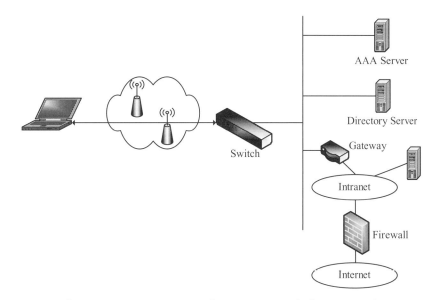

Figure 1.9: An AAA Server in an 802.11 Wireless Network.

of these three criteria. A network might allow any user possessing the correct transmitter and receiver equipment to communicate via a cell's antenna or the network could require that the user identified by the network as one of its permitted users. This concept is called *authentication* or *admission control*. Once admitted, different users may be allowed access to different levels or types of network service. This idea is called *authorization control*. For example, in a network using authorization control, user A might be authorized to perform both voice calls and text messaging, whereas user B might only be authorized to perform voice calls. The third component of AAA is *accounting*. This component is important, unless the service is delivered free of charge. Usually, when a network has accounting, the user pays for the volume of user traffic sent or the time spent on the network. In 802.11, traffic volume and network time are tracked in terms of kilobytes and minutes, respectively, and the cost increases with usage.

At one extreme, a wireless network may allow any user with the correct transmitter and receiver to connect to the access point, use the network, and use any of the services provided by the network for any amount of time or volume of traffic without charge. While this model is a theoretical rather than a practical extreme, it is closer to the Internet access model than to the traditional wireless network model. Cellular telephony represents the other end of the spectrum, where admission, authorization, and accounting, all play vital roles in providing what is almost always a commercial for-profit service.

Depending on which of the three components of AAA technology that a network needs to provide directly affects the number of active users that are allowed in a cell, the total number of subscribers permitted to access the network, and how quickly users may be able to roam from one cell to the next. From the beginning, the cellular telephone networks were designed to handle all aspects of AAA technology because they were originally conceived of as a for-profit service. Their original design entailed incorporating considerable back-end technology with each cell so that users could be identified and tracked as they entered one cell, and then roamed to the next. We will discuss this technology in greater detail in Chapter 2.

In the spirit of Internet-related technologies, the 802.11 networks evolved differently, with concerns about AAA issues remaining in the background. The original 802.11 standard assumed that 802.11 devices were trusted entities (for example, barcode scanners) and were not expected to be authenticated or charged for the service provided. Just as the Internet evolved from a not-for-profit technical experiment to what is widely considered the greatest commercial revolution of the 1990s, it soon became apparent that different aspects of AAA technology would need to be incorporated into 802.11 networks. From the perspective of a public hotspot provider such as T-Mobile or Eurospot, the 802.11 network deployed for commercial reasons has come to resemble the cellular telephone network in terms of the need to charge for the service and the concommitant reliance on AAA technology. Within the corporate enterprise, although billing for services might be rare, the notion of admitting a valid employee while denying service to the war-driver sitting in the company parking lot has become extremely important, creating a demand for the first two A's of AAA (authentication and authorization) technology even in an industrial setting.

Because the AAA requirements on 802.11 networks evolved rather than being incorporated from the beginning, there has been an ad hoc approach to adding this technology to 802.11 networks. An early brute force, nonscalable approach to admission control in 802.11 access points was to program statically an exhaustive list of every MAC address that would be permitted to use the access point. This idea seems reasonable in the *Small Office/Home Office* (*SOHO*) environment with one access point and 20 users, but if you try to scale this notion to a corporation with 500 access points and a subscriber base of 20,000 wireless users, it simply does not work. (Note that this idea does not work for another fundamental reason as well: it is trivial to make a device pretend to be the MAC address of another. This easy attack is called MAC *spoofing*.)

The approach taken in 802.11 involved leveraging AAA solutions that were already developed for wired access to the Internet. The development of these AAA solutions avoided having to reinvent the wheel. In Chapter 6, we will discuss these technologies in detail. For purposes of discussion, it suffices to say that the additional complexity of adding these AAA technologies to the 802.11 model has added considerable overhead to the process of connecting to one access point and especially roaming from one access point to the next.

The salient point related to AAA as it pertains to differences between cellular telephony and 802.11 is that 802.11 initially comprised a simpler technology due to it being decoupled from AAA requirements. Thanks to this simpler technology, the 802.11 model has been a huge success. As deployment sizes have grown accordingly, increased security concerns have, of course, emerged. Naturally, commercial entities have seen the opportunity for network access revenues to grow with larger deployment sizes. These changes led to AAA technologies being gradually added to the 802.11 model. Some of these additions have directly conflicted with the early goals of simplicity. As we proceed, we will see that back-fitting the requirements for AAA has added a complex dimension to our roaming problem.

1.7 How Fast Do We Roam on the Range?

While roaming is indeed disconnecting from one antenna and reconnecting to another, how fast you need to perform these two operations radically affects what solutions are feasible. How fast you need to or can execute this transition is influenced by a number of issues. We address these factors in the text that follows.

1.7.1 Speed of Travel

Consider a cell phone user who is traveling on an uncrowded freeway in a car and separately an 802.11-enabled laptop user who is walking down a corridor from an office to a conference room—scenarios already depicted in Figures 1.3 and 1.4, respectively. If we keep all other factors constant, since the first user is moving at a much higher velocity than the second, there is less time available for the completion of the roaming process. This fact seems quite obvious, as long as we can hold all other factors constant. In reality, a given cellular system always makes assumptions about cell size and the size of the overlapping coverage area, as well as the maximum user velocity which is to be supported. We can safely assume that a device traveling at a speed S places greater demands on a given roaming technology than the same device traveling at a speed less than S.

1.7.2 Size of Overlapping Cell Coverage Area

When a user quickly transits along an *overlapping cell coverage area*, a greater demand is placed on the roaming technology than for a slower-moving user. The overlapping cell coverage area is defined as the region where communication with two or more antennae is possible. The area labeled in Figure 1.10 denotes this region. The size of this area is a function of the proximity of the two antennae, plus other RF factors. In the case of 802.11 networks, this area may be measured in the tens of meters, whereas in cellular telephone networks, this area may be measured in many hundreds of meters or even small numbers of kilometers. Clearly, the size of the overlapping cell coverage interacts intimately with the maximum allowable user velocity in determining the operational requirements for roaming in a given network.

1.7.3 Application's Tolerance for Disruption in User Data Flow

Consider a person Deb who is driving while talking on her mobile phone and an office worker Bob who is reading his email on an 802.11-enabled laptop as he walks to an

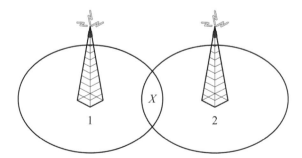

Figure 1.10: An Example of an Overlapping Cell Coverage Area X, Where a Device Could Communicate with Either Antenna 1 or Antenna 2.

upcoming meeting. Deb has very little tolerance for disruption in her data flow (less than 20 ms), whereas Bob may not notice any problem even if his laptop takes an entire minute to enter full communication with the new antenna. If we instead assume that Bob is using a VoIP phone to carry out business, as he walks from his office to the conference room, the roaming parameters for the same office network change drastically. The roaming parameters more closely resemble those for Deb, although the area of overlapping cell coverage and the velocity remain much smaller for Bob.

1.7.4 Complexity of Accomplishing the Roam

In the preliminary remarks about roaming described in this section, we purposely oversimplified the processes involved in accomplishing the roam. Of course, there is the essential basic task of switching the radio communication from one antenna to the next. The related procedures for this task in 802.11 are different from those in cellular networks. We provide details on the related 802.11 procedures in Chapters 3 and 5. The complete roam, however, involves much more than just switching the communication between the mobile radio device and two different base station antennae.

Different networks will require different authentication, authorization, and accounting processes to admit a user to the new antenna. These processes may make roaming easier or more difficult depending on the technologies involved. Roaming securely usually involves encrypting the traffic between the mobile station and the new antenna, which may involve changes in complex encryption processes. In later chapters, we will see how the initial decisions by the 802.11 designers have resulted in processes that are so complicated for secure roaming that while roaming securely in 802.11, it is still challenging to achieve the requirements in user data flow needed for minimal disruption in voice communications.

1.8 Taxonomy for Roaming

In this section, we present two geometric network cell models and discuss handoffs. These concepts will be used in later discussions about roaming.

1.8.1 Network Cell Models

Intuitively, cell size is the area where a device can communicate with an antenna. For purposes of our discussion, it is not necessary to provide a rigorous definition of cell size, rather we will merely point out a number of factors that influence cell size. Cell size is determined by the effective power of the transmitter of the cell's antenna, the transmitter at the mobile unit, and the corresponding receivers' ability to receive those transmissions. Cell size is also influenced

Chapter 1

by many other RF factors such as fading, interference with physical objects that affect the effective range of a given antenna in different directions, and other users' signals. (Note that a network in this context is simply a group of adjacent cells.) We examine two cellular network models later in this subsection: the *circular* and the *hexagonal models*.

For simplicity, in the networks we consider, cells are depicted as uniform-sized regions. We can consider each cell as containing a single antenna at its center. A user is defined to be *within the cell* if a usable signal from the cell's antenna is received, and the user can successfully transmit to the antenna. Note that because a user may be able to transmit and receive from more than one antenna in a given network at the same time, a user by this definition may be within multiple cells simultaneously.

The first network cell model we consider is the circular model. As the name implies, each cell is represented by a circle. Furthermore, the model assumes that the circles are uniform in size. Figure 1.11 depicts a network of overlapping cells. In the figure, the letters A, B, C, and D represent four different users and their locations. Note that user A is not within any cell; user B is within one cell; user C is within two cells; and user D is within three cells. Clearly, in order to obtain full coverage in the circular network cell model of an area much larger than the cells themselves, one must have overlapping circles.

The second network cell model we consider is based on hexagons rather than circles. Figure 1.12 depicts a six-cell, nonoverlapping hexagonal network cell model. As the cells are nonoverlapping, a given device can only be in one cell at a time. This corresponds to the fact that the cellular user's traffic only flows through the antenna of one cell at a time. This nonoverlapping hexagonal model is useful in network analysis involving the roaming problem, although the model does trivialize much of the complex RF analysis on which an entire network is based.

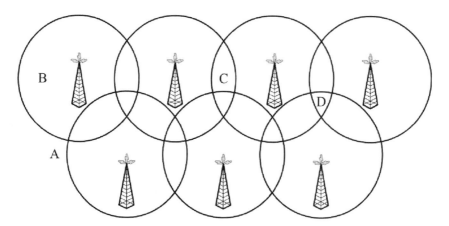

Figure 1.11: A Network Consisting of Seven Circular Cells.

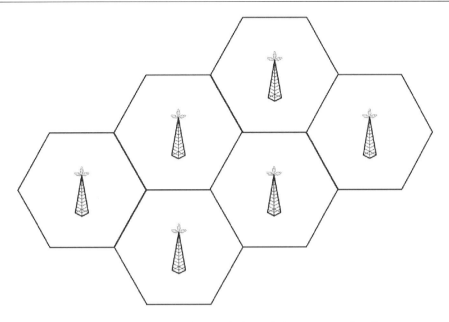

Figure 1.12: A Network Consisting of Six Hexagonal Cells.

We will draw upon both the circular and the hexagonal network cell models, as we explain how roaming works. Depending on the types of transmissions that we are interested in considering, we may even view a network using both the circular and hexagonal models simultaneously.

1.8.2 A Handoff Model

The great benefit of wireless networks is that a user can remain connected to a network even while in (fast) motion. A key concept in making sure that a user sees continuity in connectivity is the notion of a *handoff*. Our notion of a handoff consists of the processes involved in switching a user's traffic in a cellular network from one antenna to another. Ideally, a handoff occurs seamlessly without the user even being aware that a handoff has taken place. The technical challenges of this process have not yet been mastered, as evidenced by how often cellular telephone and 802.11 network users are aware that the handoff is taking place, due to interruption and sometimes temporary loss of connectivity. Nevertheless, one major goal of roaming in cellular networks is to provide a transparent handoff. In order to describe different aspects of a handoff, we need to define the following technical concepts: control traffic, user traffic from the handset to the antenna, user traffic in the network infrastructure, time period immediately preceding the handoff, and time period immediately following the handoff. As we explain these concepts, let antenna 1 (antenna 2) refer to the antenna in use before (after, respectively) the roam. It will be useful to refer to Figures 1.13 and 1.14 throughout this discussion.

Chapter 1

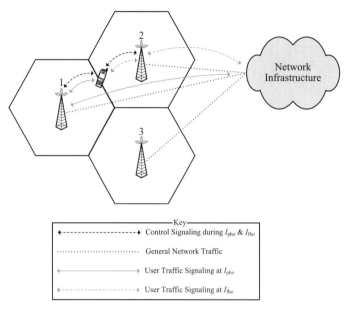

Figure 1.13: An Example of a Hard Handoff.

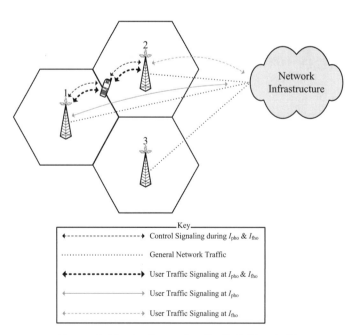

Figure 1.14: An Example of a Soft Handoff.

Control traffic This network traffic is the signaling communication, between a mobile device D and an antenna B, that is used to detect the presence of D within B's cell. Control traffic allows D to detect the presence of B's antenna, and control traffic performs all the call setup and teardown processes that are required to establish and to teardown the user connection between D and B's antenna. Note that control traffic may occur between a device and more than one antenna without breaking a user's connection. Control traffic is sometimes referred to as *control signaling*.

User traffic from a mobile device to an antenna This network traffic is the communication related to the application(s) that a mobile device enables. This communication may include voice traffic or various types of Internet-related data such as email or Web traffic. This traffic refers only to the traffic in the network between a mobile device and an antenna, and not to that same traffic as it is relayed deeper into the network infrastructure.

User traffic in the network infrastructure This traffic is the same as that described in (User traffic from a mobile device to an antenna), except that it refers to this traffic as it is relayed deeper within the network infrastructure. This infrastructure is normally a wired infrastructure, but fixed wireless infrastructure networks are also possible, for example, WiMAX. This traffic is sometimes referred to as *bearer traffic*.

Time period immediately preceding a handoff This time is the period during which the user traffic in the network infrastructure is being relayed via antenna 1. We denote this interval by I_{pho} ("pho" means pre-handoff).

Time period immediately following the handoff This time is the period during which the user traffic in the network infrastructure is being relayed via antenna 2. We denote this interval by I_{fho} ("fho" means following handoff).

It is worth noting that in this book we use the term "handoff" somewhat more liberally than certain cellular telephony specifications. A *roam* is only considered a *handoff* in certain cellular-telephony vernacular if the transition can be handled exclusively by radio-layer capabilities. When transitioning between two antennae requires intervention by more than just the radio-layer, this transition may be an instance of roaming in the cellular telephony world, but it is not properly called a handoff in that world. In this book we use the term "handoff" to mean both things. Similarly, handoffs between two antennae of the same provider do not constitute a "roam" in the specialized vocabulary of cellular telephony, whereas this transition is a "roam" within our context of this book. Finally, the term "handover" is often used in place of "handoff" in cellular-telephony documents. We will consider these two terms equivalent, for use within our 802.11 context as well.

Chapter 1

In the following two sections we provide generic descriptions of hard and soft handoffs. Since the descriptions are intended to be generic, they do not faithfully follow the handoff details of any one cellular technology.

1.8.3 Hard Handoffs

Figure 1.13 illustrates the relationships between the concepts presented in Section 1.8.2 during the course of a hard handoff. The hard handoff implies an abrupt break in the path that the user signaling takes, as it is routed into the network infrastructure. We see in Figure 1.13 that control signaling is taking place to both antennae 1 and 2 during both intervals I_{pho} and I_{fho}. The user traffic is only transmitted via antenna 1 during the interval where the mobile device is beginning to communicate control traffic with antenna 2 (I_{pho}). Indeed, the user traffic abruptly ceases to be communicated to antenna 1 at the transition between I_{pho} and I_{fho}, which is the moment that the handoff occurs. Note that an additional process must be coordinated at that moment—the process of relaying the user traffic into the network infrastructure via antenna 2 instead of antenna 1. We see in Figure 1.13 that during I_{fho}, the user traffic emanating from the mobile device is only communicated to antenna 2.

1.8.4 Soft Handoffs

In Figure 1.14 we show the transition during the course of a soft handoff. Notice the similarities and differences between Figures 1.13 and 1.14. Unlike a hard handoff, the soft handoff implies that both antennae may simultaneously be receiving the user traffic from the mobile device. When both user traffic streams are being received by the network infrastructure, the network infrastructure can relatively simply switch between the stream arriving from antenna 1 to the stream arriving from antenna 2. We see in Figure 1.14 that control and user signaling are taking place to both antennae 1 and 2 during both I_{pho} and I_{fho}.

In general, a soft handoff will provide a much greater likelihood of a seamless transition during the roam, but a soft handoff is not always feasible, either due to the particular technology in use or even due to where in the network topology the handoff occurs. In those many instances where hard handoffs are used, cellular-phone trends are currently to improve the hard-handoff technology to achieve an acceptably seamless roam in a high percentage of cases. The 802.11 technology has always been built on the principle of hard handoffs of user traffic, so this concept will be more important to the readers of this book.

1.8.5 Comparison of 802.11 and Cellular Roaming

Although the radius of the 802.11 antenna is small compared with the radius of the *Global System for Mobile Communications* (*GSM*) antenna, the fundamental problems of roaming from one antenna to another are quite similar between these two classes. In other words, for a given roaming problem, the same parameters described in Subsection 1.8.2 can be used to describe the problem for both 802.11 and cellular networks, although the values related to time and distance will differ between the two types of networks.

1.9 Organization of the Book

We cover a great deal of ground in this book; so, a high-level road map is in order. The first three chapters define important terminology about various types of roaming. We will see that these different types of roaming have a wide range of goals associated with them. While providing a historical context for the roaming challenges faced by wireless network designers during the 1980s and 1990s, we show that the roaming problem manifests itself in markedly divergent ways in different parts of the technology and business models. This situation has resulted in a host of solutions to problems that fall under the context of roaming. We examine cellular telephony in Chapter 2.

While shifting our focus to the roaming problem within the 802.11 context, we explain the interesting process of how a number of collaborative subgroups developed the 802.11 standard and how their work was ultimately ratified. Chapters 3 through 7 describe various standard approaches to the different aspects of roaming problems within 802.11. In Chapter 3 we discuss the general principles of roaming in WLANs. Chapter 4 covers the IEEE 802.11 workgroups and roaming. Chapter 5 delves into the technical details of basic 802.11 roaming. It is interesting to learn about the security model used for 802.11, and Chapter 6 provides the required background needed to understand 802.1X. Chapter 7 focuses on the technical details of improvements relevant to roaming-related security issues in 802.11.

The latter chapters of the book build on the introductory material. They discuss attempts to address some of 802.11's current shortfalls with respect to roaming, including both nonstandard solutions and proposed extensions to the 802.11 standards. In Chapter 8 we discuss some solutions that are outside of the 802.11 standard. Chapter 9 surveys other recent IEEE approaches to fast, secure 802.11 roaming, including 802.11r and the use of the 802.11k standard as a supplemental solution for fast secure roaming. Chapter 10 examines roaming between 802.11 and cellular networks, as well as other technologies. The final chapter explores current research directions. To help one cope with the avalanche of capital letters in the networking peaks, we also include an appendix of networking acronyms. At the end of each chapter, we provide bibliographic material, including references to important Websites. The reader will find our comprehensive index user friendly.

References

[1] William C. Y. Lee, *Mobile Communications Design Fundamentals*, John Wiley & Sons, Inc., 1993.

[2] D. H. Ring and W. R. Young, *Mobile Telephony—Wide Area Coverage*, Bell Labs Internal Memorandum, 1947.

[3] Dan Steinbock, Wireless Horizon, *Strategy and Competition in the Worldwide Mobile Marketplace*, New York: Amacom, 2003.

[4] *Wikipedia—Ericsson [online]*, Internet, 2006.

CHAPTER 2

Cellular Telephony: Wireless Roaming Pioneers

2.1 Introduction

2.2 The Future of Computing

2.3 Basic Concepts

2.4 Early History of Radio Telephony

2.5 The Digital Revolution

2.6 Soft versus Hard Handoffs in Various Cellular Technologies

2.7 The Quest for Convergence

2.8 Summary

References

2.1 Introduction

This chapter provides some details on the background and history of mobile computing. Given the breath of this topic, we have had to be highly selective in our choice of the material that could be included in this book. We have listed a large number of references so that the interested reader can explore these topics further.

This chapter was co-authored with Janice Stanford. Much of this material appeared in a draft of her Armstrong Atlantic State University Master's thesis titled *Current Directions in Mobile Computing and Wireless Roaming Technology*.

Modern communications and computing technologies have become so ubiquitous in our lives that it is sometimes difficult to imagine how society managed without them. The "Y2K Bug" hysteria of the late 1990s illustrated not only how completely these technologies have been integrated with nearly every aspect of our social infrastructure but also how deeply the conveniences and securities that they provide impact the daily lives of the hundreds of millions of people for whom the working of these mundane marvels remains an invisible mystery. Mobile telephones, personal computers, and Internet commerce are only a few of the more obvious examples, but certainly as important are the less obvious effects of these technologies on our ability to respond to emergencies, to diagnose and treat illness, to position resources effectively, and to achieve so many other benefits that it is not practical to list them all in this chapter.

Indeed, computing and communication technologies are considered so essential that the *digital divide*, a widening gap between those societies and classes of individuals that have easy personal access to advanced electronic resources and those that do not, was identified during the 1999 G8 economic summit as "one of the most serious problems hampering development in the Third World" [13]. As of this writing, one initiative aimed at addressing this issue is "One Laptop Per Child"—a program proposed to the United Nations by Nicholas Negroponte of the MIT Media Lab that aims to provide affordable rugged laptops with wireless Internet access and open-source software to children in every nation of the world [1]. Another development that could have a huge impact on global technology access is the launching of Inmarsat's Broadband Global Access Network—a service providing highspeed wireless communications worldwide via three satellites in *geosynchronous* orbit [11].

2.2 The Future of Computing

The initiatives mentioned in Section 2.1 and many others indicate a definite trend in communications and computing technologies. The future promises the possibility of universal affordable access to the abundant resources of the Internet and broadcast media with portable wireless *personal* communications devices. Elementary-school students in rural areas will be able to do their homework assignment (or watch their favorite cartoon) on a long bus ride home. While en route to her next appointment across town or across an ocean, a sales representative will be able to show her presentation to colleagues at the monthly staff meeting. An archaeologist doing field research in the high Andes will be able to refer to a database at the home institution. This flexibility is the goal of *mobile computing*—a convergence of communications and computing technologies that allows anyone to access anything, from anywhere, at any time.

A perusal of any technology news archive or home-electronics store will confirm that considerable progress has been made toward achieving this goal. Since their introduction about two decades ago, hand-held mobile phones have shown a steady decrease in size and cost, and an increase in functionality and availability. As a result, the number of users of mobile phones surpassed that of fixed telephone service in March 2002, and this number continues to grow at a steady rate, with the total worldwide market approaching two billion users [22]. Similarly, wireless networks primarily accessed with laptop computers and *personal digital assistants* (*PDAs*) have seen widespread deployment in private homes and businesses in the years preceding the publication of this text.

However, a considerable amount of work still remains to be done before the goal of ubiquitous mobile computing is achieved. The mobile communications networks, originally designed to carry voice communications rather than data, must be upgraded significantly before they can attain a speed comparable to that of computer networks. Although wireless computer networks have seen tremendous growth, they still have a long way to go to approach the level of market penetration of mobile phone systems. Furthermore, wide area coverage is spotty and many issues remain to be resolved to provide *seamless mobility*, or movement without requiring reconnection, between networks. Various standards organizations with memberships drawn in from the telecommunications, computer networking, and Internet communities around the world are involved in continuous research to resolve these and many other issues.

2.3 Basic Concepts

Before initiating a detailed discussion on topics related to wireless mobile computing technology, a brief overview of some of the basic concepts and terminology will be helpful. In general, we use the term *mobile computing* to refer to the process of using a portable computing device capable of wireless communication to connect with a data network, access information via the Internet, and maintain the same connection while moving to a new network. The device may be any combination of portable computer and wireless communicator: a laptop computer with a wireless modem, an 802.11-enabled PDA, a PDA with *Bluetooth*, or a mobile phone with Internet service. The important feature is that the user is free to move around without restrictions and without getting disconnected from the existing network.

Wireless communication is the transmission of information without a physical connection by means of *electromagnetic* (*EM*) radiation. EM radiation is the basic communication medium of the universe. Physical objects emit, transmit, or reflect energy in the form of EM radiation, which travels in the form of waves. Waves can be measured in several ways, but we consider only wavelength and frequency in this book. *Wavelength* can vary from thousands of kilometers to a fraction of the width of a single atom—a gamma ray [28]. The transmittance of most practical mobile-computing applications occurs via radio waves or microwaves, but wireless communication can also be accomplished via infrared and visible (laser) light waves.

The frequency spectrum is a limited natural resource that is in great demand and is, therefore, regulated carefully. The governments in each locale allocate strictly-limited frequency bands to each of the wireless communication service operators. The use of only *Frequency Division Multiple Access* (*FDMA*) to provide multiple access does not provide sufficient capacity for modern wireless communication systems; therefore, FDMA must be used in combination with one or more other methods. A *cellular* system uses FDMA in combination with *Space Division Multiple Access* (*SDMA*) to divide the coverage area into cells of many smaller geographic units. The cells are typically arranged into *clusters* where no adjacent cells use the same set of frequencies, thereby minimizing interference. These clusters of cells can then be repeated throughout the coverage area in what is called a *frequency reuse pattern*.

2.4 Early History of Radio Telephony

2.4.1 Introduction

As a convergent technology, mobile computing has roots in several prior technologies including broadcast media, telecommunications systems, networked computer systems, and the Internet. As a key progenitor technology, mobile telephony has greatly influenced the development of the basic concepts in mobile computing and continues to do so as the technologies evolve side-by-side. A majority of users of tomorrow's mobile-computing systems will most likely come from the ranks of today's mobile-telephony users. To provide a proper context for the current state of wireless roaming technology, this section provides an overview of its evolution over several generations of technological progress spanning several decades in the mobile telephone industry. Beginning with a brief review of the initial developments in wireless mobile communications, we see a progression of revolutionary concepts and evolutionary steps that have evolved and progressed through several generations and have shaped the field of wireless communications.

2.4.2 Precellular Era

The foundation of modern wireless communications was laid in the latter half of the 19th century, when in 1886 Heinrich Hertz demonstrated the wave nature of electric transmissions traveling through space, as predicted by James Maxwell's famous equations in 1864 [22]. In 1897, Guglielmo Marconi gave the first example of a *mobile* wireless communications system, when he used his new wireless telegraph machine (one of the first radios) to communicate with ships on sea in the English Channel [19].

The first public *Mobile Telephone Service* (*MTS*) capable of placing calls through the *Public Switched Telephone System* (*PSTN*) was introduced in 25 major cities in the United States in 1946 [19]. Each of these systems relied on a single high-powered transmitter that covered a 50-km service area but presented only three *half-duplex* (push-to-talk) channels because technology was not yet widely available to filter out interference in channels less than 120 KHz wide [19]. Throughout the 1950s and 1960s, technological improvements led to the reduction of channel bandwidth to 30-KHz, allowing the *Federal Communications Commission* (*FCC*) to double and then quadruple the number of available channels in the 150-MHz range without allocating any additional spectrum [19].

To place a call from a device within an MTS service area, a mobile subscriber manually scanned for an unused channel, then pressed the "talk" button to send a signal to an operator, who would connect the caller to the PSTN [10]. As the mobile units used low-powered transmitters, the system employed a number of receivers strategically placed around the service area. When any mobile went "off-hook," all the receivers opened. As the subscriber moved around the service area, the signals from any receivers that picked up the transmission were combined, amplified, and then retransmitted through the PSTN [10].

Systems with similar capabilities, but with different technology, began appearing in Europe. *A-Netz*, a manually switched network offering sixteen 50-KHz channels in the 160-MHz range, was introduced in 1958 in cities all over Germany [22]. In this system, mobile stations could place calls but not receive them. With *B-Netz*, a later version of the same system introduced in 1972, calls could be placed to a mobile from the fixed network, but only if the location of the mobile unit was already known [22]. The *Autoradiopuhelin* (car-radio phone), or *ARP*, was launched in Finland in 1971. ARP was also a half-duplex manually switched system with individual transmitters covering 30 km of service areas [26].

Although some claimed that the Swedish Telecommunications Administration tested the world's first fully automatic mobile telephone system in a 1951 Stockholm trial, it was actually the Richmond Radiotelephone Company, one of a number of small carriers competing with AT&T, that first put such system in operation in 1948 [10]. It was not until 1964 that the Bell System finally introduced the *Improved Mobile Telephone Service* (*IMTS*), which offered *full-duplex* transmission, automatic channel selection, and direct dialing [10].

One of the major problems common to all precellular mobile radio systems was the use of a single, large transmitter to cover each market. This limitation severely restricted the number of available channels, and therefore simultaneous users, because the frequencies broadcast by that transmitter could not be reused locally without causing interference [19]. A second common problem was limited mobility. Although the MTS system in the United States provided a rudimentary form of call handoff within the coverage area, it could not support intersystem handoff, which required changing frequencies. This issue seems to have presented more of an inconvenience in Europe, where the coverage areas of the large transmitters were close enough, such that they appeared similar to a unified cellular system but lacked the handoff capability to support it; for example, the German A-Netz system accounted for 80% of the coverage area of the country by 1971. However, when a subscriber went out of range of a transmitter, any call already in progress was dropped and had to be reconnected even if the mobile was already within the range of a new base station [22].

Although Figures 2.1 and 2.2 present a gross oversimplification of the situations that existed, they are useful in illustrating the denser coverage via antennas in Europe relative to the United States. The figures illustrate as to why the issue of handoffs became more pressing in Europe than in the United States. The denser coverage not only meant that handoffs were possible, but also complicated decisions of where to roam, because several choices were often available.

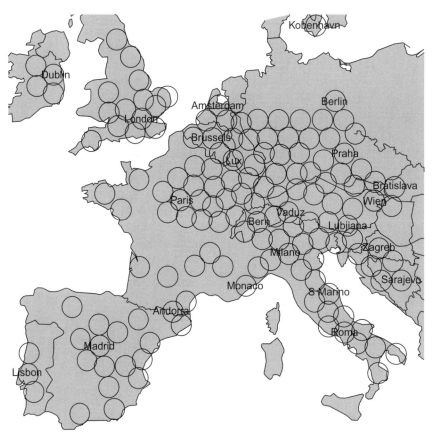

Figure 2.1: An Abstraction and Oversimplification of Antenna Placements in Europe Illustrating the Density of the Coverage Relative to the United States.

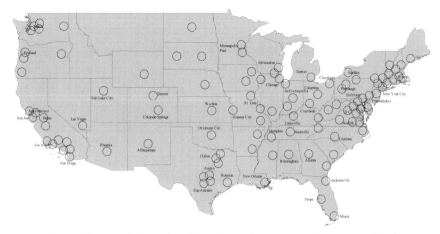

Figure 2.2: An Abstraction and Oversimplification of Antenna Placements in the United States Illustrating the Density of the Coverage Relative to Europe.

The issue of inadequate security measures was also of serious concern. The system lacked privacy, although in the MTS system, the base station and mobile frequencies were offset by 5 KHz to make it more difficult for eavesdroppers to hear both sides of a conversation [10]. Furthermore, the system became a target for toll fraud and spoofing because subscribers had to announce their mobile numbers to operators over clear channels when placing calls [10]. In later versions of the systems with automatic dialing, mobiles transmitted their identification numbers automatically using a form of *Audio Frequency Shift Keying* (*AFSK*). This type of transmission made it more difficult for third parties to spoof; however, malicious users found ways around it [10].

Despite their shortcomings relative to modern standards, there were waiting lists of up to 10 years in every city where mobile telephone services were introduced [10]. In 1976, the Bell Mobile Phone Service in New York City (an IMTS system) still offered only 12 channels and served only 543 customers with a waiting list of more than 3700 people [19]. By 1978, the FCC had allocated only 54 channels nationwide for all carriers combined [10]. It soon became apparent that regulatory agencies could not allocate enough spectrum to satisfy the growing demand. Mobile phone system technology would have to be restructured to achieve higher capacity while continuing to serve very large areas [19].

2.4.3 Advanced Mobile Phone System

The major breakthrough in mobile phone systems occurred with the advent of the concept that had been articulated as early as 1947 in a Bell Labs Technical Memorandum written by

D. H. Ring [10]. The key concept of cellular design was to replace the single highpowered transmitters in conventional mobile radio systems with a network of lower-powered transmitters, with each servicing a smaller geographic cell. Each of the smaller transmitters, or *base stations*, would be assigned a subset of the channels available to the system, with neighboring cells being assigned different subsets such that channel interference would be minimized [19]. A central *Mobile Switching Center* (*MSC*) would coordinate the activity of all the cells, connect them to the PSTN, and provide for call handoff, as subscribers moved among the cells.

In his article published in 1947, Ring focused on the general concept of increasing spectral efficiency by reusing frequencies in cells, rather than on any specific mechanisms for managing the process [10]. However, Ring did note that a method needed to be devised for managing call handoffs from one set of frequencies to another [10]. Teams from Bell Labs worked out the technical details over the next two decades, and AT&T submitted a proposal for a cellular system to the FCC in 1971. The proposal outlined the company's technical and commercial vision for the *Advanced Mobile Phone Service* (*AMPS*) [20].

The new concept proposed by AT&T provided a radical leap forward. Without the need for any major technological changes, it offered significantly higher capacity and room for continuing growth, while staying within the bounds of a limited (although initially larger) spectrum allocation [19]. The advantages of cellular system design were attributed to four key principles:

1. The service area was divided into multiple cells served by lower powered base station transmitters and was controlled using a central switching center.

2. Groups of frequencies could be reused as many times as necessary to provide adequate channel capacity to the service area.

3. The signals of calls in progress were handed off among the cells, as subscribers moved about the service area.

4. The subdivision of cells allowed system growth [9].

The first AMPS systems used large cells and omnidirectional antennas to keep the initial equipment requirements to a minimum [19], but directional antennas could also be used to further reduce interference from neighboring cells. Hexagons were used to represent cells in coverage maps because they provide a good approximation of the roughly circular coverage area of an omnidirectional antenna without overlapping or leaving gaps. After extensive testing, it was found that a seven-cell cluster *frequency reuse pattern* provided optimum performance with the AMPS 30-KHz channels and 120-degree directional antennas [19].

Communications in AMPS were managed by means of a *common air interface* that represented four different channels: a pair of forward and reverse voice channels and a pair of forward and reverse-control channels. The forward channels were used for *transmitting* signals from base stations to mobiles, whereas the reverse channels were used for receiving signals. The forward-control channels also served as beacons to advertise the presence of the base station [19]. The base stations typically had one transmitter/receiver pair for control channels and eight or more pairs for voice channels [19].

At any point of time when a mobile phone was on, it scanned for the strongest control channel signal in the area and monitored that channel until the signal strength was too weak, then the mobile scanned again. If a call came in for the mobile, the MSC sent out a request to all base stations to page the mobile over all their control channels. When the mobile received the page, it identified itself over the appropriate reverse control channel, and the base station relayed that information back to the MSC. The MSC then instructed that base station to move the call to an unused voice-channel pair. The base station signaled the mobile to change frequencies to the selected voice-channel pair, and then sent an alert over the voice channel to ring the mobile [19]. Once the call was in progress, the MSC remotely adjusted the transmitter power of the mobile and changed the channels at the mobile and base stations, as needed, to maintain call quality. This process, referred to as a call *handoff* was managed by AMPS with special control signaling.

A *Supervisor Audio Tone* (*SAT*) was applied to voice channels so that the mobiles could be controlled by the base stations and MSC during calls [19]. Each base station had an additional receiver called a *locator receiver* that was used to monitor signal strength of mobiles in neighboring cells. When a mobile call appeared to be in need of a handoff, the *Radio Signal Strength Indication* (*RSSI*) for that mobile was relayed to the MSC. The MSC compared the RSSI for the mobile from each of the locator receivers that reported it and initiated a handoff, if required [19]. For comparison, recall that the multiple receivers spaced out over the coverage area in the MTS system exhibited a rudimentary intrasystem handoff capability by performing a similar service. Locator receivers are a natural evolution of that capability.

We should point out that one of the key features that distinguish between cellular and 802.11 networks is who controls the handoff decision. As described in the previous paragraph, in cellular networks, it is the network infrastructure that makes the handoff decisions. This is in contrast with 802.11 systems' where the mobile device controls the handoff decision. We delve further into this point in Section 2.5.5.

When a mobile phone entered an area that was not part of its home service area, the phone was registered as a roamer. Periodically, the MSC issued a global command over every control channel in the system, instructing unregistered mobiles to report their identification. After receiving a reply from the new mobile, the MSC contacted the mobile's *Home Location Register* (*HLR*) to request billing information. If the mobile was authorized to roam under its billing plan, the MSC registered it as a valid roamer and automatically routed billing to the home service provider [19].

If a mobile with a call in progress began to move out of the area controlled by a particular MSC, the MSC could perform an intersystem handoff. Management of this type of handoff is more complex than that of a handoff between cells in the same system. For example, long-distance charges may be incurred in the PSTN when the mobile moves out of its home system [19]. This type of roam as well as that described in the preceding paragraph are both instances of global roaming, as defined in Chapter 1.

Due to a series of regulatory disputes (some of which led to the divestiture of AT&T in 1984), it was in 1982 that the FCC finally allocated 666 duplex channels in the 800-MHz band and began granting market licenses for cellular service. According to the new FCC regulations, two cellular providers were required for each market, with half of the channels going to the *local wireline common carrier* (usually a Bell company) and half to any other qualified carrier [20]. AT&T's first pilot of the new cellular system began operating in the Chicago market in October 1983 [20].

2.4.4 Analog Systems in Europe and Japan

Meanwhile, groups in Europe and Japan had also been working on cellular solutions. In 1969, a consortium of northern European countries began working on an open standard that was later called the *Nordic Mobile Telephone* (*NMT*) system [22]. However, *Nippon Telephone and Telegraph* (*NTT*) made it to market before NMT, deploying the first commercial cellular system in Japan in 1979. This system was proprietary, but apparently based on the AMPS technology developed at Bell Laboratories. It offered 600 duplex channels in the 800-MHz band and initially served 88 cells in the Tokyo area ([10], [19]).

Work on the first NMT system, which operated in the 450-MHz range, was completed in 1981, and, suprisingly, the system was first deployed in Saudi Arabia. Within months, NMT was operational in Sweden and Denmark and then in Finland and Norway the following year [18]. Cellular systems appeared all over Europe during the next few years. In 1985 alone, four separate cellular systems emerged. Germany launched the 450-MHz *C-Netz*—the cellular successor to the popular B-Netz system used primarily in Germany and Austria [22]. Italy and France each deployed their own cellular systems, *Radio Telephone Mobile System* (*RTMS*) and RadioCom 2000, respectively [10]. Also in 1985, the United Kingdom introduced a version of the American AMPS that was dubbed *Total Access Communications System* (*TACS*) and operated in the 850–950-MHz range [19].

An important point to note about all these European systems, indeed *all* of the analog systems, is that, although they were founded on the basic cellular system concepts developed for the original AMPS, differences in their implementation and operating frequency range made them completely incompatible with one another. The systems also had several common problems [14]. We list a few of them in the following text:

Spectral inefficiency The *first-generation* (*1G*) cellular systems relied on the same method for partitioning available spectrum into usable channels as the precellular systems introduced before them, namely FDMA. Using FDMA alone, only one call at a time could use a given pair of frequencies, which was indicative of very inefficient use of the limited spectrum resource.

Poor voice quality Interference from cochannels and environmental sources reduced the quality of cellular-voice transmissions, as compared to that of the service in the PSTN. To avoid unacceptable levels of interference, systems had a minimum channel width of 25–30 KHz, which added to the problem of spectral inefficiency.

Security and fraud Encryption was still not practical over analog radio links; therefore, privacy was a serious concern. Illegal cloning of cell phones also became a lucrative business in the black market.

Limited data services First-generation cellular systems were designed primarily for voice communications and, therefore, handled data services (fax, email, and file transfer) inefficiently. In 1993, a *Cellular Digital Packet Data* (*CDPD*) service was developed that used a full 30-KHz AMPS channel on a shared basis; voice traffic was given priority and packet traffic was dynamically assigned to various voice channels, as they became vacant [19]. However, all the previously identified issues impacted the development of fast, reliable data solutions. The lack of robust data capability limited the usefulness of wireless service for business users.

In just a few years, the AMPS system became surprisingly successful in establishing nearly ubiquitous coverage in the United States [10]. During the same period, revolutionary changes were also taking place in the design of the PSTN. With the deployment of *Signaling System No. 7* (*SS7*) beginning in the mid 1980s, the PSTN essentially became two parallel networks—one for voice traffic (bearer traffic) and one for signaling [19].

In 1988, the *Telecommunications Industry Association* (*TIA*) took up the responsibility of developing and maintaining an AMPS family of standards with the release of its *Interim Standard 41* (*IS-41*) based on SS7 [10]. Previously, mobile users arriving in a new service area had to register manually with the new system [19]. IS-41, also known as ANSI-41 after its adoption by the *American National Standards Institute* (*ANSI*), provided seamless roaming with automatic intersystem handoff, call delivery, validation, and authentication [17]. However, these features did not appear in the cellular networks until the early 1990s, nearly concurrent with the arrival of digital systems [19].

In the previous paragraph, we noted that IS-41 provides seamless roaming and "automatic handoff." This point is worth elaborating on, so that later, the reader may compare and contrast handoffs and roaming in the cellular world to those in the 802.11 framework. Although there are many different scenarios which cause roaming in the cellular world, we will focus on the one depicted in Figure 2.3. In the figure, a user of a mobile phone moves from coverage area X to coverage area Y. We assume initially that a call was placed to the mobile phone when it was in cell X. Note that cell X is served by MSC 1. As the user of the mobile phone heads toward the direction where cells X and Y overlap, MSC 1 runs an internal algorithm to decide if the call needs to be handed off. The signal strength plays a prominent role in the handoff decision. MSC 1, the station that is handling the call, will make the decision of whether or not to roam.

MSC 1 sends out a handoff message request to its neighboring stations requesting certain types of information. Once MSC 2, the MSC serving cell Y, receives the request from MSC 1, MSC 2 sends back the parameters that were requested. If the appropriate conditions are met, MSC 1 will initiate a handoff to MSC 2. The call can then be handed off to MSC 2. Of course, there may be many candidate MSCs for MSC 1 to choose from, and according to MSC 1's internal algorithm, it will select the most desirable MSC to which to handoff. Naturally, we have omitted many of the technical details. The interested readers may refer to these details through the references provided at the end of this chapter.

2.5 The Digital Revolution

2.5.1 Introduction

Even as the analog-cellular era was just beginning to take off, Europeans were already planning for a digital future. This trend was not as much due to great foresight as it was to

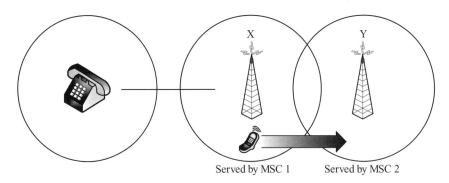

Figure 2.3: A Handoff Scenario in a Cellular Network.

strategic necessity. The AMPS system in the United States was essentially homogeneous: although automatic handoff was not available until the introduction of IS-41, subscribers could nevertheless depend on their mobiles to work almost anywhere in the country where there was cellular coverage. However, the situation was far different in Europe, where a host of incompatible conventional radio systems were spawning an equally incompatible assortment of cellular systems. The closest thing to an analog standard in Europe was the NMT, but that system was only deployed in Denmark, Sweden, Finland, and Norway [10].

In 1982, France and Germany persuaded the *Conference of European Postal and Telecommunications Administrations* (*CEPT*) to develop a truly pan-European mobile phone standard [24]. The goals of the new system would be to make use of a newly allocated 900-MHz spectrum, to allow roaming all over Europe and to offer voice and data services in a fully digital system compatible with the emerging *Integrated Services Digital Network* (*ISDN*) specifications [22]. The *Groupe Spéciale Mobile* (*GSM*) was formed to oversee the project in which 26 European national telephone companies participated [10].

Meanwhile, as coverage of the higher capacity cellular phone systems spread throughout Europe and the United States, carriers began receiving new subscription requests at a rate well beyond the most optimistic predictions [5]. In the larger cities, the demand exceeded the growth capacity of the systems. The spectrum allocated for the first-generation cellular services was fairly small, particularly in the European systems using the 450-MHz band, and the channels themselves were fairly wide (25–30 KHz) due to the limitations of the original equipment. A new version of the NMT system was launched in 1986 to take advantage of the broader spectrum available in the newly allocated 900-MHz band (the TACS system in the United Kingdom had begun using this wider band in 1985). The new system also took advantage of better technology to reduce the channel width by half, effectively doubling the number of channels available in each cell [19].

2.5.2 Global System for Mobile Communications

After the *European Telecommunication Standards Institute* (*ETSI*) was established, it took over the direction of the GSM project in 1989. The first specification for the newly renamed *Global System for Mobile Communications* was released in 1990, with the first systems being deployed the next year. As the United Kingdom had an existing cellular system already operating in the 900-MHz band, an additional GSM plan was specified in 1991 that would operate in the 1.8-GHz range. The resultant *Digital Cellular System* (*DCS1800*) began to operate commercially in 1993 [10].

Similar to their predecessors, GSM systems use FDMA for channel separation. The initial systems divided 49.6 MHz of radio spectrum into 124 full-duplex channels of 200 KHz in

Chapter 2

each direction. The division of the spectrum is illustrated in Figure 2.4. However, the channels were then further divided into eight time slots using a *Time Division Multiple Access* (*TDMA*) scheme. TDMA takes advantage of the fact that digitized voice conversations can be compressed easily. Voice conversations typically include considerable "dead air" time as well, blocks of time when there are long pauses or when one party is speaking and the other is quietly listening. With judicious use of compression algorithms and digital encoding, multiple transmissions can be parceled out into slices of time and carried over the same frequency. In the earliest versions of the GSM systems, this arrangement yielded an effective channel width of 25 KHz, which really provided no better spectral efficiency than the European analog systems that were introduced before them [2]. However, because of the digital encoding and the greater flexibility of TDMA to allocate time slots asymmetrically according to demand, it allowed for the possibility of many new features such as *Short Messaging Service* (*SMS*), group addressing, call waiting, and multiparty services [15], as well as data transmission rates up to 9.6 Kbps [22].

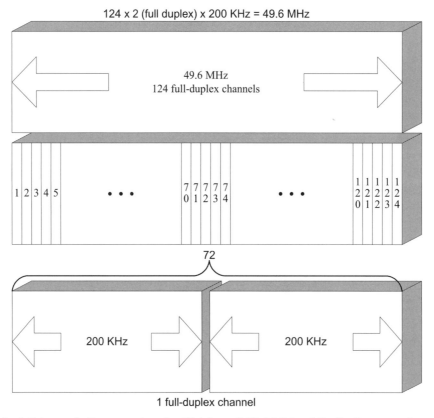

Figure 2.4: A Schematic Representing the Division of 49.6 MHz of Radio Spectrum into 124 Full-Duplex Channels of 200 KHz Each.

The elements of the GSM system architecture are grouped into three basic subsystems: the *Radio Subsystem* (*RSS*), the *Network and Switching Subsystem* (*NSS*), and the *Operation Subsystem* (*OSS*) [22]. The RSS is illustrated in Figure 2.5, the NSS in Figure 2.6, and the OSS in Figure 2.7. We highlight some of the key features of the systems in the following text and explain the various components, connections, and interfaces:

1. The *Mobile Station* (*MS*) is the subscriber telephone unit. In GSM, the MS is composed of two distinct parts: the *Subscriber Identity Module* (*SIM*) and the *Mobile Equipment* (*ME*). The ME is the physical telephone device. The SIM is a detachable smartcard that stores subscriber-specific information such as the mobile identification number, authentication codes, and lists of subscribed features and authorized roaming

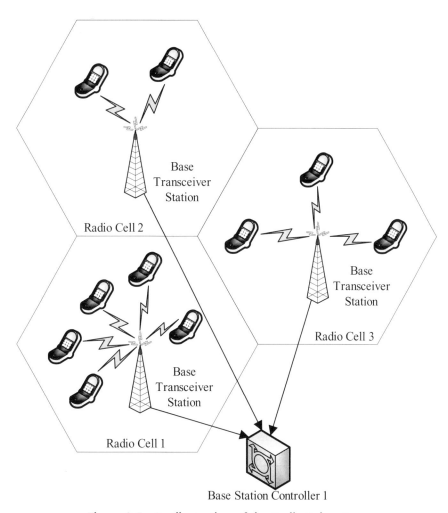

Figure 2.5: An Illustration of the Radio Subsystem.

Chapter 2

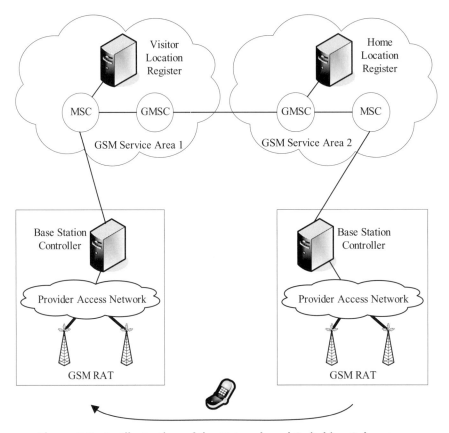

Figure 2.6: An Illustration of the Network and Switching Subsystem.

systems. The SIM may be removed from the ME and installed in a different piece of ME without affecting access to the service; however without an SIM, the ME can make calls only to emergency services [22].

2. The *Base Station Subsystem* (*BSS*) comprised a set of *Base Transceiver Stations* (*BTS*) coordinated by a *Base Station Controller* (*BSC*). The relationship between a BSC and the BTSs that it controls is shown in Figure 2.5. Mobile stations communicate with BTSs that may be either colocated with the BSC or distributed remotely and connected by microwave links or dedicated leased lines. Call handoff between two BTSs connected to the same BSC are handled locally by that BSC, thereby reducing the load on the MSC [19].

3. The NSS functions as the central nervous system of a GSM service area [22]. The primary component of the NSS is the MSC—a high-performance ISDN switch that performs call handoffs between the BSSs and to other MSCs. These two components

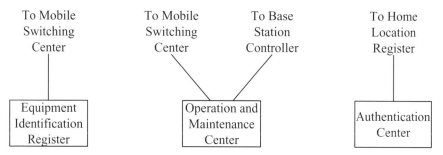

Figure 2.7: An Illustration of the Operation Subsystem.

are shown in Figure 2.6. If the MSC is a *Gateway MSC* (*GMSC*), it also connects the wireless network to the PSTN. The MSC reports subscriber activity to the *Operation and Maintenance Center* (*OMC*) in support of billing operations and performs roaming operations with the aid of the two special databases that we describe in the following text [22].

4. The HLR maintains identification and location information for each subscriber's SIM registered in a local service area. One of the key features in a GSM system is that it supports automatic localization of users. When the MS is in a GSM service area and is not switched off, the system periodically updates its location (service area) with the HLR [22].

5. The *Visitor Location Register* (*VLR*) in the service area where the MS is roaming is the entity responsible for informing the HLR of the MS's current location upon its arrival. When the MS moves to a new service area in which it is authorized to roam, the local service area VLR informs the MS's HLR of this fact, and the HLR forwards all relevant subscriber information to that VLR [22]. This process is automatic whether the new service area is operated by the same provider (local roaming) or by a different provider possibly in a different country (international roaming) [22].

6. Finally, the OSS, which comprised an OMC, an *Authentication Center* (*AuC*), and an *Equipment Identification Register* (*EIR*) is responsible for monitoring network traffic, reporting status of network entities, managing security, and handling subscriber billing [22].

In points 2 and 3 noted earlier, we dealt with two different types of handoffs—intra-BSC and inter-BSC. The intra-BSC roam is more local. In this case, the BSC will assign a new channel to the MS as it moves from one cell to another. In the inter-BSC roam, the MSC handles the roam by communications with the old BTS and BSC and then assigns a new BTS and BSC.

Specific details about how the location of a MS are maintained by GSM and the details of the various roaming possibilities in cellular networks are not within the scope of this book. For more information on these topics, we refer the reader to [22].

2.5.3 North American TDMA (IS-54)

As previously noted, the AMPS system had been quite successful in establishing broad coverage in the United States. Furthermore, by the time GSM was introduced, AMPS or one of its TACS derivatives had gained a strong presence in all the populated continents of the world [10]. However, AMPS/TACS operators were also facing looming shortfalls in system capacity. In particular, the North American network needed a capacity upgrade that would allow operators and subscribers a gradual migration path from the analog AMPS service.

Three solutions eventually emerged to address the need. Motorola developed a *narrowband* version of AMPS, called *N-AMPS* (TIA IS-88), that used FDMA to squeeze three 10-KHz voice channels into the space previously occupied by one [3]. This system was introduced in 1992 [19].

Also, TIA IS-54, or *United States Digital Cellular* (*USDC*) was introduced in 1992. Alternately known as NA-TDMA and D-AMPS because of its close relationship with AMPS, it used TDMA to support three voice calls over an existing 30-KHz AMPS channel. It could transmit on the same channels right alongside an existing AMPS; therefore, the operator could gradually upgrade the AMPS to the higher-capacity digital service without displacing existing subscribers. Subscribers could purchase a dual-mode phone that would work any location where there was AMPS service and also could have access to the extended features of USDC where it was available [10]. This strategy worked well, and USDC won out over N-AMPS in most of the markets in the United States.

IS-54 and a later version specified as IS-136 have essentially the same features as GSM networks and provide a comparable set of services. Intersystem mobility management (roaming) is specified by the GSM *Mobile Application Part* (*MAP*) in GSM [16] and by ANSI-41 (IS-41) in all TIA-specified networks [6]. The roaming process in the two protocols is remarkably similar as both are based on SS7 [6], but interoperation between the two types of networks is not possible because they use the signaling system in different ways, essentially speaking different languages [8]. The details of the intersystem mobility management are not within the scope of this book; however, in the next paragraph we give a sketch of the key concepts.

An MS constantly sends back measurement reports to its BSC. The report includes, among other items, the quality of the signal. If the signal is of good quality, no action is taken.

However, when the quality degrades, the BSC may determine that a roam is needed to improve performance. The BSC communicates with the corresponding MSC to obtain a set of candidates where the MS might be handed off to. A series of messages are exchanged among MSC, VLR, and BSC, so that eventually proper information can be returned to the MS in order to effect the roam. Various timers are set during the protocol to ensure that everything is proceeding as scheduled. Once the new channel is set up, the roam may be completed and the old channel released.

We now return to the discussion of the third solution mentioned earlier, TIA IS-95. This network was an entirely new type of network based on the *Code Division Multiple Access* (*CDMA*) scheme. It was approved in 1994 to operate in a new 1900-MHz band that had been reserved for *Personal Communication Services* (*PCS*) [4]. A version of this service also operates in the 850-MHz cellular band in the United States and in the 1800-MHz PCS band in Hong Kong and South Korea.

In 1993, the TIA adopted a system proposed by Qualcomm for its IS-95 specification. The new system, called *cdmaOne*, was based on a CDMA scheme [14]. CDMA uses a channel-assignment technique called *Direct Sequence Spread Spectrum* (*DSSS*) that assigns a unique code to each user and then spreads the transmissions of all users across the same wide-frequency band [3]. The receiver must also know the code assigned to the transmission of interest and then be able to perform the necessary calculations to distinguish the segments of that transmission from those belonging to other transmissions and from the background noise [22]. For this reason, CDMA base station transceivers are significantly more complex than those used in other systems. The channel bandwidth in an IS-95 system is 1.25 MHz; therefore, it is classified as a borderline *wideband* technology. However, it is often referred to as narrowband CDMA to distinguish it from *wideband CDMA* (*W-CDMA*)—a 3G technology in the IMT-2000 family [12].

2.5.4 Japanese Systems

In contrast to the phenomenal success of cellular systems in Europe and North America, the growth of Japan's mobile industry was stifled by the bureaucratic policies and overpriced system of the NTT monopoly [10]. In 1988, the Japanese Ministry of Posts and Telecommunications finally opened the market, thereby leading to competition [10]. Shortly thereafter, the JTACS system, equivalent to TACS of the United Kingdom, began operation [18]. However, the industry did not really begin to flourish in Japan until complete deregulation in 1994 allowed operators to offer price breaks and customers to have their own phones [10]. Around this time, a succession of new digital systems entered the Japanese market: *Japanese Digital Cellular* (*JDC*) in 1992 and a competing system called *Pacific Digital Cellular* (*PDC*) introduced by NTT in 1994 [both were based on the USDC (IS-54) standard], and *Personal Handyphone System* (*PHS*) in 1993.

PHS is a *wireless local loop* system derived from cordless telephone technologies. It is one of a class of systems collectively known as "low-tier" PCS that operates at short range (typically less then 300 m [22]), uses TDMA with *frequency hopping*, and provides basic data services. PHS can perform call handoff between *picocells* within the system and also to the PSTN or other digital networks [15]. Similar systems exist in Europe and in the United States and, due to their limited range, they are ideally suited for use in indoor applications like wireless PBX systems. In densely populated urban business centers, such systems can support up to 10,000 users per km^2 [22]. Europe's *Digital Enhanced Cordless Telephone* (*DECT*) is fully interoperable with GSM and was selected as a candidate system for the 3G IMT-2000 family [22].

One of the major improvements that digital systems offer over their analog counterparts is the ability to provide encryption. In analog systems, the voice signal is directly modulated onto the carrier wave "in the clear." Digital systems first convert the voice signal into a stream of bits, which is coded using one of the various noise-canceling algorithms, and then the bit stream is modulated onto the carrier wave [14]. An encryption algorithm may be included in the encoding process to secure the privacy of the voice conversation or of the data being transmitted.

The European GSM specification incorporated a moderate level of encryption into its mobile and base stations, which was sufficient at that time, although portable devices that are currently available can crack it. Unfortunately, the use of encryption in mobile phones is a political issue in many countries. For example, China requires operators to disable all encryption support at the base station level. In the United States, no encryption was included on the voice stream in the cdmaOne (TIA IS-95) standard at the insistence of the National Security Agency [14]. As wireless networks evolve toward carrying ever larger amounts of sensitive data, cryptographic technology and its regulation by the government is becoming an increasingly critical issue.

2.5.5 Focus on CDMA and Soft Handoff

Although CDMA is relatively new to commercial communication systems, it has a long history in military applications because of its inherent security features [22]. Between 1920 and 1940, a number of researchers filed patents that established the key concepts behind spread spectrum [14]. The first true spread spectrum patent, called "Secret Communication System," was filed in 1941 by actress Hedy Lamarr and a composer friend, George Antheil. It described a scheme for military communications that relied on frequency hopping and spread spectrum transmission to confuse radar jamming devices [14].

The first systems based on DSSS technology were built in the early 1950s [14]. The *Global Positioning System* (*GPS*) is one example of a military system that uses this technology [14]. After articles on spread spectrum research began appearing in publically available journals in the early 1980s, the FCC approved it for use in the commercial sector in 1985 and it was eventually declassified by the United States Department of Defense in 1991 [14].

All the systems discussed previously have used some combination of space-division multiplexing (where transmissions are separated by geographic distance), frequency-division multiplexing (where transmissions are separated by spectral frequency), and time-division multiplexing (where transmissions are separated by time). Code-division multiplexing separates transmissions in "code space," by assigning each transmission a code that is *orthogonal* to every other transmission [22]. All the mobile stations in each cell may transmit at any time using the same set of frequencies. This feature of CDMA substantially improves spectral efficiency by completely eliminating the need for frequency reuse planning; each frequency in the operating range is available to the entire market [19]. The data stream to be transmitted from each mobile is multiplied by its own unique code called a *chipping sequence* or *pseudonoise sequence* that appears as random noise to any receiver not privy to it [22]. The resultant combined signals are effectively spread across the entire bandwidth of the channel and then transmitted simultaneously. The 802.11 communications that are the theme of this book are an instance of this kind of spread spectrum communications. Note that the ability for multiple 802.11 transmitters to use the same set of code-division multiplexed frequencies simultaneously does not mean that two 802.11 transmitters in close physical proximity will be able to successfully simultaneously transmit on the same 802.11 *channel*. We discuss the concept of 802.11 channels and the means to contend for access to those channels in Chapter 3.

A helpful analogy to visualize how this works is a noisy party where conversations in various languages (codes) are taking place at the same time. As long as everyone is speaking at nearly the same volume, a listener can pick out the desired familiar language from the general commotion [22]. This coding is in contrast with TDMA, where at the same party each pair of speakers are only allowed to talk for short bursts in a round-robin fashion. This analogy indicates a couple of important requirements for CDMA systems. First, the base station transceivers must have knowledge of all the codes to be used. Furthermore, the transmission power levels must be precisely controlled so that transmissions from mobiles on the edges of cells can be heard over the "noise" generated by those close at hand [22]. Unlike the fixed capacity limits in FDMA and TDMA systems, CDMA cells offer *soft capacity*. As the number of users increases, the general noise level increases, thereby leading to gradual degradation of system performance [19]. As a result, CDMA cells are said to "breathe," because the effective coverage area expands and contracts with changes in the load [22].

2.6 Soft Versus Hard Handoffs in Various Cellular Technologies

We have already mentioned numerous issues related to handoffs for various technologies. In this section, we tie much of that together. The reader should keep foremost in mind that all handoffs in 802.11 networks are controlled by the MS.

In CDMA, because all base stations are operating under the same sets of frequencies, transmissions from mobile stations may be received by two or more base stations in adjacent cells. After each transmission burst, the network compares the quality of the received signals and selects the current best base station to transmit signals back to the mobile station. This method of transferring a communications link is called a *soft handoff* because the MS communicates simultaneously with all the nearby base stations [15].

In FDMA and TDMA systems, transfer of a communications link is considered to be a *hard handoff* because the MS is connected to only one base station at a time. Such handoffs may be *mobile controlled* (*MCHO*), *mobile assisted* (*MAHO*), or *network controlled* (*NCHO*) [15]. However, whether the MS or the network initiates the handoff, there is a period of time while the link is being transferred by the MSC (usually measured in milliseconds) when the conversation is physically interrupted, although normally this brief hiatus is not noticeable to the users. In CDMA, communication links with each of the base stations in range are automatic and continuous. The MSC continuously monitors the signal quality at each base station and adjusts the forward transmission path accordingly after *every* burst of the mobile transmission. As the MS moves out of the range of a particular base station, the signal simply fades into the background noise until that base station can no longer distinguish its signature.

Note that for soft handoffs, there is more than one base station receiving the MS signal; so, the system needs to decide who "owns" that MS at any given point in time. Correspondingly, the voice traffic coming from the phone should only be forwarded by the elected base station, and conversely, the network should only direct traffic destined to the phone through that elected base station. The readers interested in these technical details for cellular networks can refer to the reference section provided at the end of this chapter.

Although soft handoff offers several advantages in terms of fewer dropped calls, less interference, and reductions in required transmitter power, it also represents the biggest drain on system capacity in a CDMA system, which is attributed to the fact that within a handoff zone, multiple base stations are simultaneously handling the same call, resulting in an increased noise load for each cell that is involved. Therefore, CDMA system operators must optimize the size of handoff zones by various techniques including adjusting the transmission strength of base-station pilot (beacon) signals, prioritizing the list of neighboring cells that the MSC considers for handoff, and by minimizing environmental or system factors that contribute to the overall noise level in the cells [23].

Although second-generation digital systems showed significant improvements in capacity, security, and data handling capability over the analog systems that they replaced, the 2G systems are still primarily designed for voice communications and use a *circuit-switched* connection-oriented architecture. In a circuit-switched network, a pair of channels or time slots (a circuit) must be dedicated to two stations for the duration of their communication [27]. This arrangement can be very inefficient for most computer data transmission, which tend to come in bursts and are often asymmetrical [22]. Furthermore, with typical data rates of 14.4 Kbps or less, the 2G systems offer excruciatingly poor performance for users accustomed to computer networks where data rates of 100 Mbps are commonly available. Obviously, a new paradigm would be required before mobile communications networks could support data on the scale required for mobile computing.

2.7 The Quest for Convergence

2.7.1 Introduction

Although the second-generation networks were fully capable of handling data, they were optimized for voice communications—a market that still accounts for the largest share of revenues from mobile communication systems. When the first specifications for GSM were completed in 1991, they included a short-messaging service (up to 160 characters), group 3 fax, and data services at 9.6 Kbps; this level of service was sufficient at that time. The use of the Internet by the general public was only just beginning to bloom with the introduction of *America Online* (*AOL*) [25] for MSDOS in the same year, and the World Wide Web [31] did not make a public debut until 1993 ([25, 31]). However, such a level of data capability is completely inadequate to support image-laden Web browsing and massive file downloads that are now common.

2.7.2 High-Speed Circuit-Switched Data

The first and most straightforward solution for increasing data transmission rates was to give an MS access to multiple time slots. Recall that the basic GSM service used TDMA to divide a 200-KHz carrier into eight time slots. An MS assigned to a single time slot received a data rate of up to 9.6 Kbps (14.4 Kbps in some systems) [22]. With the introduction of *High-Speed*

Circuit-Switched Data (*HSCSD*), higher data rates could be achieved by bundling several *Traffic Channels* (*TCHs*) together [22]. Theoretically, an MS could obtain an *Air Interface User Rate* (*AIUR*) of up to 115.2 Kbps by using all eight slots in a *TDMA frame* at the maximum rate of 14.4 Kbps. However, using all the time slots simultaneously would require that the MS be able to send and receive transmissions at the same time—a capability that was not required by standard GSM [22]. Note that if the MS could not send and receive transmissions simultaneously, then during maximum outward transmission, the MS would be unable to receive any incoming messages.

In practice, ETSI specified an available AIUR of 57.6 Kbps using four slots for uplink and four for downlink. TCHs could also be allocated asymmetrically, using more slots for downlink than for uplink, as would be most beneficial to the typical profile for Web browsing and file downloads [22].

The primary disadvantage of HSCSD is that it maintained the same connection-oriented circuit-switched mechanism used by GSM. Circuit-switching requires a dedicated two-way circuit (using two TCHs) for a pair of stations to communicate, and this circuit remains dedicated until the connection is terminated whether or not any transmission actually occurs. This arrangement is an inefficient use of frequency bandwidth, particularly for data traffic, which tends to come in bursts with long pauses in between [7]. Another problem with the circuit-switched architecture was related to the network architecture. Although the introduction of a more efficient modulation scheme allowed HSCSD to triple the data rates (up to 43.2 Kbps/TCH), the ISDN interface between the MSC and the BSC limited each call to one 64-Kbps channel [7].

2.7.3 General Packet Radio Service

The next step toward achieving more flexible, efficient data transmission required a complete paradigm shift. The *General Packet Radio Service* (*GPRS*) overcame the problems imposed by circuit switching on the air interface by adopting the packet-oriented methodology used in the Internet [22]. With GPRS, time slots are shared among active users and allocated on demand as needed to transmit a *Packet Data Unit* (*PDU*) [22]. Depending on the "coding scheme" used, with no error correction, a data rate of up to 171 Kbps is possible, but the practical maximum is about 100 Kbps with realistic speeds of 40–53 Kbps [7]. Although the GPRS does not deliver the high-bandwidth services already envisioned for third-generation networks, the GPRS represents a significant improvement in terms of both user data rate and bandwidth efficiency, and it is, therefore, classified as a 2.5G.

The greatest advantage of GPRS as compared to HSCSD is that, when a user is not sending data, the time slots may be employed by other users. This flexibility provides a more efficient, and therefore cheaper service for typical Internet applications. Operators often pass these savings on to users by charging based on volume rather than on connection time [22]. The main benefits for users of GPRS, besides cheaper data service, is its "always-on" characteristic—no connection setup is required prior to each data transfer [22]. When there is no connection setup, the MS must request access to a time slot, and the network must allocate the slot before a data transfer can take place, but this process occurs so quickly that the delay is not noticeable to the user. Thus, the service appears to be "always-on." Users may also specify QoS parameters in which desired levels may be indicated for reliability, delay, user data throughput, and service precedence.

GPRS introduces a set of new coding schemes offering varying levels of error correction to help the system achieve higher-data rates and reliability. CS-1 offers the highest level of error correction and the lowest data rates, whereas CS-4 offers the highest possible data rate but has no error correction at all. CS-2 is the most commonly used coding scheme, offering reasonably robust error correction, although still allowing an improved data rate (over basic GSM) of up to 13.4 Kbps per time slot. Although CS-3 and CS-4 offer nominally higher data rates, the increased error rates that they allow require more frequent retransmission, which results in roughly the same throughput as CS-2 under typical conditions [7]. Furthermore, actual throughput also depends on the current load of the cell, as GPRS typically uses only idle time slots [22].

The capabilities of the MS further influence data rate calculations. There are 12 classes of devices with varying levels of support for using time slots to send and receive transmissions at the same time. The very best performance, typically seen only in high-end PC cards, comes from a class 12 device providing four downlink slots and four uplink slots, although only five slots may be used simultaneously [30]. A more typical GPRS-capable MS is a class 10 device supporting four slots for downlink and two slots for uplink, with simultaneous use of five slots. When used with CS-2 coding, this configuration results in a receiving rate of 53.6 Kbps and a sending rate of 26.8 Kbps [22]. Table 2.1 provides a comparison of the multislot capabilities of various classes of GPRS-capable devices.

The GPRS service operates right alongside an existing GSM service, providing packet-data services, whereas GSM continues to support voice service. Therefore, GPRS devices must be able to connect to both systems. Three classes of devices have been defined for this purpose. We describe them in the following text:

- Class A devices can be connected to GPRS service and GSM service at the same time and both can be used simultaneously. For example, on such a device, a user could engage in a conversation while viewing a Web-based presentation. Unfortunately, no such devices are known to be available at this time [30].

Table 2.1: Multislot Classes of GPRS Devices.

Device Class	Downlink Slots	Uplink Slots	Simultaneous Slots in Use
1	1	1	2
2	2	1	3
3	2	2	3
4	3	1	4
5	2	2	4
6	3	2	4
7	3	3	4
8	4	1	5
9	3	2	5
10	4	2	5
11	4	3	5
12	4	4	5

- Class B devices can be connected to both GPRS service and GSM service at the same time, but may use only one or the other at any given time. If a user needs to place or receive a call using GSM service, any currently active GPRS data session is suspended and later resumed automatically when the call is completed. Most GPRS devices have this capability [30].

- Class C devices can be connected to either GPRS service or GSM service, but only to one at any given time. The device must be manually switched between the services [30].

Although GPRS shares the same radio system with GSM, several new elements must be added to the switching network to support packet routing. The *Packet Control Unit* (*PCU*) is responsible for controlling access to the air interface and for scheduling and assembling packets. The PCU is usually physically integrated into a GSM BSC [7]. The *Charging Gateway Function* (*CGF*) maintains a record of the packet data activity of each MS and supplies this information to the billing system for volume-based billing [7].

Figure 2.8 provides a schematic of a GPRS system. Next, we elaborate on the components and relationships among them in such a system. The central element of a GPRS system is the *GPRS Support Node* (*GSN*)—a packet router that provides services in GPRS similar to those of the MSC in the circuit-switching domain including mobility management, security, and access control [7]. There is one *Serving GPRS Support Node* (*SGSN*) in each service area; usually, it is colocated with the MSC [21]. A *Gateway GPRS Support Node* (*GGSN*) connects the GPRS network to an external data network such as the Internet or X.25, and the GGSN

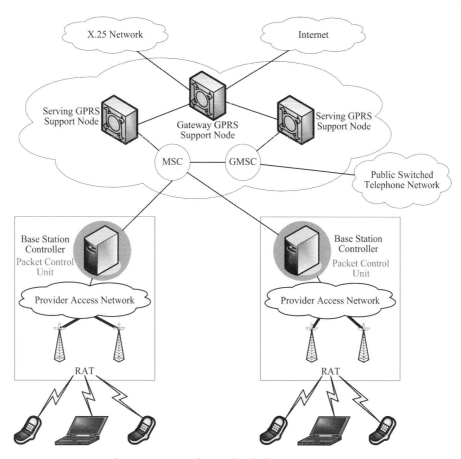

Figure 2.8: A Schematic of the GPRS System.

provides dynamic assignment of IP addresses to mobile stations through *Dynamic Host Configuration Protocol* (*DHCP*) [21]. SGSNs communicate with each other and with the GGSN over an IP-based interface using the *GPRS Tunneling Protocol* (*GTP*), and they interface with the GSM network using the SS7-based GSM-MAP protocol. An optional configuration allows the GGSN to query the HLR directly to obtain the location of an MS. This feature supports mobile-terminated data sessions, but frequently the feature is not implemented because it requires that the MS has a fixed IP address. Having a fixed IP address is often not possible given the addressing limitations of IPv4 [7].

2.7.4 Roaming for Data Applications in GPRS

This section focuses on roaming for data applications in GPRS, which is a process distinct from voice roaming. The material deals with a type of roaming different from what we have covered in earlier sections of this chapter. The service area supported by each SGSN is subdivided into *Routing Areas* (*RAs*) that are analogous to the service areas in GSM. When the MS moves from one RA into another, the MS performs a *routing area update* similar to a location update in GSM. However, unlike location updates, the RA updates can be performed during a data session, also called a *Packet Data Protocol context* or *PDP context*. In GSM, the MS cannot transmit control signals until a call has ended, but in GPRS control signals can be sent between data packets [7]. The interface between SGSNs primarily exists to facilitate forwarding of packets when a RA update takes place during a PDP context [7], but this forwarding occurs only briefly when the new routing path for the PDP context is established.

To provide the "always-on" capability, the connection to a GPRS network occurs in two stages:

- *GPRS attach*
- initiation of a *PDP context*

GPRS attach may occur when the MS is powered on, or it may occur when a certain type of application is opened (for example, a Web browser). The GPRS attach procedure moves the MS from the *idle* state to a *ready* state, and it signals the network that the MS is available to receive packet data. If a PDP context (an actual data session) is not initiated within a specified period of time, the MS may time-out and enter a *standby* state. The standby state conserves MS power and network resources by performing updates less frequently [7].

To perform an attach, the MS requests a channel and sends an attach request. The SGSN queries the MS for its authentication and ciphering code, which are then verified with the AuC. After the MS is authenticated, the SGSN updates the RA information for the MS in the HLR and finally sends an attach acceptance. During this process, the MS must request a channel each time it needs to transmit any sort of message, whereas the SGSN sends its messages directly. Although GPRS networks can be configured to provide assigned time slots, the method that we described for requesting a channel is more typical because it results in a more efficient use of channel bandwidth [7].

When initiating a PDP context, the MS sends a request to the SGSN containing several identifiers, its QoS requests, and the contents of the PDU—the PDP equivalent of an IP *datagram*. After accepting the PDP context request, the SGSN uses the MS identifiers and the addressing information within the PDU to encapsulate the PDU and *tunnel* it to the appropriate GGSN via GTP. The GGSN decapsulates the PDU and forwards it according to the protocol requirements of the external network (IP or X.25) [7]. When a PDU for an MS is received from the external network, a similar process occurs in reverse. After the PDP context is initiated, each PDU is individually passed to or from the MS. If the MS moves to a routing area served by a different SGSN during an active PDP context, it sends an RA update to the new SGSN. The new SGSN uses the *Routing Area Identity* (*RAI*) included in the update request to locate the old SGSN and sends it a *Context Request* message. The old SGSN responds using all the information required to set up the PDP context; the old SGSN stores any PDUs intended for that MS until the RA update is complete, at which time the old SGSN forwards all the PDUs to the new SGSN.

2.7.5 Enhanced Data Rates for Global Evolution

GPRS represents a revolutionary step forward in terms of data-handling capability, but is not fast enough to deliver the full range of applications that were hyped during its development. To support such applications as multimedia messaging and streaming video downloads, further enhancements were required. Specifications for the third-generation IMT2000 systems were completed in 1999, with Phase I systems beginning deployment in 2001 [22]. However, these new networks require allocation of additional spectrum and investment of billions of dollars in radio infrastructure [7].

An interim solution proposed to provide a migration path from GSM/GPRS to true 3G systems is *Enhanced Data Rates for Global Evolution* (*EDGE*). Originally proposed solely for GSM migration, EDGE was extended to provide a parallel but separate migration path for the IS-136 TDMA networks popular in North America. However, some of the largest of those network operators have already begun migrating to the GSM/GPRS platform instead [7].

The basic concept of EDGE is to use a tighter modulation scheme to squeeze higher-data rates out of the same carrier that is used simultaneously for both GSM and GPRS. The older systems use a modulation scheme called *Gaussian Minimum Shift Keying* (*GMSK*)—a type of frequency modulation in which incoming bits are converted into -1 or $+1$ symbols and passed through a filter that imparts a Gaussian (bell-curve) distribution to minimize abrupt phase changes. The frequency of the carrier wave is then shifted according to the value of each symbol. At a symbol rate of 270.833 Ksymbols/second, this results in a raw data rate of 270.833 bps over each 200-KHz carrier. EDGE introduces *8-Phase Shift Keying* (*8-PSK*), where the data bits are handled as eight different symbols of three bits each and a 45-degree segment of carrier phase is associated with each symbol value. With 8-PSK, the carrier-symbol rate remains the same at 270.833 Ksymbols/second, but a symbol is three bits instead of one, effectively tripling the data rate [7].

EDGE supports data rates of up to 384 Kbps with a theoretical maximum of 473.6 Kbps, but the actual speeds are typically lower, which is for the same reasons given in the earlier discussion of GPRS. Such data rates represent a significant improvement over GPRS, yet EDGE still does not measure up to the proposed 3G standard of 2 Mbps. Therefore, EDGE is often referred to as a 2.75G technology [7]. EDGE does offer an attractive alternative to 3G

Chapter 2

systems for existing service operators because it can be deployed simultaneously with an existing TDMA system and requires no new spectrum allocation. The increased bandwidth efficiency is achieved at the cost of requiring new MS devices and BTSs that support 8-PSK modulation, but this cost pales in comparison to the upgrades required to deploy a 3G system [7].

One small problem with EDGE is that 8-PSK states are quite close with each other and a fairly minor increase in noise level results in difficulty in distinguishing phase changes, which can lead to unacceptable error rates. To overcome the above-mentioned situation, EDGE employs nine *Modulation and Coding Schemes* (*MCS*) and can dynamically select the optimum modulation and level of error correction required for the current radio environment. As shown in Table 2.2, MCS-1 through MCS-4 use GMSK, providing nearly equivalent service as the GPRS coding schemes CS-1 through CS-4, and MCS-5 through MCS-9 use the 8-PSK modulation. The slight differences in performance result from additional error-correction bits in the EDGE header which, ensure that it can always be decoded. If a PDU sent with a higher-coding scheme is not received correctly, EDGE will retransmit it using a coding scheme with better error correction. The original PDU is also retransmitted, thereby increasing the likelihood of success. This mechanism, called *incremental redundancy*, assures the fastest possible receipt of correct data [21].

In an EDGE-enabled network, the same channels are shared by both EDGE and GPRS users [7]. To provide this capability, the software in the base station subsystems must be upgraded to

Table 2.2: Coding Schemes for GPRS and EDGE.

Service	Coding Scheme	Modulation Technique	Throughput per Time Slot (Kbps)
EDGE	MCS-9	8-PSK	59.2
	MCS-8	8-PSK	54.4
	MCS-7	8-PSK	44.8
	MCS-6	8-PSK	29.6
	MCS-5	8-PSK	22.4
	MCS-4	GMSK	17.6
	MCS-3	GMSK	14.8
	MCS-2	GMSK	11.2
	MCS-1	GMSK	8.8
GPRS	CS-4	GMSK	20.0
	CS-3	GMSK	14.4
	CS-2	GMSK	12.0
	CS-1	GMSK	8.0

support the new protocol, and new transceivers that are capable of handling both GMSK and 8-PSK modulation must be installed, but no changes are required in the core network [29]. Subscribers wanting to access the higher data-rate EDGE service must obtain an EDGE-capable MS, but subscribers with older devices can continue to use the network at the GSM/GPRS level of service. EDGE and GPRS transmissions are multiplexed over the same channels, with some time slots used to carry GPRS packets using GMSK modulation and some time slots used for EDGE with either GMSK or 8-PSK modulation. However, as the devices of both services must be able to decode the signaling to determine the time slots in which they can transmit, the signaling time slots are coded in GMSK [7].

As existing GSM/GPRS systems can be incrementally upgraded to provide EDGE service with only relatively simple improvements, many system operators may find EDGE deployment as a practicable approach for gradual migration into a 3G system. It remains to be seen what the "killer app" will be that requires migration to true 3G systems [7]. Potential technologies include IP-TV, mobile video conferencing, and the so-called quadruple play—a service that includes high-speed data, telephony, television, and wireless. EDGE is actively supported by TDMA-based operators in North America, but many operators in Europe and elsewhere view EDGE merely as a stepping stone to 3G or have skipped it altogether [29]. As of September 2005, 180 operators in 95 countries were either offering EDGE service or working on active deployment. At the same time, there were 80 commercial *Universal Mobile Telecommunications Systems* (*UMTSs*) networks (a 3G technology) in service in 35 countries with an additional 77 at various stages of development [21].

At present, there are true 3G systems that are very well defined. As Table 2.3 shows, these systems offer digital multimedia and basically came into existence at the beginning of this decade. The description of the roaming details of 3G systems is not within the scope of this chapter. However, in Chapter 10 we discuss about roaming between 802.11 and 3G technologies. The reader who is particularly interested in the concept of roaming in 3G systems can refer to the reference section provided in Chapter 10. Perhaps the section can be referred to after this book has been read, when the reader should have a much better grasp of the roaming concepts.

Table 2.3: Cellular Generations.

Name	Feature	When Deployed
1G	Analog	Late 1970s and early 1980s
2G	Digital	1990s
3G	Digital multimedia	Early and mid-2000s
4G	Movies and television	Expected 2007–

2.8 Summary

We have provided an inventory of some of the cellular telephony systems where the challenges of roaming in wireless networks were first encountered. These challenges have been addressed, to varying degrees, over the span of more than 30 years. Many fundamentally different technologies have come into play in this long process. The differences among FDMA, TDMA, and CDMA are pronounced. Hard handoffs are radically different from soft handoffs. We have seen that in some cases combinations of these technologies have been used successfully in a single system. It is out of this complex brew of technologies that 802.11 was born. In Chapter 3, we will explain from which of the fundamental technologies exposed in this chapter the 802.11 standard most closely draws its heritage. While 802.11 itself has evolved and continues to change over time, the differences between these versions of 802.11 are very minor when compared to the divergent experiments in cellular telephony that we have discussed in this chapter. This greater consistency in the world of 802.11 will allow us to dive deeply into the specific mechanics of how 802.11 confronts the many challenges of roaming securely.

References

[1] Christa Case, *A Low-Cost Laptop for Every Child [online]*, The Christian Science Monitor, 2005.

[2] Cellular Networking Perspectives, *TDMA Digital Cellular and PCS [online]*, Internet, 2006.

[3] Cellular Networking Perspectives, *The AMPS Family of Wireless Standards [online]*, Internet, 2006.

[4] Cellular Networking Perspectives, *CDMA Digital Cellular and PCS [online]*, Internet, 2006.

[5] Cellular Networking Perspectives, *Cellular Technologies [online]*, Internet, 2006.

[6] Gerry Christensen, *Mobile Networking [online]*, Internet, 2005.

[7] Daniel Collins, *3G Wireless Networks*, McGraw-Hill, 2003.

[8] David Crowe, *Global Roaming [online]*, Wireless Review, 2006.

[9] Stuart Crump, Jr. *Cellular Radio Ushers in Revolutionary Changes for Servicing Business User Needs*, Communications News, 1984.

[10] Tom Farley, *Mobile Telephone History [online]*, Internet, 2006.

[11] Inmarsat Global Ltd., *Inmarsat Announces Launch of BGAN Service [online]*, Internet, 2006.

[12] International Telecommunication Union (ITU), *About the ITU: History [online]*, Internet, November 2006.

[13] Kibum Kim, *Challenges in HCI: The Digital Divide*, Crossroads **12**(2), pp. 3–7, 2005.

[14] Alex Lightman and William Rojas, *Brave New Unwired World: The Digital Big Bank and The Infinite Internet*, John Wiley & Sons, Inc., 2002.

[15] Yi-Bing Lin and Imrick Chlamtac, *Wireless and Mobile Network Architectures*, John Wiley & Sons, Inc., 2001.

[16] *Mobile Application Port (MAP): Inter-system Handoff.* Third Generation Partnership Project 2, 3GPP2 X.S0004-200-E, Version 1.0, May 2006.

[17] MobileInfo, *Mobileinfo glossary [online]*, Internet, 2007.

[18] Petri Possi, *The history of UMTS and development [online]*, Internet, 2006.

[19] Theodore Rappaport, *Wireless Communications: Principles & Practice*, Prentice Hall, 1996.

[20] David Roessner, Robert Carr, Irwin Feller, Michael McGeary, and Nils Newman, *The Role of NSF's Support of Engineering in Enabling Technological Innovation—Phase II*, technical report, The Science and Technology Policy Program, SRI International, Washington, DC, May 1998.

[21] Peter Rysavy, *Data Capabilities: GPRS to HASP and Beyond*, technical report, Rysavy Research, September 2005.

[22] Jochen Schiller, *Mobile Communications*, 2nd edition, Addison-Wesley, 2003.

[23] Andrew Singer, *Improving System Performance—Cellular Communications Systems*, Wireless Review **15**(3), 1998.

[24] Jonathan Tarlin, *Beyond Technology: Market Forces Drive GSM—Global Mobile Communications Standard*, Telephony **219**(27), 1990.

[25] Wikipedia, *America Online [online]*, Internet, 2007.

[26] Wikipedia, *Autoradiopuhelin [online]*, Internet, 2007.

[27] Wikipedia, *Circuit Switching [online]*, Internet, 2007.

[28] Wikipedia, *Electromagnetic Spectrum [online]*, Internet, 2007.

[29] Wikipedia, *Enhanced Data Rates for GSM Evolution [online]*, Internet, 2007.

[30] Wikipedia, *General Packet Radio Service [online]*, Internet, 2007.

[31] Wikipedia, *World Wide Web [online]*, Internet, 2007.

CHAPTER 3

Roaming in 802.11 WLANs: General Principles

3.1 Introduction

3.2 Primer on the 802.11 Standard

3.3 Introduction to 802.11 Roaming

3.4 Local Roaming

3.5 Global Roaming

3.6 Mobile IP and Its Role in 802.11 Roaming

3.7 Those Pesky Laws of Physics

References

3.1 Introduction

In Chapter 2 we provided a description of the principles of roaming within the cellular telephone context. The objective of this chapter is to introduce the general capabilities and limitations of roaming, as defined in the original 802.11 standard. We illustrate the 802.11-specific aspects of local and global roaming in WLANs. A discussion of the numerous amendments and other extensions to 802.11 that enhance these basic roaming capabilities is deferred until later chapters. We conclude this chapter with the discussion of Mobile IP and its relationship with global roaming in 802.11.

3.2 Primer on the 802.11 Standard

3.2.1 Introduction

Before we delve into the specifics of roaming within 802.11 WLANs, we will provide a brief review of the basic mechanisms used within 802.11 that permit controlled sharing of an 802.11 RF channel among a number of user *Stations* (*STAs*). (Recall that in the cellular world we used the acronym MS to denote a Mobile Station.) During this introductory discussion, we focus on how access is shared between many user stations and a single access point. We do not deal with multiple access points in the same network until Section 3.4.

Certain 802.11 management frames are especially crucial to a basic understanding of establishing and maintaining connectivity between a station and an access point. These important frames are as follows:

- Beacons
- Probes
- Association requests
- Association responses

All the 802.11 management frames are prefaced with the 802.11 MAC header shown in Figure 3.1. This header has six fields contained in 24 bytes. There are two bytes for frame control and two bytes for duration. These fields are followed by six bytes representing the *destination address* (*DA*) and another six bytes representing the *source address* (*SA*). The next six bytes are devoted to the *Basic Service Set ID* (*BSSID*), which is discussed in Section 3.2.4. The last two bytes are for sequence control. All the management frames end with two bytes called the *Frame Check Sequence* (*FCS*), which are used for error control (for example, see Figure 3.2).

3.2.2 Beacons and Probes

The beacon and probe-response frames are two separate mechanisms via which a station can learn of the existence of an AP operating on a given channel. Whether beacon frames or probe frames are used depends on whether *active* or *passive scanning* (discussed in Section 3.2.5) is

Figure 3.1: The 802.11 Management Frame MAC Header.

selected. The beacons are broadcast by APs to advertise their existence on a given channel. The capability information is also included. In Figure 3.2 we show the basic components of the beacon frame. Note that similar to the MAC headers of all other 802.11 management frames, the standard MAC header shown in detail in Figure 3.1 precedes the body of the frame. We see that a beacon frame carries a time stamp of eight bytes, a beacon interval of two bytes, a two-byte capability field, and a variable-length *Service Set Identification* (*SSID*) of the AP emitting the beacon. The collection of *Information Elements* (*IEs*) that are included in a beacon will vary widely between APs depending on the capabilities and the configuration of the AP. Much of the work of the specialized 802.11 task groups, which we will discuss in detail in Chapter 4, revolves around adding new IEs to management frames in order to support the special functions that those task groups are working to implement.

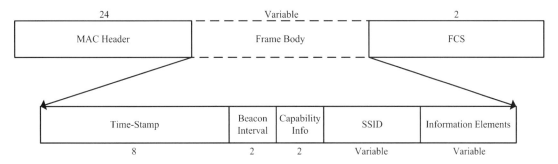

Figure 3.2: Basic 802.11 Beacon Frame.

APs only transmit beacons when they are configured to do so. In many networks, in order to render war-driving somewhat more difficult, APs are configured not to advertise themselves by beaconing. If an AP is not advertising itself via beaconing, the AP can only be discovered through the use of the probe-request and probe-response exchange. This mode is called *active scanning*, and in Section 3.2.5 we discuss examples of active scanning as well as the use of beacons in *passive scanning*. While active scanning is imperative to discover a nonbeaconing AP, active scanning is sometimes also used even in the case of beaconing APs to determine quickly what APs are available on a given channel. The probe-request frame layout is shown in Figure 3.3. The SSID field in the frame body is used to specify which SSID an STA is seeking. These probe requests may have to be sent out on many different channels until an AP listening on a particular channel can respond. The SSID field has a variable length, as does the succeeding supported rates field.

When an AP receives a probe request on the channel on which it is listening, the AP will respond with the probe-response frame whose format is shown in Figure 3.4. The reader should note that the format of the body of the probe-response frame (Figure 3.4) and the beacon frame (Figure 3.2) is nearly identical. This near duplication is because the frames serve the same purpose of advertising the capabilities of the transmitting AP. The difference is that the

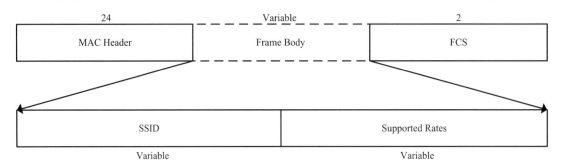

Figure 3.3: Basic 802.11 Probe-Request Frame.

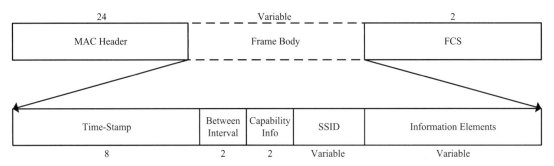

Figure 3.4: Basic 802.11 Probe-Response Frame.

beacon is proactively sent by the AP, whereas the probe response is solicited by the actively scanning STA.

Active or passive scanning for APs involves looking across a set of channels to determine on which channel a desired AP is operating. In order to proceed with the discussion of scanning for APs, we first review the concept of 802.11 channels.

3.2.3 Channels in 802.11

As we indicated in Chapter 1, a basic tenet of cellular communications architecture is to increase the overall use of a given subset of the RF spectrum by allowing multiple cells in the same network to reuse the same frequency or set of frequencies. This technique requires that cells use frequencies different from those of their immediate neighbors as well as limited transmission power. These two principles facilitate frequency reuse in cellular systems and, hence, the increased aggregate RF efficiency of these systems.

In order to provide RF signaling that is highly robust in the face of a wide array of interference factors (walls, portable telephones, microwave ovens, and so on), the 802.11 standard specifies the use of spread-spectrum transmission. This technique was developed many decades ago for military applications. As noted in Section 2.5.5, the original patent for this method was actually granted to the famous Austrian actress Hedy Lamarr and her computer friend George Antheil; they conceived spread-spectrum transmission as a means to circumvent jamming of radio-controlled torpedoes. That this very significant RF technology has its roots in an actress's pre-WWII cocktail party conversations is one of the great ironies in the annals of RF history.

There are three spread-spectrum technologies that have been used in commercially available 802.11 devices. They are listed in the following text:

1. *Frequency Hopping* (*FH*) is used in one version of the original 802.11 standard.

2. *Direct Sequence* (*DS*) is used in the 802.11b standard.

3. *Orthogonal Frequency Division Multiplexing* (*OFDM*) is used in the 802.11a and the 802.11g standards.

As we provide only a brief introduction to spread-spectrum methods in this book, we encourage the reader interested in this topic to refer to the reference section. For the purposes of this book, the concepts that we present here suffice. With FH, the user signal is actually transmitted over a set of frequencies. At any moment in time, the FH signal is only being transmitted over a subset of the set of frequencies available for the transmission. The transmitters and receivers have knowledge of the FH code; it allows a transmitter and a receiver to synchronize which frequencies are being used at a specific point in time and to know when the switch to the next subset of frequencies will occur.

With DS, the narrowband signal is spread over a relatively wide-frequency band and recovered from that wide band by the receiver. The spreading is accomplished via a mathematical transformation of the signal by applying a *numerical chipping sequence* to the original signal; the sequence is also used by the receiver to recover the signal. The numerical chipping sequence is a binary string, and it is sometimes called a *pseudorandom noise code* (*PN code*).

In the case of OFDM, the comparatively wide-frequency band used for transmission is divided into subchannels. The term operating channel is used to distinguish the channel numbers referenced by the user from the subchannels that comprise the *operating channel*. When we refer to 802.11 channels in the text, we mean operating channels. The original transmission is encoded into these multiple subchannels, which work in parallel. The aggregate transmission rate of these subchannels when recombined is the higher transmission rate perceived by the user. As there was comparatively little commercial success of the original 802.11, as compared to 802.11a, 802.11b, and 802.11g, our examples only consider the 802.11 channel assignments used for the DS and OFDM modes that are used by these three widely deployed 802.11 PHYs.

The 802.11b DS PHY has a total of 14 channels, although not all are used in any given regulatory domain. In the United States, only channels 1–11 are allowed. The particular channels allowed in Japan or various parts of Europe differ from the regulatory domain of the United States. Each channel is 5 MHz wide, and they are all in the 2.4-GHz band. As channel 10 is available in all the main regulatory domains, it is often the default operating channel. In a multi-AP environment, the channels in neighboring cells should be nonoverlapping to minimize *cochannel interference* (*CCI*). This requirement normally dominates the selection of an AP channel when the cell topology is laid out.

In the case of the DS used in 802.11b, only three nonoverlapping channels are available: channels 1, 6, and 11. In most examples cited in this book, we will show roaming topologies based on adjacent APs being assigned channels from this set of three channels. Although the 802.11g PHY is based on OFDM similar to 802.11a, it operates in the same 2.4-GHz band as 802.11b and its channel assignments are the same as 802.11b. The use of channels 1, 6, and 11 for 802.11b deployments has evolved into a de facto standard, which as of this writing is often referred to as *traditional 3-color deployments*. This reference is to a computer science graph of the cell topology that can be colored using only three colors, (channels) where no two adjacent cells share the same color.

The 802.11a OFDM PHY has 12 nonoverlapping channels; they are numbered 36, 40, 44, 48, 52, 56, 60, 64, 149, 153, 157, and 161. Each channel has a bandwidth of 20 MHz and is in the 5-GHz band. The availability of dual-mode 11a/11g and even tri-mode 11a/11b/11g chips has effectively increased the channel selection in laying out the cell topology beyond the limit permitted by any one of these single PHYs.

3.2.4 Basic Service Set

The 802.11 *Basic Service Set* (*BSS*) is a group of STAs and an access point that are communicating with each other. In an *Independent BSS* (*IBSS*), which is also referred to as an *ad hoc 802.11 network*, stations directly communicate with each other rather than through an access point. Such networks are also sometimes called *peer to peer networks*. The 802.11 networks covered in this book are operating in *infrastructure mode*. Infrastructure mode networks, which dominate all familiar 802.11 applications and installations, consist of infrastructure BSSs. We do not deal with IBSS 802.11 networks here, as such networks are not cellular-based networks and, hence, do not fit into the roaming paradigm which is the focus of this book. As we exclusively refer to infrastructure BSSs in this text, our use of the term BSS will imply infrastructure mode.

A BSS denotes a single AP and all of the user stations currently communicating with it. The AP and all the stations in the BSS communicate on the same 802.11 channel. The MAC

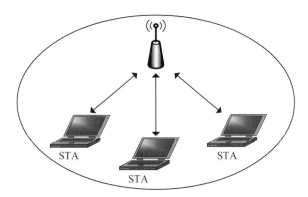

Figure 3.5: Basic Service Set.

address of the AP is used to identify that BSS and is called the *Basic Service Set Identification* (*BSSID*). A sample BSS is shown in Figure 3.5. The figure shows three STAs communicating via one AP. Note that it is common practice to drop the "B" in BSS and refer to the "Basic Service Set" as the "Service Set." The Service Set ID (SSID), which we saw earlier in an 802.11 frame, is the name given to a group of access points that jointly provide wireless access to a given IP subnet.

3.2.5 Active and Passive Scanning

In passive scanning, an STA will iteratively listen on all channels that are available to it. While listening, an STA will receive beacons from SSIDs. The STA records the SSIDs that it identifies. It is typical for an STA to receive beacons from APs advertising different SSIDs as well as different APs advertising the same SSID. In Figure 3.6 we show two access points beaconing: AP1 and AP2. The example does not specify whether these APs are advertising the same SSID or different SSIDs. The process of beaconing continues by the APs independently of the state of the STA. In the example, we see that the STA begins its passive scan midway through the diagram. For example, the start of the passive scan might correspond to a user enabling an 802.11 card on a laptop. This scan does not directly result in the transmission of any frames, hence the name passive scan.

During a passive scan phase, an STA collects information from the beacons it receives. In our example shown in Figure 3.6, the STA receives two beacons: one on channel 6 and the other on channel 11. On receiving these beacons, the STA learns the channels being used by AP1 and AP2, respectively, the SSID that they are advertising, plus capability information from their capability fields and information elements. If the STA wants to join the BSS of one of these APs, the STA will follow the association procedure which will be described in Section 3.2.6. In Figure 3.6 the STA wishes to associate with the BSS of AP2, which results

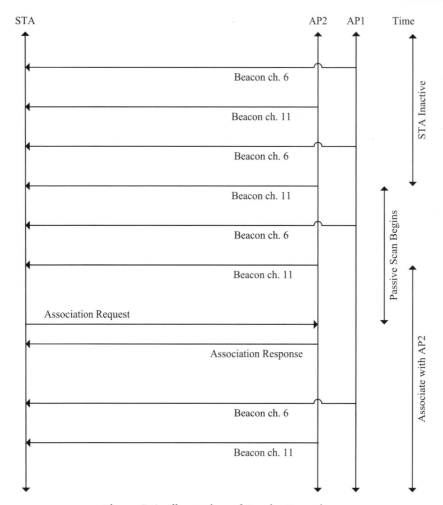

Figure 3.6: Illustration of Passive Scanning.

in the association request/response exchange as shown at the lower portion of the figure. Note that although the STA has associated with AP2, the beaconing continues from both access points as they are proactively advertising to other potential STAs. An STA can only receive a beacon if it is listening on the channel on which the beacon is sent. Thus, determining when to perform the passive scan is a fertile area for optimization in the STA. We deal with this topic again in Section 3.4.2.

Whether an STA opts to join a BSS can be a complicated decision. At the very least the STA must want to communicate with that AP's SSID. As shown in Figure 3.6, if only one of the two access points has a desired SSID, then only that AP is a viable candidate. Another criterion that impacts the decision of which BSS to associate with is capability information—such as what level of security is offered by the AP. If both AP1 and AP2 are advertising the same SSID and have the same capabilities, then a basic 802.11 STA will likely associate with the AP showing a stronger received signal. We use the caveat "basic 802.11 STA" because an enhanced 802.11 STA may apply other criteria to the decision beyond simple signal strength. These other criteria are introduced in the discussions about the 802.11 task groups in Chapter 4.

The process of active scanning differs from passive scanning in that the STA actively searches for an SSID to which the STA wishes to connect. For each channel that is actively scanned, the STA must first tune the 802.11 radio to the specific channel, listen in order to avoid a collision, send a probe frame, and finally listen for a prescribed period of time for any probe responses or beacons. In order to discover an AP using a particular SSID of interest, probe requests are transmitted on the channels that are available to that STA until an AP having that SSID responds or the STA gives up. Note that the AP responds to a probe request only if the AP has the SSID that matches the probe request.

A probe request may specify a broadcast SSID, in which case it is a *wildcard* probe, to which any AP on that channel may respond. Wildcard probes are also called *undirected probes*, in contrast with *directed probes* that specify a particular SSID. We show the process of active scanning in Figure 3.7. The first five probes do not result in a matching SSID. The probe request on channel 6 results in a match on AP1, and a probe response is sent back from AP1 in

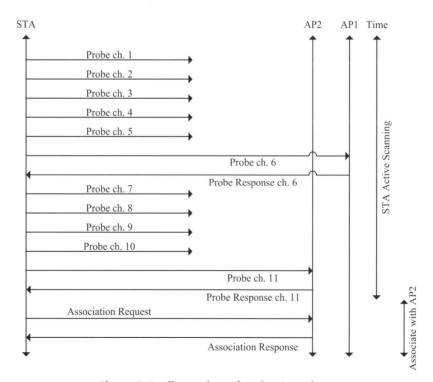

Figure 3.7: Illustration of Active Scanning.

this case. Probe requests on channels 7–10 do not result in matches. The probe request on channel 11 results in a match on AP2, and a probe response is sent back from AP2 in this case. The STA then requests to associate with AP2. AP2 sends the STA an association response.

Unlike beaconing, which is performed proactively by the access point, probe responses cease when no active scan is occurring by an STA in range of that access point. The example shown in Figure 3.7 assumes that there are no other stations doing active scans; so, no further probe responses are shown after association.

3.2.6 Association

Once the STA has identified an acceptable AP, the STA will tend to associate with that AP's BSS by sending an association-request frame to that AP. The association-request frame format is shown in Figure 3.8. The frame has two bytes of capability information, two bytes for the listen interval, a variable-length SSID field, and a variable-length supported-rates field.

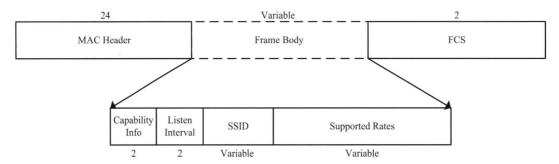

Figure 3.8: Basic 802.11 Association-Request Frame.

The reassociation-request frame shown in Figure 3.9 serves a function similar to that of the association-request frame. The reassociation request is used when an STA was already associated with an AP with the same SSID as the AP with which the STA is now trying to associate. The reassociation request contains additional information, notably six bytes for the current (old) AP address. This information is intended to facilitate the handover of the connection from the old AP to the new AP. This 802.11 process is an instance of the concept of local roaming introduced in the earlier chapters. We delve into the details of local roaming later in this chapter in Section 3.4.

When an AP receives the (re)association request, the AP will choose either to accept or to reject the request. If the AP chooses to accept the request, the AP will do so with an association response or a reassociation response as appropriate. The format of the (re)association-response frame is shown in Figure 3.10. There are two bytes of capability

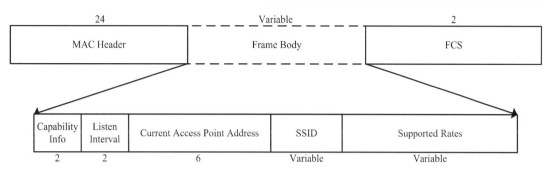

Figure 3.9: (Re)Association-Request Frame.

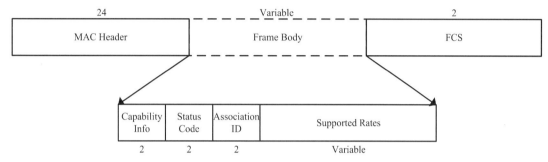

Figure 3.10: (Re)Association-Response Frame.

information, two bytes for status code, two bytes for the association ID, and a variable-length field for supported rates.

In all the protocol exchanges described in this chapter, we show the association-request/-response exchange without any preceding authentication frames. In reality, before an STA can associate with an AP, the STA must authenticate with the AP, which involves a minimum of a two-frame exchange in the open-authentication mode. This exchange needs to precede association; however, the exchange does not need to precede association directly. That is, the authentication can occur at any time before the association.

The number of authentication frames exchanged prior to association depends on what authentication mode is in use. If open authentication is selected, there is a two-frame exchange. If *Wired Equivalent Privacy* (*WEP*) authentication is used, a four-frame exchange takes place. As we will explain in Chapters 5 and 6, at the time of this writing, *neither* of these two types of authentication is considered acceptable for true authentication in 802.11 implementations. The most common situation, in implementations in place as we write this book, involves the use of the two-frame open-authentication exchange to maintain compatibility with the original 802.11 specification, all the while acknowledging that the true authentication will not take place until the more complicated exchanges described under 802.11i have occurred.

Due to the confusing nature of the authentication-frame exchange and in order to simplify the figures for the sake of clarity, we have omitted this exchange from the protocol diagrams presented in this chapter. While these authentication frames have little to do with actual authentication in secure 802.11 networks, we will see in Chapter 10, that these frames may be exchanged multiple times and the exchange occurs well before an association, thereby making them an ideal vehicle for communicating information in anticipation of a possible 802.11 roam.

3.2.7 Contention-Based Access

Both 802.3 (*Ethernet*) and 802.11 use contention-based media-access mechanisms to share access to a broadcast medium. One of the reasons for the rapid uptake of 802.11 technology is that despite the fundamental differences between wired and wireless media, the MAC and PHY layer-abstractions mask these differences and allow the same upper-layer protocols and paradigms to work transparently over both. Thus, the evolution of protocols that has made Ethernet increasingly easy to configure and use has directly accelerated 802.11's adoption. Discovery protocols like *Address Resolution Protocol* (*ARP*) for MAC addresses and *Dynamic Host Configuration Protocol* (*DHCP*) for IP addresses function in the same manner over both 802.3 and 802.11.

Ethernet is based on the *Carrier Sense Multiple Access/Collision Detect* (*CSMA/CD*) access scheme, whereas 802.11 is based on *Carrier Sense Multiple Access/Collision Avoidance* (*CSMA/CA*). The difference between the collision detection of Ethernet and the collision avoidance of 802.11 is that the 802.3 transceiver listens to the medium when it is transmitting and can *detect* the additional energy of another transmitter which is operating simultaneously. When multiple transmissions are attempted simultaneously, a *collision* occurs. Transmitters involved in a collision follow the same paradigm. As each transmitter has detected a collision, the transmitter invokes a *random back-off algorithm* that is part of the media-access scheme.

This algorithm randomizes how long each transmitter delays before attempting to transmit again, thereby reducing the likelihood of a repeated collision. In all cases, each transmitter verifies whether the medium is free—no other transmitter is detected as currently active—before beginning to transmit. This check also reduces the likelihood of collisions. The network smoothly recovers from collisions because although the colliding frames are not received, the upper-layer protocols have mechanisms to detect and recover from lost frames. This concept is the basis for the operation of the CSMA/CD paradigm. There are fundamental issues with 802.11 that force the use of CSMA/CA instead of CSMA/CD. We explore these issues in the following text.

The first issue is that it is much more expensive to build RF transceivers which can simultaneously transmit and receive. The second issue is that it is very possible that two STAs in the same BSS may be so distant from one another that they cannot hear each other's signals, although the AP that they share can communicate with each of them. This is called the *hidden node problem*. For these two reasons, the 802.11 standard uses algorithms that avoid collisions rather than relying on detecting collisions. *Virtual carrier sensing* uses the *Network Allocation Vector* (*NAV*) that is carried in the duration field of 802.11 data frames. The NAV indicates the duration in microseconds of transmittance by the transmitter. The NAV thus allows a waiting 802.11 transmitter to know a priori how long it should wait before attempting to gain access to the medium. Then, much like Ethernet, before the 802.11 transmitter starts to transmit, the transmitter listens to the medium to see if another transmitter is active. This process is called *physical carrier sensing*. Both virtual carrier sensing and physical carrier sensing are used in 802.11.

For considerably congested 802.11 environments, where the probability of collisions is high, the 802.11 standard supports a *Request To Send/Clear To Send* (*RTS/CTS*) protocol to gain guaranteed access to the medium. The use of this protocol is only appropriate when comparatively large data frames are queued to transmit, which is attributed to the fact that the use of the RTS/CTS exchange itself is not immune to collisions, and the shorter the frame length, the lower the probability of a collision. If a short data frame is queued for transmission, the likelihood of the frame suffering a collision is similar to that of a collision occurring during the RTS/CTS exchange; so, no network efficiency is gained from the additional RTS/CTS overhead.

When collisions do occur in 802.11 (and they are indeed possible), the collisions are detected by the 802.11 MAC layer itself. The 802.11 MAC layer can detect collisions because it incorporates a mechanism where each successfully received frame is acknowledged to the transmitter. This situation is in contrast to 802.3. To appreciate the comparative underlying complexity of 802.11, as compared to Ethernet, it is worth juxtaposing the transmission of an 802.11 data frame against the transmission of a similar data frame in 802.3. As the physical-collision-detection mechanisms of 802.3 are more robust than those of 802.11, the 802.3 standard will have a high probability of successful transmission of a large frame by the simple transmission of that single frame without any additional control-frame overhead. In 802.11, transmission of that large frame may entail four frames:

- Transmission of the RTS
- Waiting for the CTS
- Transmission of the data frame
- Waiting for the received *Acknowledgment* (*ACK*) frame

This considerable extra 802.11 overhead is justified to approximate the level of reliability and performance to which we have become accustomed to with 802.3, despite 802.11 operating

Chapter 3

over a much less reliable medium. It should be noted that the RTS/CTS exchange is often skipped in a typical 802.11 operation.

This section has provided a very high-level review of the contention-based access mode of basic 802.11. This access mode resulted in the initial success of 802.11. A *contention-free* (*CF*) access mode was defined in the original 802.11 but was not generally implemented. In Chapter 4 we will discuss details and extensions to the contention-based access mode, as well as a new contention-free access mode, which was defined in the recent amendments to the 802.11 standard.

3.2.8 Rate Adaptation in 802.11

If all other factors are held equal, a lower bit-rate transmission will more likely be error free than a higher bit-rate transmission. The 802.11 standard incorporates this reality by dynamically reducing the transmission bit rate in the face of transmit errors. These errors are manifested by the need to retransmit frames, which, in turn, is prompted by the lack of acknowledgments to the initial transmission of those frames. This feature of 802.11 is called *rate adaptation*. Rate adaptation only applies to data frames because 802.11 control information is *always* sent at low data rates.

When transmission errors persist, the transmit bit rate for data frames is ratcheted down on a stepwise basis in accordance with the fundamental transmission rate hierarchy of 802.11. With 802.11b, this would start at 11Mbps and gradually stop at 1Mbps. Since the decision to adjust the transmission rate dynamically rests solely in the hands of each individual transmitter, different transmitters in the same BSS can all be transmitting at different speeds. Thus, it is necessary for the receivers to be able to detect the underlying transmission rate of a frame dynamically while the receiver is receiving the frame. This apparent conundrum resolved by encoding the bit rate of the payload of an 802.11 frame in the PHY header of that frame. The PHY headers are received in the same way regardless of the differing bit rates possible for the payload.

The 802.11 rate adaptation allows STAs and APs to continue effective communications even under circumstances where the RF environment would make operation at the nominal bit rate impossible. The rate adaptation also extends the effective radius of the 802.11 cell by allowing STAs that are very distant from the AP to communicate with that AP at reduced bit rates. These two points indicate very real and tangible benefits of the 802.11 rate adaptation, but there are potential drawbacks as well when many stations and APs are involved, which is the typical situation that we consider when dealing with secure roaming in 802.11.

One such drawback is related to the ability to achieve frequency reuse in 802.11-topology planning. Specifically, APs are normally positioned so that no two adjacent APs use the same channel. If we again consider 802.11b as an example, the topology will be created using assumptions that the cell sizes are a function of the configured transmit power of the APs and of the effective transmission range of 11 Mbps 802.11b frames. As the distance from the AP increases much beyond that range, such frames appear in neighboring cells as noise and at some distance from the original AP, these frames actually do not interfere with simultaneous transmissions on the same channel in the other cell. When 802.11 rate adaptation is being used and data frames are transmitted at lower bit rates, these slowed-down frames may be received without error at increasing distances from the original AP. This circumstance effectively increases the cell size beyond that accounted for in the original network topology, which in turn increases the number of potential collisions.

Because errors due to collisions often resulted in the initial invocation of rate adaption, this can result in a downward spiral of transmission rates. A related contributing effect is the fact that as the average transmission speed of the data frames degrades, there is relatively more time spent occupying the medium (the air) with the data that is taking longer to transmit, which therefore makes collisions more likely, and so perpetuates the cycle. There is continued debate about how serious these rate-adaption issues are, but this issue is certainly a complicating factor in the congested multicell 802.11 environments where roaming often occurs.

3.2.9 Other 802.11 Frames

In the previous sections, we provided an overview of the important 802.11 frames which are relevant to our subsequent discussion about roaming. The frames covered so far do not constitute an exhaustive list. A more complete list of 802.11 frames appears in Tables 3.1 and 3.2. In these tables, the 802.11 frames are grouped based on type. The first group, which shares type value 00, consists of management frames, and is shown in Table 3.1. The second group, which is shown in Table 3.2, includes control frames; they have a type value of 01. The final group, data frames, has a type value of 10; these frames are shown in Table 3.2. It should be noted that the type value 11 is reserved for future use; so, there are no table entries for this value.

The last column of Tables 3.1 and 3.2 shows the class to which the frame belongs. To understand the concept of frame class in 802.11, we must first understand the concepts of the 802.11 frame transmission, association, and authentication state. There are three states as follows:

1. The STA is not authenticated and not associated.

2. The STA is authenticated (basic 802.11 authentication—not *secure* authentication) but not yet associated.

3. The STA is authenticated and associated.

Chapter 3

Table 3.1: The 802.11 Standard's Frames Based on Type and Class: Part 1.

Frame Name	Type	Type Code	Subtype	Class
Association request	m	00	0000	2
Association response	m	00	0001	2
Reassociation request	m	00	0010	2
Reassociation response	m	00	0011	2
Probe request	m	00	0100	1
Probe response	m	00	0101	1
Beacon	m	00	1000	1
Announcement traffic indication	m	00	1001	1
Disassociation	m	00	1010	2
Authentication	m	00	1011	1
Deauthentication	m	00	1100	3
Action	m	00	1101	3

(m = Management)

Table 3.2: The 802.11 Standard's Frames Based on Type and Class: Part 2.

Frame Name	Type	Type Code	Subtype	Class
Power-save (PS) poll	control	01	1010	3
Request to send (RTS)	control	01	1011	1
Clear to send (CTS)	control	01	1100	1
Acknowledgment (ACK)	control	01	1101	1
Contention-free (CF) end	control	01	1110	1
CF-end and CF-ACK	control	01	1111	1
Data	data	10	0000	3
Data and CF-ACK	data	10	0001	3
Data and CF-Poll	data	10	0010	3
Data and CF-ACK and CF-Poll	data	10	0011	3
Null data	data	10	0100	3
CF-ACK	data	10	0101	3
CF-Poll	data	10	0110	3
Data and CF-ACK and CF-Poll	data	10	0111	3

2	2	6	6	6	2	0–2,312		
Ctl	Dur	Destination Address (DA)	Source Address (SA)	Address 3 (Filtering)	Seq-Ctl	Address 4 (Optional)	Frame Body	FCS

Figure 3.11: An 802.11 Unicast Data Frame.

It should be noted that not all frames may be transmitted in every state. All three classes of frames may be transmitted in state 3. Both classes 1 and 2 may be transmitted in the second state. Only class 1 frames may be transmitted in state 1.

Figure 3.11 shows an 802.11 unicast data frame. The figure shows that the frame consists of two bytes for frame control, two bytes for duration, six bytes for the DA, six bytes for the source address, six bytes for filtering, two bytes for sequence control, six bytes for an optional fourth address, 0–2,312 bytes for the frame body, and four bytes for error checking.

Data frames are not only for user traffic. Frames that carry higher layer control protocols, such as ARP and DHCP, are transmitted as 802.11 data frames, although they themselves are control protocols that support the 802.* family. However, in a typical busy BSS, it is probable that the vast majority of frames to and from the AP will be unicast data frames carrying IP packets. The IEEE fastidiously maintains that the 802.* family protocols are independent of any particular layer-three technology. Nevertheless, because of the near ubiquity of IP as the network-layer protocol in actual 802.11 networks, we have focussed on IP in the examples and discussions throughout this text.

3.3 Introduction to 802.11 Roaming

3.3.1 Extended Service Set

A group of Basic Service Sets, each one individually identified by its BSSID, is called an *Extended Service Set* (*ESS*) and is identified by its SSID. As mentioned earlier, the MAC address of the AP has the dual role of serving as the BSSID of that AP's BSS, whereas the SSID does not correspond to the MAC address of any particular AP of the ESS and is merely a name specified in the AP's configuration. Such an ESS is shown in Figure 3.12. Each of the three APs in the figure advertises the same SSID in its beacon and probe-response frames. It should be noted that some high-end enterprise APs support multiple SSIDs in the same AP. This multiplicity allows a single AP to behave as multiple *virtual APs*. This feature is normally used in an environment where the different virtual APs would each offer distinct security, QoS, or *Virtual LAN* (*VLAN*) characteristics.

The IP subnet corresponding to an SSID is often similar to the subnet of the 802.3 wired-Ethernet ports to which all the APs sharing that SSID are connected. Resources, including a DHCP server and other servers, as well as a gateway router providing access to

Chapter 3

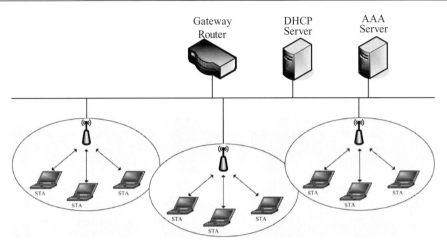

Figure 3.12: An Example of an Extended Service Set.

other subnets and the public Internet, are also likely to be connected to the same Ethernet and belong to the same subnet. *Distribution Service (DS)* is the 802.11 term that describes the network linking the infrastructure side of the APs in an ESS. Whereas wired Ethernet is the most prevalent DS in current deployments, other wired LAN, MAN, or WAN technologies are possible. It is also possible that the DS be wireless based. The 802.16 standard, also known as WiMAX, which is a wireless-MAN technology, is often proposed as the back-haul technology linking the infrastructure side of the APs in an ESS.

3.3.2 Example of Multiple ESSs in Operation

Figure 3.13 shows the configuration of a real network with multiple ESSs in operation in the same geographical location. The ESSs displayed on the user-interface screen are named den_hidden_1, den_hidden_2, den-ra2, and gateway-AP. The names of these ESSs are actually the names of the SSIDs shared by their member APs. As gateway-AP is currently selected on the panel, the display shows the BSSID of the three APs belonging to that ESS, their signal strength, and their security level. As the security levels of the member APs are not homogeneous, the aggregate security of the ESS is shown as mixed.

When a connection to one AP is broken and a connection is made with another AP in the same ESS, we consider this local roaming. For example, in Figure 3.13 a local roam would occur in switching from AP 000c41f57d89 to AP 00e0b87608d8. For the sake of clarity, when we are connecting to a new AP with an SSID different from the last one which we connected to, we consider this global roaming, even if the new AP is geographically colocated with the AP in which connection is broken. So, for example, in Figure 3.13 a global roam would occur in switching from gateway-AP to any AP in one of the ESSs den_hidden_1, den_hidden_1, or den-ra2.

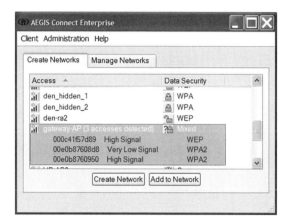

Figure 3.13: The Configuration of a Network with Multiple Extended Service Sets in Operation.

3.3.3 Phases of 802.11 Roaming

There are many factors that actually contribute to the roaming delay as perceived by the 802.11 user. Because each of the various 802.11 enhancements presented in this book focus on one or at most a few of these factors, in this section we discuss about these individual components of roaming delay and their relationship with one another.

Figure 3.14 shows a sample of roaming-delay measurements in the context of the IEEE 802.11t task group. The problem with providing straightforward measurements for 802.11-roaming delays is that this delay may differ in its meaning based on context. One industry anecdote is that measuring roaming delay is like answering the question "How long does a divorce take?" One valid answer is that the couple goes to court and the divorce is usually completed within a day. However, a more common understanding of how long a divorce takes is that it takes years from the time a couple begins to realize that the marriage is not working, begins considering alternatives, makes a decision to proceed with the split, goes through an extended legal proceeding, and then arrives in court to complete the divorce. Similarly, after the divorce is legally in place, the period of recovery before one actually resumes normal life can be very long.

This long process for decision-making in divorce prior to the actual divorce is like the STA staying connected too long to its current AP, and then when the STA realizes that it must change, it takes too long to select a new AP. At this point, the quick "court-proceeding" analogy of the handoff is apt, which is why the delays measured in Figure 3.14 seem short. The delay measured by the 802.11t efforts shown in Figure 3.14 corresponds to the time consumed by the STA to reassociate after the client decides to roam. Similar to the postdivorce recovery, after the layer two roam is complete, there may be a prolonged process before the user actually perceives the roam to be complete.

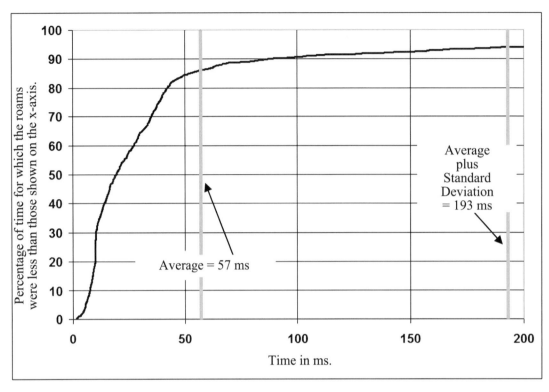

Figure 3.14: Example of Roaming Delay Times. (Courtesy: Azimuth Systems.)

One complication is with regard to deciding when to initiate a roam. Intuitively, one would guess that a good approach would always be to associate with the AP that currently presents the strongest signal. However, this strategy does not work in practice. The RF domain is one that is subject to many variations, even when the STA is stationary. When we add user mobility to the mix, signal strengths from nearby APs can vary widely. In early 802.11 networks, the most common practice in place in 802.11 STAs was to defer a roam until signal strength has dropped to a nearly unusable level. As of this writing, more sophisticated 802.11 STA implementations have become available that are able to use multiple criteria in deciding when to initiate a roam. At that point, the target AP is selected among those that display the strongest signal strength and the strongest match to other implementation-dependent criteria. For discussion purposes in this chapter, we simply emphasize that it is important to make a roaming decision and follow it for a reasonable period of time; otherwise, the STA runs the risk of entering a state of oscillation—constantly switching back and forth between two APs. This *ping-ponging* results in a very bad aggregate level of service while in search of the optimal one.

We now take a broader look at the various factors that contribute to the roaming delay, as perceived by a user. We turn the reader's attention to Figure 3.15. Along the top of the figure,

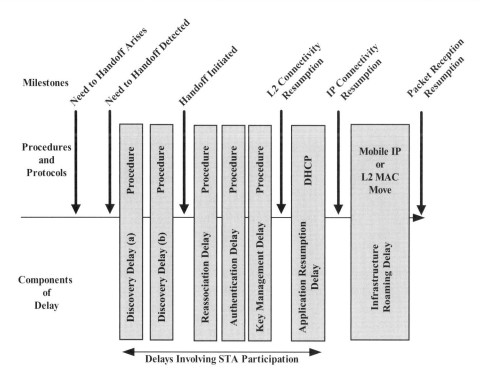

Figure 3.15: Phases of 802.11 Roaming Delay.

we show several milestone events that occur during the time when the user is affected by the 802.11 roam. In the figure, the progress of time is shown as from left to right. Below and between the milestones are a set of shaded boxes. The top part of each box is labeled with the procedures and/or protocols that are active and lead up to that milestone. The bottom of each box is labeled with the name of the delay type that precedes the achievement of the next milestone. We identify the components of secure-roaming delay in the following text, and in the discussion of each component, make reference to the procedures and protocols that are used during the phase of delay. We encourage the reader to refer to Figure 3.15 throughout the following discussion. Note that Figure 3.15 is a generic portrayal of the phases of roaming delay. Specific STA implementations may diverge somewhat from the diagram. Also, while the relative widths of the relay components in the figure are intended to be meaningful, in what follows we explain that wide variances in the actual durations of the delays may occur, depending on the given circumstances.

Discovery delay The discovery process consists of scanning and other measurement processes required for the STA (a) to determine that it needs to roam and (b) to determine the best AP to roam to. Part (a) is essentially a determination that indicates that the current AP is unacceptable. There are different metrics that can be used to make this determination. These include the following:

- Observing an increase in retry counts due to transmission errors

- A downward shift of the transmission rate due to the inability to communicate at higher speeds

- Too many missed beacons

- Too many frames received with errors

- A direct measurement of decreasing Radio Signal Strength, as measured by the RSSI parameter

Different implementations may require one or more of these metrics to trigger a roam. Because all these are directly or indirectly a function of signal strength, we will refer to them generically as *signal-strength-based roaming triggers*. The firing of these triggers corresponds to the milestone "Need to Handoff Detected" in Figure 3.15.

It should be noted that the delay between "Need to Handoff Arises" and "Need to Handoff Detected" shown in the figure is attributed to the common scenario that the triggering of the need to handoff actually occurs later than the ideal moment to begin handoff procedures. This gap is real, but-probably unavoidable, because the decision to handoff must be made on real discrete metrics and the ideal time to begin the handoff often only becomes clear with hindsight.

Reassociation delay This period is the time required to complete the association with the new access point. It involves the exchange of a pair of 802.11 authentication frames and an association-request/-response exchange. The beginning of the reassociation delay marks the definitive end of application data being passed through the current AP.

Depending on which triggers are used in part (a) of discovery delay, application data may be unable to be passed through the current AP even before the start of the reassociation delay but certainly not after it begins. It should be noted that the data given in Figure 3.14 is based on measuring the time taken by the roam after the client decides to roam. This duration would include part (b) of the discovery delay and the reassociation delay. Also, the measures in Figure 3.14 are based on the assumption of weak or no security; so the following two delay types are not shown in the figure.

Authentication delay This time is the delay attributed to the authentication conversation between the STA and an AAA server. Depending on what version of 802.11 security is in place, this exchange may include several frames, often 13 or more, and theoretically hundreds of frames when network-admission control is running. We provide details on this authentication conversation in Chapters 6 and 7.

Key management delay   This period is the delay caused by the four-way exchange of key management frames that are used to derive the keys that will be used to encrypt the wireless link. As in the case of authentication delay, we provide full details on this authentication conversation in Chapters 6 and 7. At the end of the key management delay phase, the milestone "L2 Connectivity Resumption" shown in Figure 3.15 is reached. It should be noted that "L2" represents "layer two."

Application resumption delay   Once the 802.11 chips have been programmed with the keys derived during the key management delay phase, there is some additional internal delay when the driver sends a LINK UP event to the upper-layer protocols and before they react to allow the application traffic to resume. Modern STA implementations will send an ARP frame to the default gateway to determine if the IP subnet has changed. In the event that there was a change of IP subnet, the delay involved in using DHCP to obtain a new IP address is part of the application resumption delay. This point is an important part of what makes global roaming distinct from local roaming. We discuss local and global roaming in detail in Sections 3.4 and 3.5, respectively. At the end of the application resumption delay phase, the milestone "IP Connectivity Resumption" is achieved, as shown in Figure 3.15.

Infrastructure routing delay   This last phase of delay captures delays that may occur in the infrastructure after the AP and the STA are completely ready to resume user-data flow. Figure 3.15 shows a bifurcation in this phase between the local roam case and the Mobile IP case. These two represent the extremes of infrastructure complexity with respect to this additional delay.

In the case of a local roam, even though the AP and STA are completely ready to resume data flow, in order for traffic destined from the network to the STA to reach the STA, it has to be routed via the new AP. Until the infrastructure reacts to this move, there is still a *loss of connectivity to the network* from the application's point of view. What we have described thus far has not explained how the infrastructure learns that it must change this routing. In the most basic approach, the normal paradigm of a *layer-two MAC move* would apply. Under this paradigm, the infrastructure actually does not learn that a layer-two device has moved until a frame has been received on a new port for that device. Thus, part of the infrastructure routing delay may occur as a result of application timeouts, culminating with the STA sending a packet toward the infrastructure. Receiving this frame from the new port on which the new AP is connected will stimulate the infrastructure switch to learn the new location of the AP. Similarly, if the infrastructure gives up first and attempts to locate the STA, it will broadcast frames on all ports until it receives a response from the STA, at which point the learning occurs. Clearly, it is possible to construct an AP to infrastructure-signaling mechanism that proactively informs the infrastructure of the association of an STA to that AP. Such a signaling protocol could essentially eliminate the infrastructure routing delay phase for local roams, in which case Figure 3.15 exaggerates the duration of the delay. An example of how this may be accomplished is to send an XID frame into the network, achieving a very fast layer-two MAC move.

In the event that this is a global roam using Mobile IP technology, considerably more complexity is involved. We show this as another possible component of the infrastructure routing delay in Figure 3.15, in which case the relatively greater amount of time depicted in the figure for this delay is appropriate. Although Mobile IP is a very complex layer-three technology whose interacting components may be located in different continents, Mobile IP is in some ways the layer-three analog of layer-two MAC move just described. Just as the layer-two MAC move repairs the STAs loss of connectivity to the network, the Mobile-IP procedures can keep the STA's static IP address reachable to the Internet despite the physical move. The user's applications remain disconnected from the network until the Mobile-IP procedures are completed. We will defer further remarks on Mobile IP until Section 3.6.

3.4 Local Roaming

3.4.1 Introduction

Local roaming for WLANs is similar to one's mobile phone connecting to the "next" cellular antenna provided by the same provider to which one is currently connected. Figure 3.16

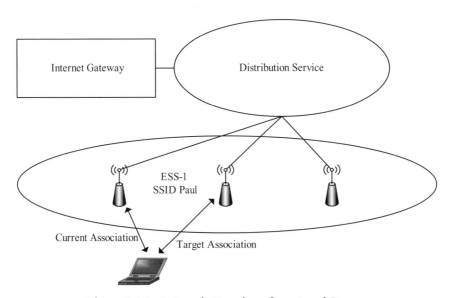

Figure 3.16: A Sample Topology for a Local Roam.

provides an example of topology for an 802.11 local roam. The device is currently connected to the left-most antenna, and the device will roam and associate with the middle antenna. Roaming in 802.11 networks within a set of APs sharing a common SSID is the counterpart to roaming within a single zone provided by a cellular provider: there are normally no issues of different billing, accounting, or security parameters between these cells. In the case of local roaming in basic 802.11, the stimulus to roam to the new AP is just that the signal strength-based roaming triggers have fired, and another AP within the same ESS has a stronger signal. In 802.11 networks, all handoffs are hard handoffs; the connection to the current AP is broken before the application-packet stream can resume to the new AP. The handoff time can vary considerably due to the stochastic nature of the 802.11 scan and the reassociation procedures. When 802.11i security is in use, the variability of the handoff times can become extreme, even possibly several seconds.

From a purely 802.11 perspective, roaming from one AP to another AP in the same ESS involves only a very simple frame exchange. In Figure 3.17 we see that the STA first joins the BSS of AP1 by associating with AP1, and then after a brief period of lost connectivity, joins the BSS of AP2 by associating with AP2. The apparent duration of the lost connectivity in the basic 802.11 case, without consideration of security or QoS issues, may be quite brief, say of the order of a few milliseconds. However, in reality the duration of lost connectivity depends on a plethora of factors, as we have explained in Section 3.3.3.

3.4.2 Scanning Tradeoffs

It is reasonable to wonder how the STA knew to reassociate with AP2. This choice depends on the driver and higher layer software implementations on the STA. If the STA is either actively or passively scanning for other APs while communicating with AP1, then the STA may already

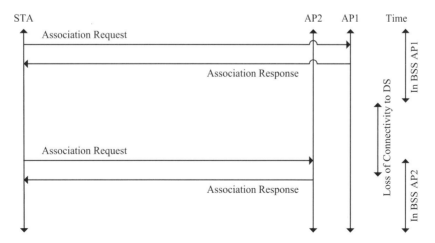

Figure 3.17: An Example of an 802.11 Management-Frame Exchange in a Local Roam.

know that its next roaming target should be AP2. In this situation, the reassociation may commence immediately after breaking the connection with AP1. However, such scanning does not come without cost because listening on other channels does require that the STA switch its radio from the current channel to the other channels in order to listen for beacons or initiate probe-request/-response sequences. In the case where the scanning was interleaved during the period of association with the BSS of AP1, the *loss of connectivity to the DS* period will be short as the target AP has already been identified. If the scanning process was deferred until the decision to roam was made, the duration of the loss of connectivity will be longer. Details and tradeoffs regarding maintaining scan lists will be discussed in more detail in Chapter 5.

3.4.3 Assumptions about Local Roaming and IP Subnets

It is certainly possible to have different ESSs on the same IP subnet. In theory it should be possible to roam between these different ESSs without forcing a change in IP address. What actually happens to the IP address of the STA in this kind of roam is implementation dependent, and as such a configuration is to a certain extent unusual, we will assume in the examples that different ESSs are indeed on different IP subnets.

Global roaming can have a much greater impact on a user than local roaming can have. In order to appreciate this impact, we need to consider some of the higher layer protocols that operate over 802.11. We look at how changes in the underlying 802.11 connections will affect those protocols and the applications that rely on them to a greater or lesser degree. To attain this broader perspective, it is instructive to look at a graphic representation. Figure 3.18 shows the same local roam that we saw in Figure 3.17, but now the higher layer DHCP and *Transmission Control Protocol* (*TCP*) protocols have been added. For simplicity, we do not attempt to depict accurately each frame corresponding to these protocols, as the diagram would become cluttered to the point of becoming unreadable. We do provide sufficient details to illustrate the effects that the 802.11-layer events have on the higher layer protocols, the applications themselves, and ultimately the end-user's experience.

From Figure 3.18 it can be seen that despite the association of the STA with a new AP, the underlying subnet did not change, the IP stack in the STA needed to be reset, and the TCP connection proceeded with at most a brief delay. It should be noted that there is no need to reestablish the TCP connection. In the next section, we focus on global roaming.

3.5 Global Roaming

3.5.1 Introduction

The examples in the previous section dealt with situations where the choice of the next AP was straightforward. In Figure 3.19, a situation where the selection of the target AP is not

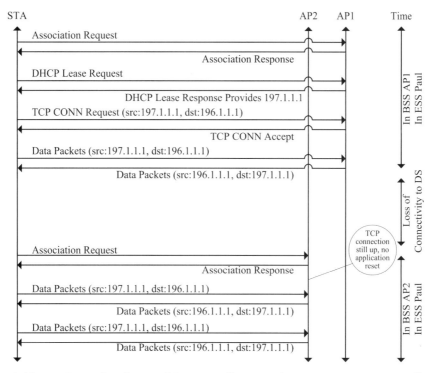

Figure 3.18: An Example of a Local Roam to Illustrate the Impact on a User Application.

necessarily obvious is shown. In this figure, the STA is currently associated with AP3. It is assumed that the signal from AP3 is losing strength, perhaps because as the STA moves, a physical obstacle appears between the STA and AP3. Let us first assume that the STA can hear the beacons of AP2 and AP6. In this case, with all other factors being held constant, the optimal roaming target will be AP2, as roaming to AP2 does not result in the change of the SSID, whereas roaming to AP6 does.

From the layout of the access points shown in Figure 3.19, one would expect that if the STA is receiving a weak signal from AP3, then the signal from AP2 is likely even weaker. Under this assumption, AP6 may be the only roaming candidate. As AP6 is in a different ESS from that of the STA's current association, associating with AP6 constitutes a *global roam*. Global roaming is sometimes also called *nomadic roaming*.

When we roam to a new SSID, there are additional issues to consider other than just signal strength. The accessible user needs to understand whether the new SSID is offering any services; that is, whether the SSID is a private organization for which the user has access rights or a public hotspot with which the user either has or is willing to have a service agreement. The user may not be able to comply with the minimum security requirements of the

Chapter 3

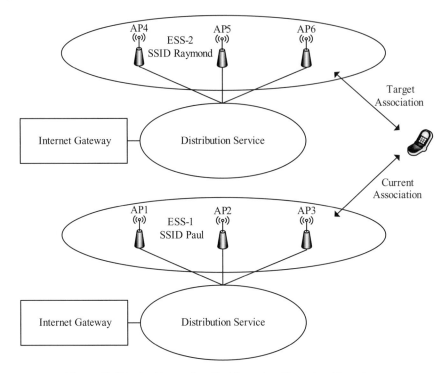

Figure 3.19: An Example of a Complex Roaming Topology.

new SSID (for example, the user's device might be *WPA* compatible, but the new network might require *WPA2*[1] compatibility).

Conversely, the user's device may be administered under an information technology policy prohibiting the connection to a network with the lax security displayed by the new network (for example, the new network is completely open and does not encrypt the traffic over the 802.11 RF hop). The 802.11 *client* function that we discuss in Chapter 5 is responsible for considering these issues when roaming.

3.5.2 Multiple Alternative SSIDs

Another scenario that has to be considered is the possible existence of multiple alternative SSIDs, all with acceptable signal strengths. In this case, the user must choose from among the SSIDs, the one that best meets the desired access, billing, and security requirements. Finally, when we do a global roam, we will change our IP address upon association with the new AP, and thus will lose our TCP/*UDP*[2] connection. Of course, in this situation our applications will

---

[1] *WPA* is *Wi-Fi Protected Access*; *WPA2* is *Wi-Fi Protected Access 2*.
[2] *User Datagram Protocol.*

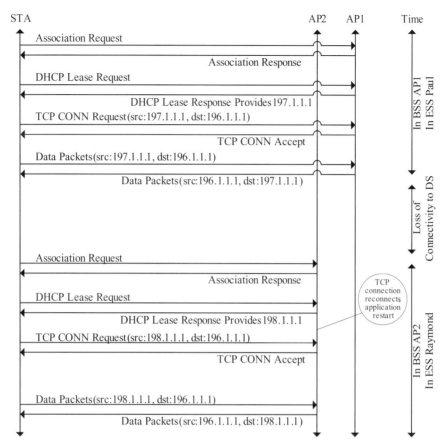

Figure 3.20: An Example of Global Roaming Showing the Impact on User Data.

have to restart. We illustrate this scenario in Figure 3.20. This figure differs from Figure 3.18 only in that in Figure 3.20 AP1 and AP2 are now in different ESSs (which is what makes this roam global). We can see that when the second DHCP-lease request is issued, the IP address that is granted is different from the original one; so the TCP session is torn down and reestablished, forcing an application reset.

Although the underlying technology differs significantly, it is important to draw a parallel between the concepts of global roaming that we have explained for cellular telephony in the first two chapters and that we have just discussed for 802.11. We defined global roaming for cellular telephony as a roam that resulted in a change of providers. In this section, we have presented global roaming for 802.11 as a change of ESS and IP subnets. Assuming that we are roaming between 802.11 networks, which either restrict access for the purpose of charging for their access or merely to keep unwanted visitors out of their network, switching IP subnets is likely to trigger similar admission, authorization, and accounting processes as the global roam of cellular telephony. These processes are the focus of Chapter 6.

3.6 Mobile IP and Its Role in 802.11 Roaming

3.6.1 Introduction

This section is devoted to Mobile IP [4]. As we described in Section 3.5, while roaming from one SSID to another SSID, it is normal to be assigned a new address belonging to the target subnet. This results in your application having to make a new TCP/UDP connection with your new IP address. To the extent that your device is "known" to other network entities, by its IP address being changed each time the device roams to a new location, this readdressing disrupts those entities. *Mobile IP* attempts to address this issue. With Mobile IP, you retain the same static IP address even though you are connecting to the Internet from different geographic locations. As of this writing, this technology is complex and, therefore, has not been widely implemented on a commercial basis. In this section, we will show how the use of Mobile IP can mitigate some of the negative user-application impact of a global roam. First, we introduce the basic concepts of Mobile IP.

3.6.2 Review of the Mobile-IP Architecture

Mobile IP is an IETF-proposed standard solution for IP mobility. It is described in IETF RFC 3220 [3]. Since the proposed standard was first published in November 1996, the commercial uptake of the technology has been limited. One of the principal reasons for the limited uptake has been the comparatively high tolerance for handoff-caused disruption on the part of the traditional mobile-Internet user. For example, if a business traveler arrives at a hotel, boots up a laptop, and has email and Web connectivity within a couple of minutes, this timing is acceptable. Other reasons for the limited deployment of Mobile IP include the facts that it has not been integrated into the popular operating systems and the paucity of neighboring networks where the benefits of Mobile IP can be seen. An important exception is the *3GPP2* technology that we discuss in Chapter 10, where Mobile IP is used for roaming in CDMA networks.

In general, the Internet-user community has come to expect that IP-application connectivity will be lost if there is a significant change in location. This paradigm is shifting with the availability of 802.11 access through a growing number of public hot-spot providers. Now, when a business traveler walks from the airport coffee shop to the waiting lounge at the gate, the person is less likely to be tolerant of the loss of an email session. If the email application provides access to archived folders that are on a server on the person's corporate network, restarting the email application can easily take a minute or two. This period will definitely try the patience of our executive who has only moved 50 m from the coffee shop. Mobile-IP technology can alleviate this problem. In order to make the discussion more concrete, we will now consider a simple example of Mobile IP in operation.

Roaming in 802.11 WLANs: General Principles

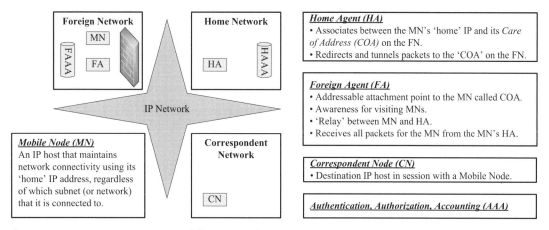

Figure 3.21: Some Important Mobile-IP Terminology and Its Relevant Locations in an IP Network.

Figure 3.21 introduces basic Mobile-IP terminology. The *Mobile Node* (*MN*) corresponds to the STA in our 802.11 nomenclature, although Mobile IP is independent of any particular layer-two access method. The MN has a home network where its native IP address resides. The native address in this context means an externally visible and externally routable Internet address, not a private address hidden behind a firewall. There is a *Home Agent* (*HA*) located in the home network that serves as a relay between the MN and any *Correspondent Node* (*CN*) which sends and receives packets with the MN. The CN can be any IP device in the Internet that communicates with the MN via its native address. The HA utilizes a database called the *Home Location Register* (*HLR*). User credentials sent locally by the MN to the *Foreign AAA Server* (*F-AAA*) are relayed back to the MN's home network where they are verified by the *Home AAA Server* (*H-AAA*).

The HLR contains information about the MN and also information about how the MN can be reached at the current time. When the Mobile-IP connection is operational, the MN is reachable via a *Foreign Agent* (*FA*) which includes an Internet-addressable attachment point for the MN. This attachment point is called the *Care Of Address* (*COA*). The reader should note that supporting FAs and HAs requires that the network-infrastructure equipment in the Foreign Network and the Home Network have Mobile-IP capability and that this capability is enabled. As of this writing, this situation is generally not the case in the Internet, although many manufacturers claim that their equipment supports Mobile IP. Also note that while the traditional instantiation of the FA function is in the network-infrastructure equipment, the Mobile IP protocol supports a special mode of operation called *co-located Foreign Agent* whereby the FA function resides on the MN.

Before an FA and an HA can collaborate to relay packets between a Correspondent Node and the MN, a Mobile-IP registration process must occur. We show the basic steps of this process in Figure 3.22. Before any Mobile-IP specific exchanges occur, the MN must gain basic connectivity in the Foreign Network. If the Access Network is an 802.11 network, achieving connectivity will entail joining a BSS. If the MN is running Mobile IP, the MN then solicits the local FA. The local FA will advertise a COA for the MN and registers the COA with the HA, which updates a binding table in the Home Network's HLR. When this registration process is complete, the MN establishes an IP tunnel from its COA (10.0.0.1 in Figure 3.22) back to the HA.

The tunnel created between the Foreign Network and the Home Network based on the example given in Figure 3.22 is shown in Figure 3.23. When a CN sends a packet to the MN, the CN sends the packet directly to the MN's Home Address (207.168.1.1 in Figure 3.22).

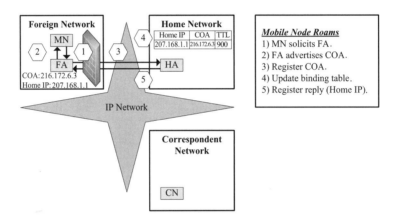

Figure 3.22: The Basic Steps in Mobile-IP Registration (IPv4).

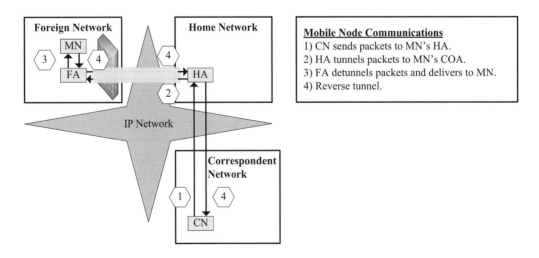

Figure 3.23: An Illustration of Mobile-IP Data Flow (IPv4).

When these packets are received by the HA, it sends these IP packets in the tunnel to the COA (216.172.6.3 in Figure 3.22). Upon receipt of these packets, the Mobile-IP application in the MN decapsulates the original packet as sent from the CN. The MN delivers the decapsulated packet to the bottom of the IP stack in the MN, which is working with the same IP address (207.168.1.1 in Figure 3.22) that the MN was using before it roamed into the Foreign Network. The packets sent from the MN to the CN are first encapsulated and sent via the tunnel back to the HA, where they are decapsulated and then relayed to the CN.

There is a more recent layer-three mobility protocol from the IETF's *IKEv2 Mobility and Multihoming Working Group* (*MOBIKE*) that provides both mobility and multihoming extensions to the *Internet Key Exchange* (IKEv2) protocol [1]. This protocol provides the user the layer-three mobility just described for Mobile IP while enjoying the protection of passing application traffic through an IKEv2 tunnel. For example, MOBIKE can provide a security-conscious MN with the ability to keep the connection with the *Virtual Private Network* (*VPN*) gateway active while roaming from one IP subnet to another.

3.6.3 802.11 Global Roaming with Mobile IP

In earlier sections, we defined global roaming as a roam entailing a move to a new ESS. It has been slated that such a move would require obtaining a new IP address and, as such, disrupt IP-based applications. The material covered in Section 3.6.2 about Mobile IP illustrates that if Mobile IP is in use, then it is not true that the local IP stack will be reset in the event of a global roam. In Figure 3.24, we illustrate how the impact on the TCP connection is reduced in an 802.11 global roam when combined with Mobile-IP technology. The figure shows that the application's TCP session is maintained across the roam. The figure also shows the complex protocol exchange taking place in order to mask the roam from the TCP-application session, and the extended time line illustrates the period when TCP communications, although not disconnected, are temporarily delayed.

Thus, the business traveler's walk from the coffee shop to the waiting lounge at the gate does not necessarily result in restarting the email application if the traveler's employer supports Mobile-IP infrastructure, the separate coffee shop and airport networks both implement Mobile-IP FAs, and all parties agree to the operation. The last step seems innocuous, but since these different entities bear the cost for these services, billing and accounting will be involved before the Mobile-IP registration is allowed to complete. The net result of all of this behind-the-scene activity is that while our executive may not lose an email session, the handover will definitely include a disruption in service, probably of the order of several seconds. Thus, while Mobile IP can mitigate some of the drastic effects of a global roam across 802.11 networks, as of this writing, Mobile IP does not offer a realistic solution for

Chapter 3

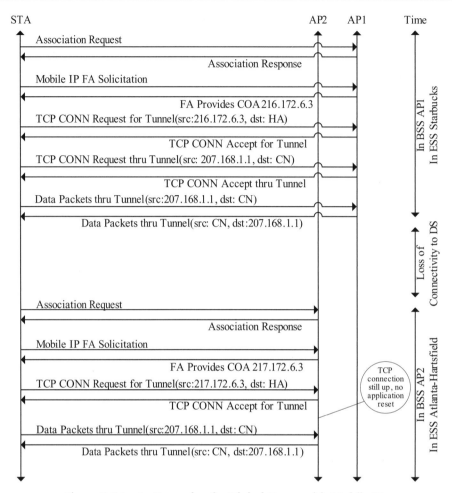

Figure 3.24: An Example of a Global Roam with Mobile IP.

transparent global roaming for time-sensitive voice. It remains to be seen whether this technology will play its promised role in IP-based roaming, or if it will remain, as some industry veterans believe, "A technology in search of a market."

3.6.4 Alternatives to Mobile IP

While Mobile IP provides a mechanism to maintain the IP address of an MN across geographical moves of unlimited scope, there are more simple, manufacturer specific solutions available. In such solutions, if the MN roams among APs that are all interconnected through a set of controllers from the same manufacturer, the IP address of the MN can be maintained. We discuss this in more detail in Chapter 8.

3.7 Those Pesky Laws of Physics

We have stated that this book attempts to cover the protocols and digital technologies which are the basis for fast secure roaming in the 802.11 standard. The substantial length of this text bears testimony to the fact that this topic is sufficiently broad, even without incorporating serious discussions of the RF realities underlying all these protocols and systems. In this section, we briefly touch upon a few of these RF realities that most blatantly interfere with our digital-only perception of how this process works.

3.7.1 Picocells—A Double-Edged Sword

Given a normal cell size utilizing the maximum transmit-speed capability of the AP, the further you move away from the AP, the more 802.11 will automatically reduce the speed in order to increase the likelihood of error-free transmission—a process known as *rate adaptation*. This fact suggests that it is better to reduce the cell sizes by increasing the density of the access points and thereby reducing their average transmit power. This paradigm of AP placement is called *picocells*. The picocell network-design approach theoretically provides higher bandwidth and better site coverage. Its proponents claim that with the picocell architecture, one can use lower power levels and still maintain high data rates due to the limited distances from the transmitters to the receivers.

While the small cell sizes purportedly minimize CCI, the 802.11 standard's control information is *still* sent at low data rates, which means that it will propagate over longer distances and thus still cause some CCI with nonadjacent APs on the same channel. This issue is tempered by the fact the percentage of the time that 802.11's control information is being transmitted by an AP is so low that the effect may be negligible in the bigger picture. Another drawback of the picocell environment is that when you reduce the transmit power of the AP, you reduce the signal to noise ratio, which renders the detection of the signal more difficult, as environmental RF noise grows. Finally, the denser spacing of APs will result in more frequent roams for the mobile user. While the number of roams will increase, the picocell approach is an effective way to increase capacity. In order to accomplish this while minimizing the drawbacks of picocells, it is important that the STAs reduce their transmit power in accordance with the reduced transmit power of the picocells' APs.

3.7.2 Limitations on Avoiding Channel Overlap

We stated in Section 3.2.3 that neighboring cells should be allocated to nonoverlapping channels. This logic is sound, but it is naive to believe that just because we separate two APs transmitting on the same channel by an intervening nonoverlapping channel AP that we can completely eliminate CCI. RF energy does not suddenly decline just because it crosses an arbitrary cell boundary drawn on a network topology; it continues to weaken and only gradually becomes indistinguishable from background noise. The notion of the three-color topology described in Section 3.2.3 will still yield some channel overlap once you have a fourth AP even though it is placed nonadjacent to an AP sharing its channel. Some of the energy of the signal from the same-channel AP will reach the new AP, causing some degree of interference. As just explained, this interference is exacerbated by the extent to which rate adaptation is in effect.

Optimal control of transmit power of an AP implies two things:

1. Assuming normal noise levels, the transmit power will be strong enough that a frame transmitted by that AP will be able to be received by all clients assigned to it

2. The transmit power is weak enough to minimize the number of receive errors on other devices that simultaneously transmit to it.

If another AP or client close by and on the same channel does in fact transmit at the same time as the AP in question, there will indeed be a collision with a consequent receive error. However, by regulating the transmit power, it is possible that the signal be kept weak enough to perturb a minimum of neighboring devices on the same channel when both of them transmit simultaneously. Some manufacturers offer dynamic control of the transmit power where a centralized management determines the optimal transmit-power distribution between the APs in order to achieve full coverage while minimizing CCI.

Such automatic *Transmit Power Control (TPC)* algorithms rely on a power capability information element communicated at association time and a TPC report that may be required at any times while associated. The TPC report includes statistics about the transmit power of the report frames itself, as well as the link margin as measured by the sender of the TPC report. Automatic TPC algorithms can adjust transmit power dynamically as a function of changing conditions, such as distance from the access point. Ideally, transmitters would be dynamically adjusted to transmit at a power level just above that required for successful reception. This carries multiple benefits including increased battery, life in the MN, as well as minimizing co-channel interference.

For further details, the reader can refer to one of the following references: [2] or [5].

References

[1] Pasi Eronen, IKEv2 Mobility and Multihoming Protocol (MOBIKE), Internet Engineering Task Force, 2005.

[2] Matthew S. Gast, *802.11 Wireless Networks, The Definitive Guide*. O'Reilly, 2002.

[3] Charles Perkins, IP Mobility Support for IPv4, RFC 3220, Internet Engineering Task Force, 2002.

[4] Jim Solomon, Applicability Statement for IP Mobility Support, RFC 2005, Internet Engineering Task Force, 1996.

[5] Charles Perkins, *Mobile IP: Design Principles and Practices*, Prentice Hall, 1997.

CHAPTER 4
Dynamics of 802.11 Task Groups

4.1 Introduction

4.2 Evolution of an IEEE Standard

4.3 Battle for Speed, Cost, and Market Dominance

4.4 The 802.11 Standard's Physical Layer

4.5 Fast Secure Roaming Task Groups

4.6 802.11i Security

4.7 802.11e Quality of Service

4.8 802.11k Radio Resource Measurement Enhancements

4.9 802.11r Roaming

4.10 Other 802.11 Subgroups

4.11 Wi-Fi Alliance versus IEEE 802.11

4.12 Patents and Intellectual-Property Issues with Standards

References

4.1 Introduction

As of this writing, the IEEE 802.11 WLAN standard is by far the most widely deployed wireless local area network technology in the world. Work on this standard began in the early 1990s. At that time, the 802.11 standard was seen largely as an untethered alternative to the

Chapter 4

802.3 Ethernet cable that connected most of the world's enterprise users to the Internet. The concept of supporting QoS-intensive features such as voice was not one of the priorities of the design. Although since the year 2000, it has become evident that 802.11 WLANs will be used for applications demanding QoS. Similarly, in the initial design, relatively little attention was paid to the security-related areas of authentication and encryption. It is now taken as a given that securing the first hop of 802.11 connections is indeed a prerequisite for widespread enterprise deployment of the technology. In parallel with these changes, the physical layer of 802.11 has undergone a number of major updates, and in the process has greatly increased its speed and range of technology. These concurrent factors of increased speed, greater range, QoS requirements, and security have combined to create a plethora of challenges for 802.11 implementations, and indeed for the IEEE 802.11 committee itself.

In this chapter along with a review of how the different 802.11 subgroups operate within each of these areas, we provide an explanation of how the IEEE 802.11 committee develops and ratifies standards through these subgroups. We believe that without an understanding of the machinations of the 802.11 standardization process, the reader will lack appreciation for the fact that so many years and so many evolutionary steps were required to transform 802.11 from its precepts to the state-of-the-art secure roaming that we describe in Chapter 9. We contrast the IEEE 802.11 situation to that of the *Wi-Fi Alliance*—a body for industry standards that mirrors 802.11 standards. The Wi-Fi Alliance moves more quickly than the IEEE in order to satisfy the commercial needs of its members. The Wi-Fi Alliance makes compromises and ratifies interim standards, when necessary.

4.2 Evolution of an IEEE Standard

4.2.1 Introduction

Two imperative principles guide the work of any IEEE standards group. These principles center around how the IEEE works as an organization, the far-ranging impact that the IEEE standards have had over the years, and the IEEE's largely successful effort over decades to avoid being co-opted by the selfish interests of a small segment of industry. These two principles are *due process* and *openness*. Due process means following publicly available procedures to the extent that the procedures are already defined by an IEEE entity hierarchically superior to one's task group, which means following those rules. If more-detailed procedures are established by one's task group, it means making those guidelines public.

Due process dovetails with the principle of openness. All written materials produced by a group, including minutes of meetings, are documents of public record; these documents can even be used for legal purposes. Task-group meetings are open to materially and affected parties, which in practice means anyone willing to attend the meetings. This second characteristic of openness clearly distinguishes the IEEE 802.11 process from the Wi-Fi Alliance process, which we will describe in Section 4.11.

To gain voting membership in 802.11, the prospective member must attend two of the four most recent plenaries (One interim can be substituted.), and must inform the Chair of the intention to become a voting member. The provision of membership begins at the start of the next plenary that is attended. In the sliding window of four plenaries, membership in 802.11 is maintained by attending two (One interim can be substituted.) of these, and by responding to two out of the three most recent *Working Group/Task Group* (*WG/TG*) ballots.

The IEEE allows for three types of standards balloting as follows:

1. Traditional balloting with one vote per individual.

2. Mixed individuals and nonindividuals, such as companies and organizations.

3. Nonindividual balloting only.

The 802.11 workgroup uses the traditional individual-based balloting.

4.2.2 New Standards

The IEEE wants to ensure that any new standardization effort follows the proper rules and procedures for IEEE standardization, and that the scope and nature of the technical content of the standard are precisely defined to avoid overlap and confusion with any other efforts underway in the IEEE. The IEEE-SA Standards Board is responsible for the first of these policing functions. The second function is performed by the sponsor for the standard.

Identifying the organization within the IEEE that will sponsor a new standard is the first step in the standardization process. The sponsor may be the IEEE board itself, one of the IEEE societies, or an active committee in one of those societies. In Figure 4.1, we see that the identification of the sponsor is the step immediately following the conception of the idea of the standard itself. The next two steps in the process are the submission and the approval of the *Project Authorization Request* (*PAR*).

The PAR is a very detailed but relatively short document (at least compared to the size of the normal document for the IEEE) that precisely describes the industry and/or market reasons why the corresponding project is needed, and additionally, what the project will do. A standardization effort is not official until its PAR is approved. IEEE standards often become legal requirements when they are adopted by governments or other international authorities. Because the PAR is in essence the root of the tree of documents for a particular

Chapter 4

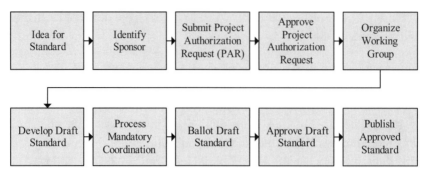

Figure 4.1: The Development Flow Chart for IEEE Standards.

standardization effort and because the PAR is formally approved at a precise moment in time, the PAR plays a critical role in these legal situations. A PAR may be formally extended after it has been approved.

4.2.3 Chairs and More on Balloting

Each IEEE working group must have a chairperson and may have other officers such as a vice chair, secretary, treasurer, or technical editor. Task groups are committees assigned by the working group to be the author of either a standard or amendments via a PAR approved by the working group. Task groups have individual chairs. Naturally, the role of the chair is pivotal. The standardization process inherently seems to show a trend toward deadlock, as different camps hold fast to their biased opinions. The chair has the essential and challenging role of ensuring that progress is continually made at the task-group meetings. Decisions on how to move ahead at these meetings are not made by the cumbersome and formal process of balloting, instead, a consensus of the meeting attendees gradually refines the material being encoded into their draft. It is imperative that this consensus be genuine and that participants really agree to compromise for the progress of the standard to move ahead. The reason why this consensus is so important is that when a series of meetings finally evolves such that the document reaches the point of a formal vote, a large majority of the task group must be happy with the document in order for it to be approved. The consensus system for advancing the document usually results in the formal ballots being cast to approve the document in question. This fact is important as the process in itself is unwieldy and sufficiently slow, thereby alleviating having to go back to the drawing board after a failed vote. A failed vote entails lengthy delays.

When a draft standard is being finalized, the group must determine the constituents who will actually cast the votes in the approval process. Unlike the task group itself, where anyone may freely participate, more care is exercised in the selection of the ballot group. Task-group members are not automatically included in the ballot group. Similar to the task group, the ballot group may be a mixed balloting group consisting of both individuals and non-individuals (for example, companies or universities), individuals only, or nonindividuals only. In order to achieve fairness in this important function, an effort should be made to construct a balance by ensuring that the ballot-group members represent fundamentally different interest groups, usually falling into the categories of producers, consumers, government, and others, without a preponderance of any one of these factions.

Coordination with other functions within the IEEE normally occurs during the development of the draft standard and prior to the approval by ballot. The process is overseen by the IEEE editorial staff. They ensure that rules regarding formatting and definitions have been followed. The goal is to provide some level of uniformity and consistency in IEEE standards.

The process of conducting the ballot in the IEEE task group is an effort to formalize the horsetrading and compromise, which resembles those regularly taking place in the upper and lower houses of governments around the world. The approval of an IEEE ballot is defined as a positive response to the ballot from at least 75% of the responders, where at least 75% of the ballot group must respond within the time limit of the ballot (normally 30–60 days) in order for the vote to be considered valid. As IEEE task groups cannot afford to haggle daily for months (as government officials do), the IEEE has constructed an iterative mechanism in which the debate is controlled and moves toward the 75% threshold needed to approve the document under consideration. This 75% approval is the formal definition of consensus within the IEEE context.

In a given IEEE ballot, there may be positive responses with comments. A negative response must come with a comment, which implies that if the comment can be addressed, the negative vote will be changed to a positive vote. This counting system is quite effective at stymieing intransigents who do not want to show flexibility in order to move the standard ahead. Abstentions are allowed.

The task-group chair works arduously to attempt to address the comments that were included in the negative votes, and the chair strives to convert a sufficient number of the negative votes to positive ones so that the ballot recirculation will achieve consensus, and the standard will move forward. A ballot recirculation is not a brand new vote; votes may only be changed due to modified language since the last vote was cast. For the process to progress efficiently, the duration of a ballot recirculation is usually kept much shorter than the original ballot itself. Assuming that after a number of re-circulations the ballot is ultimately approved, the standard will then undergo a final thorough editing by a professional IEEE standards editor. This revision will then be a completed standard.

Chapter 4

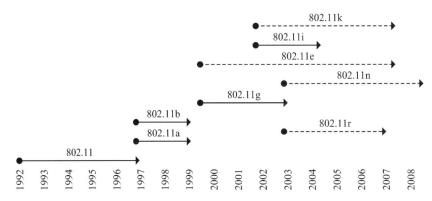

Figure 4.2: Selected 802.11 Task-Group Timelines.

4.2.4 Timeline of 802.11 Task Groups

Figure 4.2 shows the 802.11 task group timelines. A black circle indicates when the PAR was approved. A solid line followed by an arrowhead indicates when the standard was ratified. A dashed line indicates that the standard is still under progress and has not been ratified yet. The figure indicates, for example, that the PAR for 802.11 obtained approval in 1991 and that the standard was ratified in 1997. Several other task groups, for example, 802.11a and 802.11b achieved shorter ratification times. It should be noted that the efforts of the 802.11r and the 802.11n task groups are ongoing. As we delve into the individual standards in later chapters, the reader may want to refer back to the timeline shown in Figure 4.2.

4.3 Battle for Speed, Cost, and Market Dominance

4.3.1 Market Dynamics

The period from 1995 to the publication of this text has witnessed the evolution of 802.11 *Physical Layer Technology* (*PHY*) from the original 1-Mbps 802.11 standard, then to the 11-Mbps 802.11b standard, and ultimately to the 54-Mbps 802.11a and the 54-Mbps 802.11g chips available. While such an increase in speed is dramatic, in this age of high technology we have become inured to the drama of breakthroughs and have become accustomed to expect them. In parallel with this industry-wide effort to push back technical boundaries, other epic battles are being waged among the semiconductor companies, as they vie to succeed in this high-stakes game.

There are certain fundamental dynamics of the semiconductor industry that have revealed themselves and will continue to reveal themselves in the 802.11 chipset markets. For example, a company that excels in low-cost, high-volume manufacturing using established technology will not want to push ahead the bleeding edge of technical performance, and indeed may try to slow down its adoption. The reason is that very rapid adoption of the next generation of chips reduces the large volumes of the older technology that such companies depend on to make profits. A company that excels at technical innovation will want to continue to push out the latest technology, allowing it to maintain a high-profit margin on its continual innovation. For this kind of company, as soon as the price of the now-commoditized chipsets drops, it will exit that market, leaving that market to the high-volume commodity manufacturers who generally struggle with low-volume profit margins.

With the advent of commercial shipments of 802.11b chipsets in 1999, the initial markets for these chips were manufacturers that built these chips into *Personal Computer Memory Card International Association* (*PCMCIA*) and other adaptor form factors that could be added to existing laptops. By 2002, the 802.11b chipsets were offered as standard equipment on high-end enterprise laptops. This enterprise-class product is one that is normally ordered in large quantities by an IT manager who is willing to pay a premium for features like 802.11 support. The SOHO customers of the lower-end laptop still needed to upgrade their laptops with an 802.11 card (usually PCMCIA, although *Contention Free* (*CF*) and *Uniform Serial Bus* (*USB*) models were also common) during this phase. By 2005, the industry was reaching a turning point where on-board 802.11 support was increasingly considered standard, similar to what had happened with 802.3 support years earlier. This situation demanded a decreasing price premium; so this direction opened the door for the lower cost, lower margin, later to market, less-differentiation designers/manufacturers who were often Taiwan based. Examples of such companies include Realtek and Inprocomm, thriving on extremely high volume, razor-thin margin chipset sales.

4.3.2 Innovation

While the presence of onboard 802.11 in the laptop has declined from being a premium feature to a relative commodity, leading chipset manufacturers have temporarily retained market share by being pioneers in the latest 802.11 PHY technology. This scenario has been played out with each stage of the evolving 802.11 PHY technologies. An innovating semiconductor manufacturer, most-often North American or European, will grab an early lead in that particular technology. They are able to maintain high-profit margins, as long as the presence of that technology demands a significant premium over comparable devices, which are absent in that technology. This fact is true whether the 802.11 chipset is sold in a PCMCIA format or as a built-in adapter in a laptop. As long as such a company can be innovative either in having the latest PHY technology or in having some value-added software feature bundled with that chipset, they can succeed in maintaining higher profit margins. Manufacturers sustain these inevitably temporary situations as long as possible before giving up the battle to lower-cost manufacturers or to Intel.

Intel successfully used its enormous laptop manufacturer penetration (for example, HP, Dell, and Toshiba) as a launch platform for its Centrino chipset, Intel's marketing term for its wireless communications devices. Intel's marketing budget for Centrino was reputedly 250 million dollars for promotion alone! Intel is the juggernaut that ultimately leaves some of the early innovators looking for new markets in which to innovate, and the lower cost chipset manufacturers fighting over the lower end of the laptop market, where the customer is not likely to pay a premium for the Intel Centrino brand. This very sizable lower-end market is a low-margin one, making it of less interest to the original innovators.

4.3.3 Recent History

The five years leading up to 2005 witnessed the early innovators like Intersil and Atheros gradually losing ground in the laptop market, first to Broadcom, and then ultimately all the three companies losing out to Intel and to low-cost Asian competitors. The process that we see happening with 802.11 chipsets happened in a similar fashion with 802.3 Ethernet chips that went from high-margin devices in the early 1990s to commodity items just a short time later.

While the initial inroads for the 802.11 chipsets were with laptops, the 802.11 technology rapidly spread to handheld devices ranging from VoIP phones, to dual-mode phones, to consumer devices such as TIVO and combined MP3 player phones, first released by large Japanese consumer-electronics manufacturers. These markets are very different from the IT-dominated laptop market; so it is not surprising that a different set of semiconductor vendors first innovated and now dominate in this segment of the 802.11 market. In particular, Texas Instruments, a large semiconductor manufacturer which was never particularly successful in penetrating the 802.11 market for laptops, is likely to be a major vendor in the 802.11 handheld space due to its high penetration of cellular phone handset manufacturers. These handsets form a natural platform to expand into dual-mode 802.11/GPRS or 802.11/3G devices.

4.4 The 802.11 Standard's Physical Layer

4.4.1 Introduction

In the previous section, we examined several market forces that drive the course of 802.11 evolution. In this section, we change gears and deal with the 802.11 Physical Layer.

The basic 802.11 PHY standard was ratified in 1997. The relatively slow 1–2 Mbps theoretical-speed limitation of basic 802.11, as compared to wired 802.3 Ethernet, ensured that work would begin on higher speed variants of the standard as soon as this initial effort was complete. In Table 4.1, we summarize the salient differences between the original 802.11 PHY and the variants that have quickly followed its footsteps, and we list the various 802.11 standards, dates of ratification, and other important properties of the standards.

The term "transfer mechanism" in Table 4.1 describes the modulation technique(s) used by a particular PHY. *Direct Sequence Spread Spectrum* (*DSSS*) and *Frequency Hopped Spread Spectrum* (*FHSS*) were two alternatives for the original 802.11 PHY standard. The 802.11a standard uses *Orthogonal Frequency Division Multiplexing* (*OFDM*). The *Complimentary Code Keying* (*CCK*) used in 802.11b is an extension of DSSS. The 802.11g standard uses CCK for its slower speed backward-compatibility mode with 802.11b and OFDM for its high-speed mode. The 802.11n standard proposes to use both OFDM and *Multiple-Input Multiple-Output* (*MIMO*).

4.4.2 Deployments and the Players

As of this writing, because 802.11n is far from being commercially deployed, the choice is between two 54-Mbps alternatives: 802.11a and 802.11g. The 802.11g standard currently enjoys greater commercial success because it represents an easy migration from the widely deployed 802.11b and shares 802.11b's longer ranges, as compared to 802.11a. In Table 4.2,

Table 4.1: The PHY Characteristics of 802.11.

Name	Ratification Date	Frequency Band (GHz)	Theoretical Speed (Mbps)	Average Range	Transfer Mechanism
802.11	1997	2.4	1–2	–	DSSS FHSS
802.11a	1999	5.8	54	25 m	OFDM
802.11b	1999	2.4	11	55 m	CCK
802.11g	2003	2.4	54	55 m	CCK OFDM
802.11n	TBD	2.4 and 5	540	TBD	OFDM MIMO

Note: 802.11a has twelve nonoverlapping channels; 802.11b and g each have three nonoverlapping channels; 802.11n's number of nonoverlapping channels has not been determined yet.

Table 4.2: The PHY Advantages and Disadvantages of the Various 802.11 Versions.

Name	Advantage	Disadvantage
802.11	1. No significant advantage	1. Very slow
802.11a	1. No interference problem like 802.11b had 2. Larger number of nonoverlapping channels 3. High speed	1. Not backward compatible with 802.11b 2. Shorter range
802.11b	1. Widespread adoption 2. Longer range	1. Potential interference with microwave ovens, cordless phones, and Bluetooth
802.11g	1. Backward compatible with 802.11b 2. Longer range 3. High speed	1. Same interference problems as 802.11b
802.11n	1. Very high speed	1. Not yet known

we compare and contrast 802.11, 802.11a, 802.11b, 802.11g, and 802.11n. However, it is unlikely that 802.11g will completely eclipse 802.11a.

The small number (three) of nonoverlapping channels of 802.11g compared to 802.11a (twelve) increases the likelihood of cochannel interference and thus can be a severe limitation with regard to AP placement. The ability to place more APs in a given area helps provide better aggregate bandwidth to the users in that area, as well as improved ability to avoid dead spots. As more and more mobile users expect to roam through their facilities without passing through dead zones and thus dropping their voice call, the 802.11a standard might grow in popularity. RF-spectrum allocation policies in different countries can make the 2.4 GHz 802.11g either more or less problematic than the 5 GHz-based 802.11a; so this issue may also affect a trade-off between the two.

Proxim was one of the early companies to launch commercial products based on the original 802.11 PHY standard. Proxim's decline is an example of the fact that being a technical pioneer does not always ensure long-term survival. The reality is that the speed limitations of the original 802.11 restricted its adoption, and the real explosion of 802.11 did not begin until the commercialization of 802.11b chipsets.

Initially, Agere and Intersil were dominant players in 802.11b. Their respective Orinoco and Prism chipsets were the ones used in early driver implementations on Linux platforms, as early on, they were the most stable and readily available hardware. At first Atheros dominated in the 802.11a market. Undoubtedly, the market forces and trends will continue to be favorable to some companies and destructive to others.

4.5 Fast Secure Roaming Task Groups

4.5.1 Introduction

We focus on the work of task groups 802.11i, 802.11e, 802.11k, and 802.11r in this section and also later in this chapter, as the work of these groups most directly influences the form that fast secure roaming in 802.11 will take. We need the focus of 802.11i for security, the measurement and reporting focus of 802.11k to make intelligent decisions about which AP to roam, and the QoS provisioning capabilities afforded by 802.11e to provide for that most common justification for enhanced 802.11 roaming—the 802.11-enabled VoIP phone. The 802.11r standard will attempt to combine the roaming-related work of these other groups into a single standard.

4.5.2 Basic Architectural Services

First, we list the 15 basic architectural services defined for the 802.11 standard. They are as follows:

- Authentication
- Association
- Deauthentication
- Disassociation
- Distribution
- Integration
- Confidentiality
- Reassociation
- *MAC Service Data Unit (MSDU) delivery*
- *Dynamic Frequency Selection (DFS)*

Chapter 4

- *Transmit Power Control (TPC)*
- Radio Measurement
- Higher Layer Timer Synchronization
- QoS Traffic Scheduling
- BSS Transition Services

We provide a mapping of the 802.11 task groups to these basic architectural services in the following section.

4.5.3 Workgroup Foci

The work of task group 802.11i focused on authentication and confidentiality, all of which fall under the general umbrella of security [6]. Radio Measurement falls under the purview of 802.11k. The 802.11e standard deals with QoS Traffic Scheduling and Higher Layer Timer Synchronization. The BSS Transition Services are specified in 802.11r.

While some improvements to roaming performance can be achieved via combined use of 802.11i, 802.11e, and 802.11k, the industry consensus is that the long-term 802.11 roaming solution will indeed be based on 802.11r, which itself builds on the 802.11i, 802.11e, and 802.11k technologies.

It is also noteworthy that while each of these projects produces an enormous-sized specification, these documents are amendments to the 802.11 specification and, thus, ultimately result in changes and additions to the base standard rather than existing as separate stand-alone documents. As the work of these groups evolves independently and often in parallel, it is quite challenging to understand the combined impact of these amendments until they are all incorporated into a version of the full standard. In the subsequent sections we briefly summarize the work of each of these four task groups.

4.6 802.11i Security

4.6.1 Introduction

The 802.11i task group began in 2001 with the goal of providing safe secure access to 802.11 networks [2]. This effort started as a result of the market's very negative reaction to the perceived vulnerability of WEP, which at that time was the standard means for protecting an 802.11 network connection. The 802.11i amendment was the result of three years of intense debate and compromise, and it was finally ratified in 2004. The economic consequences of getting this particular part of the standard done correctly and quickly were enormous. If 802.11 could not "get security right," the expected wide adoption of the technology would never occur.

The terms *Robust Secure Networks* (*RSN*) and *Safe Secure Networks* (*SSN*) have both been used publicly by 802.11i to describe its general goal, although the finally ratified amendment uses RSN. The 802.11i amendment defines RSN procedures that occur during the association phase; these allow the client and AP to determine the security context of their particular association. Figure 4.3 illustrates how the basic 802.11 association procedures are expanded to include the RSN information. The figure shows the requests and responses between a client and a wireless AP. The security context, at the most basic level, will establish whether the *personal* or the *enterprise security* mode is to be used.

The personal security mode is called *Preshared Key* (*PSK*) mode in the 802.11i standard parlance and is intended to provide easy-to-use, yet adequate security for the SOHO markets. The enterprise security mode uses the IEEE 802.1X standard for authentication and key exchange. This mode is a much more robust security technology than personal mode, but, a much deeper knowledge of security technologies is required in order to deploy enterprise security growth effectively. A more complex authentication infrastructure is also required. This complexity is the reason why the term "enterprise" is being used.

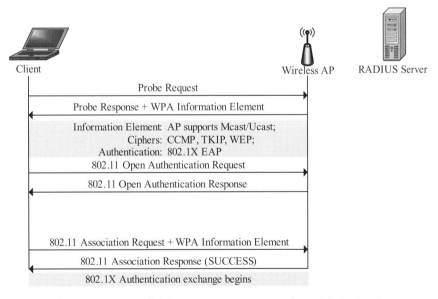

Figure 4.3: Establishing an 802.11 Connection with 802.11i.

4.6.2 WEP's Limitations

One of WEP's greatest limitations is that the symmetrical encryption of the data frames was based on a particularly vulnerable way of using *Rivest's Cipher 4* (*RC4*) encryption. A number of vulnerabilities in RC4, as implemented in WEP, were revealed early in WEP's deployment. These vulnerabilities served to underscore the immediate need for task group i to ratify a solution that addressed these problems. In fact, during the early years of the task group's work, the discovery of a new possible exploitation of WEP seemed to be such a frequent occurrence that many eventually considered WEP to provide no protection at all. In order to address this issue, task group i developed two new ciphers: *Temporal Key Integrity Protocol* (*TKIP*) and *Counter Mode CBC-MAC Protocol* (*CCMP*).

TKIP actually uses the RC4 cipher in an improved way that circumvents the vulnerabilities exposed in WEP. CCMP is based on a fundamentally different and more robust cipher, *Advanced Encryption Standard* (*AES*), that was adopted by *National Institute of Standards and Technology* (*NIST*). One advantage of TKIP is that because it was based on RC4, users could upgrade their WEP-based 802.11 hardware implementations to TKIP, whereas with CCMP it was necessary to upgrade to new 802.11 hardware in order to use chip-based AES encryption.

4.6.3 802.11 Cipher

Regardless of whether personal security mode or enterprise security mode is used, and regardless of the cipher used, an 802.11i cipher needs two keys: a *Pairwise Transient Key* (*PTK*) and a *Group Transient Key* (*GTK*). The PTK is used to encrypt unicast traffic both from the STA to the AP and from the AP to the STA. The GTK is used by the AP to encrypt broadcast/multicast traffic sent to all STAs currently in the BSS, and the STAs need this key in order to decrypt the traffic. The basic operation (refer to Figure 4.4) is as follows:

1. The STA associates and then negotiates the security parameters used with the association.

2. The AP authenticates the user in enterprise security mode. This step does not exist in personal security mode.

3. A four-way key validation protocol is executed such that the PTK becomes available in the STA and on the AP.

4. The agreed-upon temporal keys are programmed into the local 802.11 chipset, using the negotiated cipher, and subsequent frames are encrypted.

In the final version of the 802.11i standard, corresponding to WPA2, the encrypted GTK is sent with message three of the four-way handshake just described. In the interim version that corresponded to WPA, the GTK was communicated in a two-way handshake dedicated to communicating the broadcast key. When the GTK is updated by the AP for all STAs in the BSS, the two-way handshake is still used by itself, both in WPA and WPA2.

Step 3 is identical in both personal and enterprise modes in that the four-way handshake is used to derive a PTK from a *Pairwise Master Key* (*PMK*) in both cases. The two modes differ with respect to the source of the PMK. This process for personal security mode is shown in Figure 4.4. In the figure, the PMK appears to be already present on both the STA and the AP before the association. This situation is in fact the case because the PMK is part of the Preshared Key, which is a static pass-phrase encoded on the AP and any STAs expecting to associate with it. In the case of enterprise security mode, the PMK is dynamically derived through the authentication process that occurs in step 2. This method adds a greater degree of entropy to the enterprise mode process since the PMK itself is fresh for each session.

In Figure 4.5, we depict the full process of deriving the PMK during the 802.1X authentication and then deriving the PTK from the PMK using the four-way handshake. However, both personal and enterprise modes in 802.11i do exhibit some degree of key freshness because regardless of how the PMK was derived, the PTK is always fresh for each new association. This statement is not true with static-key WEP, which exposed it to *dictionary attacks* on the key, because the key was the same across all stations in the BSS and the same from session to session over extended periods of time. The steps outlined here for 802.11i have fortunately mitigated these vulnerabilities.

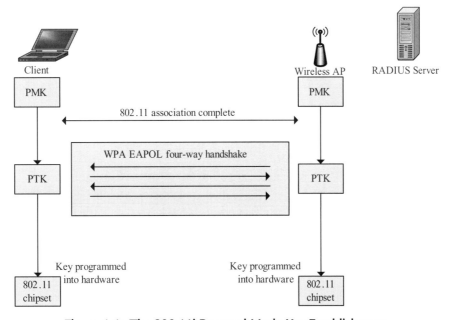

Figure 4.4: The 802.11i Personal Mode Key Establishment.

Chapter 4

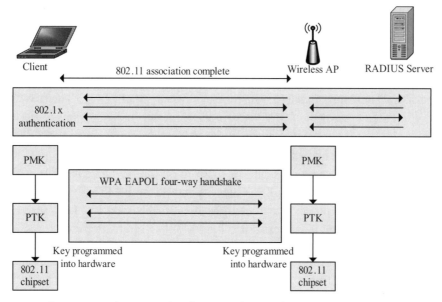

Figure 4.5: The 802.11i Full Enterprise Mode Key Establishment.

4.6.4 Preauthentication

One of the last major features added to 802.11i was *preauthentication*. This feature is highly relevant to the discussion on roaming because the security gained from the task group i work is achieved at the expense of additional complexity. This complexity manifests itself in terms of latency between associating and being able to use 802.11 for effective user communications. If this lag occurs only once at the beginning of a long session, the additional delay, ranging from hundreds of microseconds to a few seconds, may be acceptable. In a mobile-roaming environment, a brute-force implementation of 802.11i security would result in this penalty being paid at the start of each and every roam. Task group i recognized this solution as unacceptable for a technology that is increasingly being considered as a medium for the mobile VoIP user. Such a user cannot be subjected to multisecond interruptions in phone calls while wandering down the corridor.

Preauthentication lessens latency by caching some of the keying material derived during the authentication (step 2 described earlier) in neighboring APs to which the user is likely to roam. This advanced work is accomplished by performing the authentication for the candidate APs through the current AP and distribution service, usually a wired LAN infrastructure, connecting it to the candidate AP. This technique permits control dialog between the STA and roaming candidate APs without disruption of the current data flow, as would be the case in basic 802.11, which presumes that the STA needs to associate with the candidate AP in order to communicate. This exchange would in fact be necessary were the candidate APs not connected via the same DS, but this situation is rarely the case. We will see later in Section 4.9 as to how 802.11r takes advantage of this infrastructure-side connection as well, in what is called *Over-the-DS* (*OTD*) mode. As of this writing though, support for 802.11i preauthentication has been dropped from the 802.11r draft.

Preauthentication, including its limitations, will be discussed in a more detailed manner in Chapter 7. The fact that preauthentication lessens the latency but does not really reduce it sufficiently for reliable roaming for voice calls made this topic of great contention with regard to the task group and resulted in this part being optional in the standard.

4.7 802.11e Quality of Service

4.7.1 Introduction

The 802.11e task group deals with standardizing aspects of 802.11 that relate to providing QoS guarantees about loss and delay in the inherently unpredictable 802.11 environment [3]. In addition to a historical overview of the task group's work, we also discuss their current directions and how these plans bode, ill or well, for the support of QoS-sensitive applications on 802.11.

The original 802.11 standard included some basic concepts related to QoS. These ideas were based on dividing the interval between two beacons into two separate phases: the *Contention Period* (*CP*) and the *Contention-Free Period* (*CFP*). Access to the medium during the CP is arbitrated by the *Distributed Control Function* (*DCF*), which is based on the collision-avoidance principle of CSMA/CA introduced in Chapter 3. Access during the CFP is based on the *Point Coordination Function* (*PCF*) in the AP. During this period, when the PCF decides that the STA should be allowed to transmit, the PCF polls the STA via a CF-Poll frame. The fact that the PCF is centralized in the AP allows for the required deterministic access to the medium which is necessary for QoS guarantees. The original 802.11 standard did not provide details on how this access should be arbitrated. Although both DCF and PCF were specified in the original 802.11 standard, the commercial sector only implemented the DCF and 802.11-QoS guarantees remained merely a concept. The 802.11e task group expands upon those basic concepts to provide richer and better specified QoS mechanisms for 802.11.

The 802.11e standard can provide QoS in either of two basic access mechanisms. The first is called *Enhanced Distributed Channel Access* (*EDCA*) and is based on an extension of the DCF called the *Enhanced DCF* (*EDCF*). As some QoS priority is allowed even in contention-based (best-effort) mode, this direction is a significant departure from basic 802.11. This mechanism defines different traffic queues for different traffic classes such that the higher priority traffic will get better access to the wireless medium during the CP. However, as the process is still contention-based, there are no true QoS guarantees in this mode. Despite the lack of hard guarantees, many believe that the higher priority for VoIP frames allowed by this mode will provide an acceptable level of service for voice. This mode is sometimes called *Wi-Fi Multimedia Mode* (*WMM*).

The second access mechanism is called *Hybrid Coordinator Function Controlled Channel Access* (*HCCA*). This mechanism is based on an enhancement of the PCF called the *Hybrid Coordination Function* (*HCF*). One of the principal enhancements to 802.11's PCF is that the *Hybrid Coordinator* (*HC*) of 802.11e can grant access during the CFP in any order it chooses, not just round robin, as was the case with the PCF. This flexibility gives the HCF more capability in providing fine-grained QoS guarantees. A second major enhancement is that, similar to the EDCF, the HCF provides for different traffic classes. The fact that the HC may allow the STA to send multiple packets in a row for a total time slot stipulated by the HC to the STA is the third significant difference from 802.11. This feature would be useful if the HC knows that the amount of QoS-sensitive traffic queued at the STA requires a longer time slot in order to meet its QoS guarantees. A fourth noteworthy enhancement is that the HCF may even interrupt the CP to grant immediate access to an STA via a CF-Poll frame. Although this ability violates the strict division between the contention-based and contention-free periods, it does augment the HCF's capability to deliver QoS guarantees. This mode is sometimes called the *WMM Scheduled Access* (*WMM-SA*) standard. At the time of publication of this text, there is no active work going on by this subgroup due to a lack of interest from vendors.

We show the EDCF and the HCF in a *QoS BSS* (*QBSS*) with two *QoS STAs* (*QSTAs*) in Figure 4.6. This figure shows, that from the single AP, frames are exchanged with the two STAs via both the EDCA and the HCCA access methods. Frames may only be sent from the STAs under the HCCA method after a CF-Poll has been received from the AP's HCF. Similar to the original 802.11's PCF, commercial adoption of HCCA is very poor. There is better commercial adoption of EDCA, as we will see in the discussion of the Wi-Fi Alliance WMM certification in Section 4.11.2.

4.7.2 Mixed Environment Terminology

The 802.11e task group anticipates that these new functions will be implemented in hybrid environments consisting of devices that are 802.11e enabled and others that are not 802.11e enabled. It is thus necessary to define expected behavior in such a mixed environment. To this end, the 802.11e standard defines the following terminology:

- QBSS: A BSS capable of operating in 802.11e mode.

- QSTA: An STA capable of operating in 802.11e mode.

- QAP: An AP capable of operating in 802.11e mode.

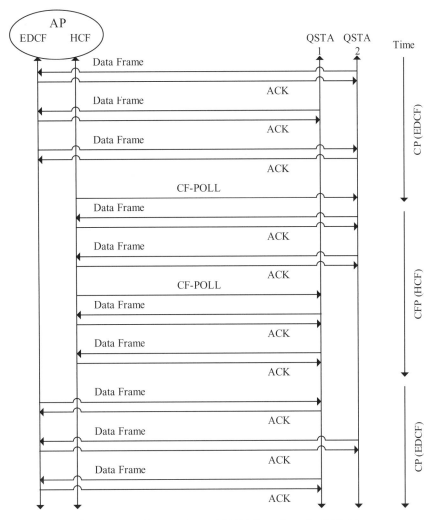

Figure 4.6: The EDCF and HCF Functions in 802.11e.

- nQBSS: A BSS not capable of operating in 802.11e mode.

- nQSTA: An STA not capable of operating in 802.11e mode.

- nQAP: An AP not capable of operating in 802.11e mode.

An nQSTA implements a subset of QSTA functionality; therefore, a QSTA may associate with an nQAP serving in an nQBSS, and thus the QSTA can receive non-QoS data service from an nQAP. As of this writing, it appears that this "Q∗" and "nQ∗" terminology will be formally dropped from future specifications. We continue to use this terminology in this text as it provides a useful shorthand.

4.7.3 Miscellaneous Issues

The work of task group e reflects the IEEE principle of avoiding the invention of new standards when existing ones can be reused. This fact is evidenced in the parts of 802.11e dedicated to priority queues and traffic classes. These concepts are the same ones used in earlier IEEE and IETF standards related to QoS (see [9]). For instance, the 802.11e standard uses the priority values 0 to 7, which are identical to the IEEE 802.1P priority tags. Task group e uses the definition of *Traffic Specification* (*TSPEC*) for traffic classes identical to those used in IETF RFC 2216 [10]. The 802.11e standard can distinguish the QoS requirements of different applications either by priority tags or by TSPECs.

Traditional mechanisms for carrying QoS-sensitive digital traffic, such as *Time Division Multiplexing* (*TDM*) equipment, had automatic clock synchronization built into the transmission medium. Asynchronous transmission media like 802.11 do not have this inherent synchronization that is very beneficial to QoS-sensitive traffic. The 802.11e standard specifies the use of a multicast synchronization "pulse" frame sent to all the QSTAs in a QBSS which the QSTAs can use to synchronize their transmission schedules.

4.8 802.11k Radio Resource Measurement Enhancements

4.8.1 Introduction

The IEEE 802.11k standard or *Radio Resource Measurement* (*RRM*) adds measurements that extend the capability, reliability, and maintainability of WLANs through measurement communications primitives involving the MAC, the *Station Management Entity* (*SME*), and *MAC Layer Management Entity* (*MLME*) layers [4]. The information gathered is made available via a standard *Management Information Database* (*MIB*) interface.

When 802.11k is implemented, the STA will maintain lists of known STAs and APs that are visible to the requester. The list entries will include link statistics, channel utilization, data rates, and link margin, as measured and reported by those STAs and APs. The specification also optionally supports asynchronous reports to a network management system of association and authentication events.

To support these statistics, there are a number of reports defined as part of the 802.11k standard, which are as follows:

- Beacon Report
- Frame Report

- Station Statistics Report
- Channel Load Report
- Medium Sensing Time Histogram Report
- Noise Histogram Report
- *Location Configuration Information (LCI)* Report
- QoS Metrics Report
- Link Measurement Report
- Neighbor Report
- Measurement Pause

Each of these reports, except for the Measurement Pause, actually defines a Request/Response pair where information is solicited in the request and returned in the response. The information is sometimes returned immediately, and at other times, the request initiates a measurement interval at the conclusion of which the information gathered is returned in the response. In these cases, when a measurement interval of zero is stipulated, this dictates that the responding STA should provide the instantaneous value of the requested statistics. The Measurement Pause differs from other reports in that rather than soliciting information, it merely stipulates the start and stop of a measurement pause in which no measurement request is honored.

All the requests in the preceding list, except for the Link Measurement Report and Neighbor Report, are communicated via a Measurement Request Frame, and the corresponding reports are returned in Measurement Report Frames. A type field in the Measurement Request Frame stipulates the type of report requested. The two exceptions differ in that there are dedicated Neighbor Report Request and Link Measurement Report Request frames. These two cases are treated separately because they do not formally request that a measurement be initiated, merely the reporting of the information already held by the recipient of the request.

The Beacon Request is used by an STA to learn what channel(s) is (respectively, are) seen to be active by another STA. The request can be made in three different modes: passive, active, and beacon-table mode. In both passive and active modes, the request initiates a measurement interval of a specified length. At the end of this interval, the information gathered by the STA from beacons or probe responses is reported to the requester. The difference between active and passive modes is that in active mode the STA will initiate a probe request in order to solicit the information from the AP. In passive mode, no probe request is transmitted when the

802.11k Beacon Request is received, and the statistics returned will be the result of beacons and probe responses unsolicited by this Measurement Request. In both cases, the information returned is for the channel(s) and BSSID specified in the Beacon Request. In beacon-table mode, there is no measurement interval, and the STA returns information already stored in its beacon table.

The Frame Request solicits summary information about the frames received by the STA that receives the request. The STA that receives the request returns a report summarizing the frames received from each different transmit address from which it has received frames. For each such transmit address, a quadruple consisting of received frame count, the average *Received Channel Power Indicator* (*RCPI*), BSSID, and transmit address is included in the report.

An STA sends a Link Measurement Request to another STA in order to request that it measure transmit-power information and an estimation of the link margin. This information enables understanding the instantaneous capabilities of a link for QoS-type requirements.

4.8.2 Station Statistics Request

When an STA receives a Station Statistics Request, it responds with a Radio Measurement Report Frame containing statistics for the interface over which the request was received. These statistics consist of the following counts:

- Transmitted fragments
- Multicast transmitted frames
- Transmit failures
- Retries
- Multiple retries
- Duplicate frames
- RTS successes
- RTS failures
- ACK failures
- Received fragments

- Multicast received frames

- FCS errors

- Transmitted frames

4.8.3 Additional Reports Important to Roaming

The remaining six request/responses are particularly significant for the role that they can play in enabling fast secure roaming. A Medium Sensing Time Histogram Request provides a flexible means for the requesting STA to retrieve a variety of complex reports from the receiving STA. As the name implies, the request provides the required parameters for a flexible histogram report, namely, bin offset, bin duration, and number of bins. There are four types of histograms that can be requested. They are as follows:

- *Received Power Indicator* (*RPI*) Time Histogram

- *Clear Channel Assessment* (*CCA*) Idle Time Histogram

- CCA Busy Time Histogram

- *Network Allocation Vector* (*NAV*) Busy Time Histogram

All the four histograms mentioned previously provide an indication of the relative usability of the RF medium, as perceived by the reporting STA.

A Channel Load Report returns the proportion of the specified measurement interval during which the reporting STA determined the specified channel to be busy.

The Noise Histogram Report consists of a nine-bin histogram measuring non-802.11 energy present in the medium. These energy readings are obtained by sampling the specified channel only when CCA indicates that the medium is idle. Each of the bins contains a count of the RCPI samples that fall into the power range (measured in dBm) corresponding to that bin.

The LCI Report pair returns a requested latitude, longitude, and altitude. The requesting STA may request the location information corresponding to itself or to the reporting STA. If the reporting STA lacks the ability to provide the information, the data is returned as zeros.

The QoS Metrics Request stipulates a Peer QSTA and traffic class for which transmit queuing delay is to be measured. The Peer QSTA is identified by a destination MAC address to which the reporting QSTA is currently transmitting frames of the given traffic class. The traffic class is specified as a number from 0 to 15 indicating a traffic priority (0–7) or TPSEC (8–15). The queuing delay is reported via a histogram.

Neighbor Reports contain information from the MIB Table *dot11RRMNeighborReportTable* that contains information concerning neighbor APs. The 802.11k standard does not stipulate as to how the information in this table is derived by the AP. An STA requests a Neighbor Report by sending a Neighbor Report Request Frame to the AP with which it is associated. This information can be used by the STA to determine potential roaming candidates. In Chapter 9 we discuss how this information can be used by the STA to improve its scanning efficiency by only scanning on those channels where there are neighbor APs, and thus reducing the length of the discovery phase.

The preceding six report types can greatly enhance the roaming capabilities of 802.11. With the deeper knowledge of the current network topology, RF, and QoS states, it is possible to make more informed and, therefore, better roaming decisions. As 802.11k can play a significant role in improved roaming, we provide additional information of this task group's work in Chapter 9.

4.9 802.11r Roaming

4.9.1 Introduction

The original 802.11 standard does incorporate roaming in a simple form. However, the limited speed and security conferred by the original standard makes it essentially of no use for secure-voice applications. The 802.11r task group deals with the standardizing aspects of 802.11 that will reduce the transition time during the roam, referred to as a BSS transition in 802.11r [5]. With regard to the principal 802.11 amendments required for fast secure roaming for VoIP applications, the 802.11r standard is less mature than the others. By less mature, we mean that the other standards are either already ratified or closer to ratification than 802.11r. Chapter 9 builds upon the security, QoS, and reporting capabilities defined by 802.11i, 802.11e, and 802.11k in order to provide details about an all-encompassing *Fast BSS Transition* (*FBT*) architecture. In this sense, it was necessary for the other groups to develop their solutions before 802.11r could integrate them.

A complete solution to Fast BSS Transition will rely upon 802.11e overloading information about an AP's QoS state in its beacons and probe responses. Similarly, the 802.11k standard provides Neighbor Reports and other measures that the roaming STA can use to make informed decisions about which AP to choose as its transition target. While use of such information is clearly part of improving the roaming experience, how the STA chooses the transition target is beyond the scope of 802.11r. The goal of 802.11r is to minimize the time that the STA loses connectivity to the DS. In the most common cases, the DS is the wired LAN infrastructure connecting the APs. Losing connectivity with the DS equates to stopping the flow of user application traffic, something to be minimized with real-time traffic, notably voice.

The 802.11r standard defines the term *mobility domain* as the set of fast-transition-capable APs to which an STA can roam to at the moment. All the APs in a mobility domain are interconnected via a single DS. In order to describe 802.11r-capable STAs and APs, the amendment introduces the terms *Fast Transition Enabled STA* (*TSTA*) and *Fast Transition Enabled AP* (*TAP*).

4.9.2 Basic Service Set Transition Pre-802.11r

Based on pre-802.11r techniques, a BSS transition for a secure QoS-sensitive application required the following five stages:

1. Scanning for target APs.

2. Open 802.11 authentication. (As we pointed out in Chapter 3, this exchange is for compatibility with the original 802.11 specification and achieves no true user authentication. Such true user authentication occurs in step 4.)

3. Reassociation.

4. PTK derivation. The complexity of this step varies depending on whether key caching, preauthentication, or a new complete 802.1X reauthentication is involved in providing the PMK at the new AP. Even in the abbreviated cases of key caching and preauthentication, a four-way handshake is required to derive the PTK. (Key caching is an alternative to preauthentication that, similar to preauthentication, makes the PMK available in roaming candidates. In Chapter 8 we provide the complete details of key caching.

5. QoS admission control with the new AP.

Considering these five serialized stages, even if we assume the shorter latency of key caching or preauthentication, the total time for the BSS transition will likely be measured in the many tens of milliseconds or even more, which will disrupt a voice conversation. In addition, the fact that the QoS admission occurs only at the end of this lengthy process implies that admission may fail and the roaming STA will have to start over and explore another potential candidate. The next section focuses on how the 802.11r standard compresses these steps.

4.9.3 How 802.11r Handles the BSS Transition

The 802.11r standard compresses the five stages described in Section 4.9.2 in two major ways—collapsing the four-way handshake into the 802.11 authentication/association exchange and QoS-resource pre-allocation. New Information Elements are added by 802.11r. The

Mobility Domain Information Element (*MDIE*) provides information that identifies the current mobility domain. The *Fast Transition Information Element* (*FTIE*) enables the advertisement of network-infrastructure resource-reservation information and security-policy information.

Both the MDIE and the FTIE are present in beacons, probe responses, association requests, and association responses sent from TAPs and TSTAs. In the later stages of task group r's work, the FTIE was extended to include fields from the 802.11i EAPOL-Key messages. By adding this security-related information to the 802.11 authentication request and response and also to the association request and response, the 802.11r standard effectively piggybacks the 802.11i four-way key exchange on top of four existing frames. This piggybacking can significantly reduce the latency involved in the BSS transition because step 4 of the process described in Section 4.9.2 can be overlaid onto steps 2 and 3 without increasing the round-trip frame counts of those steps. This method is the first of the two primary ways in which the 802.11r standard reduces BSS transition time. The second reduction is called *prereservation*, which we elaborate in the following text.

Prereservation allows the TSTA to perform the function of step 5 described in Section 4.9.2 before steps 2 and 3. There are two mechanisms specified for prereservation: *Over the DS* (*OTD*) and *Over the Air* (*OTA*). Using OTD, the TSTA communicates to the target TAP via the DS passing through the AP with which it is currently associated. This traffic flow is similar to that used by 802.11i for preauthentication with roaming candidate APs. OTD is the preferred prereservation mechanism under 802.11r as OTD provides the prereservation capability without interrupting the current traffic flow. The optional OTA method requires that the TSTA perform an 802.11 authentication request/response exchange with the target TAP. This exchange is enhanced under 802.11r to include the QoS reservations requested by the TSTA. Because the OTA method allows the TSTA to make the reservation without breaking its association with the current TAP, this method provides an alternative for preserving resources.

The 802.11r standard additionally allows for fast BSS transitions with QoS provisioning without performing prereservation and instead delaying the reservation of resources until the association-request/-response exchange occurs at the time of the transition. This form of QoS provisioning is appropriate when the TSTA detects that the target TAP is lightly loaded and likely to have the resources to accept the reservation when the request is made. The TSTA may determine this lightly loaded state from information in beacons or probe responses from that AP or from 802.11k Neighbor Reports. When hard guarantees are required, QoS provisioning of this latter sort is not appropriate because there is a time lag between the report of the lightly loaded state and the new reservation. Another TSTA could potentially have reserved those resources in the meantime, thereby resulting in a failed roam.

In this section we have attempted to provide a brief review of the major import of the 802.11r standard. As 802.11r is central to fast secure roaming in 802.11, we provide additional in-depth coverage of the standard in Chapter 9.

4.10 Other 802.11 Subgroups

In addition to the a, b, g, i, e, k, and r standardization efforts discussed earlier, in Table 4.3, we briefly summarize the other 802.11 task groups for the sake of generating a complete reference. The table shows the d, F, h, j, n, p, s, T, u, v, w, and y task groups as well as the status of their work. Standard 802.11d was ratified in 2005, whereas the next three standards were ratified in 2003. The remaining standards listed in the table are active.

The 802.11t standard provides the precise sort of measures exemplified in Figure 3.14 and referenced in Chapter 3 in our general remarks about the phases of roaming. A graph such as the one shown in Figure 3.14 is only of real value if the test criteria are clearly defined and

Table 4.3: Summary of Other 802.11 Task Groups.

Task Group	Status	Description
802.11d	Ratified 2001	Operation in additional domains (country-to-country roaming)
802.11F	Ratified 2003, withdrawn 2006	*Inter Access Point Protocol* (*IAPP*); originally thought to have relevance to roaming, not used in practice
802.11h	Ratified 2003	5-GHz spectrum, *Dynamic Channel/Frequency Selection* (*DCS/DFS*) and *Transmit Power Control* (*TPC*) for European compatibility
802.11j	Ratified 2003	Extensions for Japan; 4.9–5 GHz operation
802.11n	Active	High throughput
802.11p	Active	*Wireless Access for Vehicular Environment* (*WAVE*) such as passenger cars
802.11s	Active	Mesh networking
802.11T	Active	Test metrics
802.11u	Proposal	Interworking with non-802 networks such as cellular network
802.11v	Proposal	Wireless network management
802.11w	Proposal	Protected management frames
802.11y	Proposal	3.65–3.7 GHz operation in United States

uniformly applied to all equipment being evaluated for roaming performance. Establishing such definitions form a large part of the work of the 802.11t task group. Such formal measures of roaming times can be very useful to system designers. They need to assign probabilities to a roam which is taking so long that a voice call is dropped or to a dejittering buffer being under run during a roam, among other things. There are several ongoing research studies related to 802.11, and this work will continue for many years in the future.

4.11 Wi-Fi Alliance Versus IEEE 802.11

This section provides a comparison of the Wi-Fi Alliance and IEEE 802.11.

4.11.1 Introduction to the Wi-Fi Alliance

The Wi-Fi Alliance was founded in 1999 as an industry consortium intended to promote the successful commercialization of 802.11 products. In 2005, the Wi-Fi Alliance comprised more than two-hundred 802.11 technology companies. This group publishes standards and performs interoperability tests between products of members. The Wi-Fi Alliance was originally known as the *Wireless Ethernet Community Alliance* (*WECA*), but the name was officially changed in 2002. While dealing with precisely the same technology as 802.11, the motivations and time constraints of the Wi-Fi Alliance participants are quite different from those of 802.11 committee members. The IEEE is certainly in favor of the successful commercialization of the technologies that it standardizes, but there are at least two drawbacks that limit the IEEE's ability to pursue this goal aggressively:

1. The IEEE does not offer certification services.

2. The IEEE processes are driven such that they primarily result in the best technical standard possible. These processes are not likely to be compromised due to commercial pressures to ship products that the market demands.

For these two reasons, there were many prestandard 802.11a, 802.11b, and 802.11g chipsets sold in the market. The fact that these devices reflected one manufacturer's snapshot of a PHY standard meant that there was always the risk of interoperability problems with other vendors' products and even with devices that conform to the ultimately ratified standard. It was with this backdrop that the Wi-Fi Alliance was formed. Most importantly, the Wi-Fi Alliance offers commercial certification and related branding services that allow companies to first demonstrate and then market their product's interoperability with other Wi-Fi-compliant devices. Passing the basic interoperability tests in one of the approved test labs allows member companies to ship their product with the trademark *Wi-Fi CERTIFIED*. This same scenario is currently being played out with 802.11n, where prestandard commercial implementations are beginning to ship and the likelihood growing of a preIEEE certification program by the Wi-Fi Alliance.

4.11.2 Wi-Fi Alliance Certification

The Wi-Fi Alliance does not attempt to create standards that deviate from 802.11. In some cases, they might develop a Wi-Fi standard that represents an interim snapshot of a full IEEE standard or a proper subset of one. *Wi-Fi Protected Access* (*WPA*) and *Wi-Fi Protected Access 2* (*WPA2*) are examples of an interim snapshot and a ratified version, respectively, of the 802.11i standard. WPA was announced in October 2002. The decision to ratify a standard when work was still in progress at the IEEE was considered to be a result of the uproar about the growing list of vulnerabilities with the then-current 802.11 security and because the IEEE had already solved some of the most egregious problems. The facts that the IEEE still had work to do to improve upon these solutions and that the IEEE still had to address some new problems did not deter the Wi-Fi Alliance from wanting to promote commercialization of those initial solutions. These initial improvements that formed the basis for WPA lay in the following three basic areas:

- Encryption cipher
- Key management
- Authentication

WPA specifies the use of the TKIP encryption cipher. TKIP uses RC4 as does WEP, but TKIP virtually eliminated the possibility of the easy attacks for which WEP was then known. Key management was extended so that the idea of key freshness was now a fundamental precept that could be relied upon. Lastly, in the WPA enterprise mode, incorporation of 802.1X port authentication allowed for robust mutual authentication between the user and the network. WPA2 was ratified in June of 2004, and extended WPA to incorporate the advances included in the ratified 802.11i. These improvements included use of the CCMP-AES cipher and preauthentication.

The Wi-Fi Alliance provides a number of certification labs around the world where hardware manufacturers can have their devices stamped as Wi-Fi CERTIFIED, WPA capable, or WPA2 capable. (As of the writing of this book, WPA certification is no longer offered on new devices.) Although many of the 802.11i specifications are implemented in software, it is the current Wi-Fi policy that only hardware is certified. This issue poses a challenge to vendors of Wi-Fi-compliant technology building blocks, as it is not their product that bears the certification stamp, but only their product as implemented on a particular end-user device. These devices are most often APs, laptops, PCMCIAs, and other removable 802.11 adapters, and increasingly 802.11 handheld devices.

The QoS work done by the IEEE 802.11e task group represents a broad collection of methods for providing QoS guarantees over 802.11 networks. The effort is so broad that it represents more of a "QoS toolkit" than a mandatory set of features for which compliance will be mandated. To this end, the Wi-Fi Alliance's *Wi-Fi Multimedia* (*WMM*) group is standardizing a subset of IEEE 802.11e that will be comprehensive enough to allow for commercial multi-vendor implementations of VoWIP but sufficiently limited to permit a certification program that ensures 100% compliance with WMM. WMM-EDCA is the part of 802.11e that has strong support in WMM and will be part of the certification. WMM-EDCA defines an alternative to the DCF access mechanism.

WMM *Power Save* is a set of features for Wi-Fi networks that help conserve battery power in small handheld devices. This certification is available for access points and for client devices enhancing the original 802.11 power-save mode using mechanisms from the IEEE 802.11e standard. It is expected that other new Wi-Fi certifications related to areas such as roaming will be forthcoming, as standards which offer scalable solutions that seem likely to lead to widespread commercialization emerge from the IEEE.

Unlike the IEEE, membership in Wi-Fi is exclusively by company, not by individual, and membership requires a significant investment in annual dues ($15,000/year in 2005). In the IEEE, leadership of an important task group or even membership on an important board does not imply that your organization has significant commercial clout. Many academics or individual consultants with great technical pedigrees reach these levels in the IEEE. The chairing of a Wi-Fi committee, and particularly board membership, connotes that your company is a significant player in the Wi-Fi industry.

4.11.3 IETF Versus IEEE 802

For many years, the IETF and IEEE 802 have cooperated in developing the standards on which the Internet has been based. In general, it is appropriate to say that the IEEE 802 has specified most of the OSI layer-one and layer-two technologies for the LAN components of the Internet, whereas the IETF has specified the network protocols at layer three and above. From the outset, despite this layer-wise separation, cooperation between these two standards bodies has been necessary.

The earliest collaborations were in the areas of MIBs and AAA applications related to IEEE standards because both MIBs and AAA applications were traditionally standardized under the auspices of the IETF, yet present in IEEE standards. The IETF, on the other hand, found itself needing to intrude upon the traditional domain of IEEE 802 in the specification of layer-two

Ethernet types. When the overlap was limited, the IETF and IEEE 802 members found ad hoc mechanisms to resolve issues. In many cases, an individual would act as an informal liaison by serving as a participant in both bodies.

As the level of dependency between the IETF and the IEEE has rapidly grown, more systematic rules about liaison activity are now being formalized [1]. One of the strongest motivators for this synergy was the IEEE 802.11i's experience in specifying EAP technology, as a key part of its security solution. This experience highlighted the fact that most often the IETF documents are not written to the level of clarity required by the IEEE 802 standards process. The review of new EAP methods and specific applications thereof was greatly enhanced by the additional review of the EAP methods to be used with IEEE 802.11i by both the bodies.

Under the new policies, the IEEE is now granting the IETF better access to interim specifications. One of the recommendations is that the IEEE task groups follow the IETF MIB guidelines. The IEEE may formally request assignment of an IETF MIB review. As the areas for collaboration rapidly expand beyond the traditional overlap areas of MIBs and AAA, these new rules of collaboration will help these two great organizations continue to work independently, yet cooperate efficiently and draw on the other's expertise as necessary. Some of these rules have already been codified and others remain as recommendations. An example of this liaison activity is discussed in Chapter 10 with respect to IEEE 802.21.

For further details, the reader can refer to one of the following references: [7] or [8].

References

[1] Bernard Aboba, The IEEE 802/IETF Relationship, `draft-iab-ieee-802-rel-05.txt`, Internet Draft (work in progress), December 2005.

[2] Draft Amendment to Standard for Information Technology—Telecommunications and Information Exchange Between Systems—LAN/MAN Specific Requirements; Part 11: Wireless Medium Access Control (MAC) and Physical Layer (PHY) Specifications: Specification for Enhanced Security, IEEE Std 802.11i/D3.0, November 2002, New York.

[3] Draft Amendment to Standard for Information Technology—Telecommunications and Information Exchange Between Systems—LAN/MAN Specific Requirements; Part 11: Wireless Medium Access Control (MAC) and Physical Layer (PHY) Specifications: Amendment: Medium Access Control (MAC) Quality of Service (QoS) Enhancements, IEEE P802.11e/D13.0, January 2005, New York.

[4] Draft Amendment to Standard for Information Technology—Telecommunications and Information Exchange Between Systems—LAN/MAN Specific Requirements; Part 11: Wireless Medium Access Control (MAC) and Physical Layer (PHY) Specifications: Amendment 7: Radio Resource Measurement, IEEE P802.11k/D2.2, July 2005, New York.

[5] Draft Amendment to Standard for Information Technology—Telecommunications and Information Exchange Between Systems—LAN/MAN Specific Requirements; Part 11: Wireless Medium Access Control (MAC) and Physical Layer (PHY) Specifications: Amendment 8: Fast BSS Transition, IEEE P802.11r/D2.1, May 2006, New York.

[6] James Edney and William Arbaugh, *802.11 Security: Wi-Fi Protected Access and 802.11i*, Boston, Addison-Wesley, 2004.

[7] Matthew S. Gast, *802.11 Wireless Networks, The Definitive Guide*, O'Reilly, 2002.

[8] William Lidinsky, IEEE Standard P802.1D Information Technology—Telecommunications and Information Exchange between Systems—Common Specifications—Part 3: Media Access Control (MAC) Bridges: Revision 24.11.1997, 1997.

[9] Scott Shenker, Craig Partridge, and Roch Guerin, Specification of Guaranteed Quality of Service, RFC 2212, Internet Engineering Task Force, 1997.

[10] Scott Shenker and John Wroclawski, Network Element Service Specification Template, RFC 2216, Internet Engineering Task Force, 1997.

CHAPTER 5

Practical Aspects of Basic 802.11 Roaming

5.1 Introduction

5.2 The Driver and Client in an 802.11 Station

5.3 Detailed Analyses of Real-Life Roams

5.4 Dissection of a Global Roam

5.5 Dissection of a Local Roam

5.6 Access-Point Placement Methodologies

References

5.1 Introduction

This chapter provides a detailed tutorial on the mechanism of how an 802.11 station switches from its current AP to a neighboring AP, assuming that the network has no security measures in place. Regardless of whether we consider the now obsolete security of static WEP or the state-of-the-art WPA2, all these security measures add greatly to the factors that play a role in a roam. As security is no longer an option but a requirement and as 802.11 style security adds significantly to the complexity of the roam, we treat these items separately in Chapters 6 through 9. Even ignoring the question of security, the basic 802.11 roaming process is sufficiently complicated to merit a chapter of its own. Detailed examples of 802.11 association procedures are also provided here.

Normal 802.11 roaming is prompted by the firing of the *signal strength-based roaming triggers* that we introduced in Chapter 3, and the standard roaming does not contemplate other important factors, such as AP *load balancing* and QoS considerations. We will provide examples to show how scan lists are maintained internally in the 802.11 software drivers, how these drivers monitor the signal strength from the current AP, and how, based on the drop in signal strength below a prescribed threshold, the 802.11 software attempts to roam to the best neighboring AP in its scan list. As we will see in later chapters, the concept of the *best* neighboring AP is subjective and open to wide interpretation. In order to keep this chapter's practical examples and tests easy to follow, there will be only one roaming candidate available.

5.2 The Driver and Client in an 802.11 Station

5.2.1 Introduction

Figure 5.1 depicts the software implementation in a typical 802.11 STA. The driver directly interfaces with the 802.11 hardware and exposes a number of primitive interfaces to the next higher layer of software—the 802.11 client. There are many variations with regard to how these specific components may be implemented on a given device. Due to the prevalence of PC implementations of 802.11, our examples will show variants that correspond to actual implementations that run on standard PC hardware. While we restrict ourselves in this chapter to conventional implementations, it is possible to implement some of the driver features in user space and some of the client features in kernel space.

Some drivers may be partly realized in firmware coresident with the 802.11 chipset rather than being implemented entirely as host-based software. These variants do not alter the fact that there are fundamentally separate functions of driver and client that must be implemented in some fashion. Because each of these two functions has become significantly more complex than they were initially envisioned to be, the division in functionality is becoming more entrenched. The reason is partly due to the establishment of standard interfaces, which allow for the mix and match of 802.11 software and hardware components from different vendors. There is a growing body of vendors that specialize in developing and marketing reusable software that implements client-specific and driver-specific functionality. It is now less common to find 802.11 hardware vendors who independently attempt to implement all this rapidly increasing complexity.

5.2.2 Driver Functions

In this section, we present the standard 802.11 driver interface exposed by a Microsoft Windows *NDIS*[1]-compliant driver to the client. This interface between the driver and the client is shown in Figure 5.1. Here it would be convenient to use Linux Wireless Extensions as examples due to the openly available source code of Linux device drivers. However, the reality is that Linux drivers lag behind Windows drivers with respect to some of the features that are

---

[1] *Network Driver Interface Specification.*

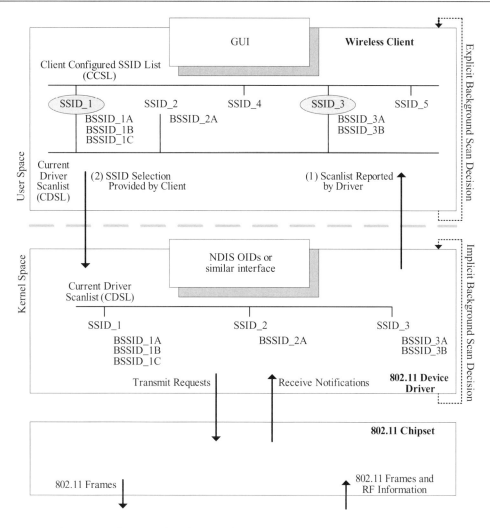

Figure 5.1: Software Implementation in a Typical 802.11 Client Showing the Driver and Interface on a PC.

most relevant to roaming, such as control of background scanning. These windows features are discussed in [6].

In Tables 5.1 and 5.2, we inventory the 802.11 NDIS 5.1 *Object Identifiers* (*OIDs*) that are mandatory for Windows XP implementations. In addition to the mandatory OIDs, WPA and WPA2 support on Windows XP requires the OIDs listed in Tables 5.3 and 5.4, respectively. Because the features enabled by the WPA and WPA2 OIDs are security oriented, we defer further treatment of them to Chapter 6. We list the optional 802.11 NDIS 5.1 OIDs in Tables 5.5 and 5.6 to provide a complete reference. While the list of mandatory OIDs is

Chapter 5

Table 5.1: Listing of Mandatory NDIS 5.1 OIDs: Part 1.

OID	Description
ADD-WEP	This allows for specification of the WEP key.
AUTHENTICATION-MODE	This interface sets the IEEE 802.11 authentication mode.
BSSID	This allows for the specification of the BSSID of the AP with which to associate.
BSSID-LIST	This returns the list of all BSSIDs stored in the driver's database, which were accumulated during the past scans.
BSSID-LIST-SCAN	This requests that the driver initiates a scan using active- or passive scanning techniques, as specified.
CONFIGURATION	This interface either gets or sets radio parameters associated with the 802.11 interface. They may only be set when there is no current association.
DISASSOCIATE	This requests that the driver disassociates from the currently associated AP and then turn off the radio.

Note that all items in column 1 start with the prefix "OID-802-11-."

Table 5.2: Listing of Mandatory NDIS 5.1 OIDs: Part 2.

OID	Description
INFRASTRUCTURE-MODE	This allows for specification of either Infrastructure mode, IBSS mode, or automatic mode (automatic switching between the two as a function of the 802.11 peers discovered; not recommended).
NETWORK-TYPES-IN-USE	This returns the 802.11-PHY type in use.
RELOAD-DEFAULTS	This requests the driver to reload the default WEP keys.
REMOVE-WEP	This forces the driver to remove the corresponding WEP key from the specified index for the current association.
RSSI	This gets the *Received Signal Strength Indication (RSSI)*.
SSID	This allows for specification of the SSID with which to associate.
SUPPORTED-RATES	This returns the data rates that the 802.11 interface supports.
WEP-STATUS	This has been deprecated to OID-802-11-ENCRYPTION-STATUS.

Note that all items in column 1 start with the prefix "OID-802-11-."

somewhat lengthy, there are really only a few basic functions provided by the driver to the client that are really important to roaming. These are as follows:

- OID-802-11-BSSID-LIST-SCAN

- OID-802-11-BSSID-LIST

Table 5.3: Listing of WPA NDIS 5.1 OIDs.

OID	Description
ADD-KEY	This allows specification of the pre-shared key for WPA or WPA2.
ASSOCIATION-INFORMATION	This gets the IEs in the last association response from the AP, or sets the IEs for the next association request to the AP.
ENCRYPTION-STATUS	This returns the current encryption status. A response indicating that the key is absent implies that no group keys are available for the NIC to encrypt data. *(Note: Same as the now-deprecated OID-802-11-WEP-STATUS.)*
REMOVE-KEY	This requests that the driver remove the key at the specified key index for the current association.
TEST	This is used to request that the driver initiate a test. This is not intended to be used as part of normal operation.

Note that all items in column 1 start with the prefix "OID-802-11-."

Table 5.4: Listing of WPA2 NDIS 5.1 OIDs.

OID	Description
CAPABILITY	This returns the authentication and encryption capabilities that are supported by the 802.11 interface and driver.
PMKID	This is used to query and set the PMKID in the PMK cache.

Note that all items in column 1 start with the prefix "OID-802-11-."

Table 5.5: Listing of Optional NDIS 5.1 OIDs: Part 1.

OID	Description
DESIRED-RATES	This gets or sets the preferred data rates for the 802.11 interface.
FRAGMENTATION-THRESHOLD	This is used to specify the fragmentation threshold such that packets that are larger than the fragmentation threshold are split before they are transmitted.
NETWORK-TYPES-SUPPORTED	This returns an array of all PHY network subtypes, which IEEE 802.11 NIC and the driver support.
NUMBER-OF-ANTENNAS	This returns the number of antennae on the 802.11 radio.
POWER-MODE	This gets or sets the current 802.11 power mode.
PRIVACY-FILTER	This allows specification of the privacy filter mode.
RSSI-TRIGGER	This gets or sets the trigger threshold for an RSSI event. The RSSI passing the threshold results in the event.

Note that all items in column 1 start with the prefix "OID-802-11-."

Table 5.6: Listing of Optional NDIS 5.1 OIDs: Part 2.

OID	Description
RTS-THRESHOLD	This allows the caller to specify the RTS threshold. Only packets that exceed this RTS threshold will trigger the RTS/CTS mechanism. An RTS threshold value of zero indicates that the NIC should transmit all packets using RTS/CTS.
RX-ANTENNA-SELECTED	This returns the number identifying the antenna used for receiving on the 802.11 interface.
STATISTICS	This requests the driver return the current statistics for the 802.11 interface.
TX-ANTENNA-SELECTED	This returns the number identifying the antenna used for transmitting on the 802.11 interface.
TX-POWER-LEVEL	This sets or gets the transmit-power level. Specified in milliwatts.

Note that all items in column 1 start with the prefix "OID-802-11-."

- OID-802-11-SSID

- OID-802-11-BSSID

- OID-802-11-RSSI-TRIGGER

The OID-802-11-BSSID-LIST-SCAN interface requests that the driver initiates a scan using active- or passive scanning techniques, as specified on a channel list that is supplied by the caller. This technique is the NDIS 5.1 standard method for initiating a scan. As this request may be initiated at any time, this may invoke a background scan. However, because the requestor is not sensitive to the current transmit/receive state on the active channel, overuse of this command could result in too much time being spent away from the currently active channel, which could in turn result in inadvertent disassociation. To avoid this situation, an NDIS-compliant driver must not spend so much time responding to an OID-802-11-BSSID-LIST-SCAN request that it loses the association. We will see in Section 5.2.7 that not all drivers live up to NDIS standards though.

NDIS also specifies that OID-802-11-BSSID-LIST-SCAN will discard the current driver scan list and that the new scan list may not be available for up to six seconds. The delay of six seconds is to allow the driver time to complete the scan without unduly sacrificing attention to the active channel. However, we have observed that this delay may not be uniformly implemented in NDIS-compliant drivers. In particular, when more than one client application is making overlapping calls to OID-802-11-BSSID-LIST-SCAN, if the drivers were actually discarding the list as specified, the concurrent applications should interfere, clearing the lists at inopportune moments, resulting in empty or incomplete scan lists. As we have been unable to observe this happening, there is some serious doubt as to whether or not the scan list is cleared in accordance with the specification.

The OID-802-11-BSSID-LIST-SCAN should be distinguished from vendor-specific interfaces that exploit more intelligent scanning decisions by putting the decision about when to initiate the background scan into the driver and, hence, making better decisions about when and how long to stay away from the currently active channel. Vendor-specific interfaces are only likely to be used by clients that are specific to that 802.11 hardware, so such features are not commonly exposed in the more familiar generic 802.11 clients.

The OID-802-11-BSSID-LIST is called by the client to force the driver to return the current list of BSSIDs found on the driver's recent survey of potential APs. We refer to this list of BSSIDs as the *current driver scan list* (*CDSL*). This list is shown inside the driver component in Figure 5.1. The figure also shows that the information on this list is present in the client as well. This information was transferred from the driver to the client as a result of the client invoking OID-802-11-BSSID-LIST.

A call to OID-802-11-SSID causes the driver to associate with a nonspecific AP belonging to the ESS denoted by the SSID parameter that is provided. The decision of which AP in that ESS to associate with is left to the driver. Normal driver implementations will associate with the AP showing the strongest signal strength and automatically roam to another AP in the ESS should the association with the first be lost. Invoking OID-802-11-BSSID requests the driver to associate with the AP designated by the BSSID parameter. In this case, if association with the AP is lost, the driver will *not* automatically roam to another AP in the ESS. This interface is useful to a client application that wishes to apply additional intelligence to the STA to AP mapping.

An example of this type of intelligence would be to force the STA to associate with a more lightly loaded AP that shows a weaker signal strength than a heavily loaded AP in the same ESS. Such clients are not yet commonly available. Another example that does occur in clients is to ensure that the STA only associates with a member AP that meets defined minimum security levels. A client can use OID-802-11-RSSI-TRIGGER to set the signal strength threshold. When the RSSI passes this threshold, a RSSI event is sent from the driver to the client. If the client is exercising BSS level control of association, and therefore of roaming, this OID provides the client with the fine-grained information that is necessary to decide when to leave one AP and join another. Thus, in the case of a NDIS-only driver interface, the *signal strength-based roaming triggers* are simply this RSSI trigger.

5.2.3 Client Functions

The 802.11 client functions have not been as formally standardized as the device–driver interface. The reason for the greater uniformity of driver interfaces is that there are many different manufacturers of 802.11 interfaces, most of which make their product available on general-purpose operating systems like Windows and Linux. In these operating systems, it is expected that the driver conforms to a standard interface specification such as NDIS or the Linux Wireless Extensions, so that the different hardware interfaces and their corresponding drivers can be seamlessly integrated into a functioning and complete system. This situation contrasts with the case of the 802.11 client where the interface is a *Graphical User Interface* (*GUI*) designed for human interaction and, thus, understandably differs from one client to another. While different clients may offer varying feature sets, there are certain common features that are offered by most 802.11 clients. These include the following:

- The client may maintain a list of SSIDs that are known to be acceptable to the user. We call this configuration information the *Client Configured SSID List* (*CCSL*), and it was shown in Figure 5.1. The client must take the intersection of this set with the current driver scan list. This intersection is denoted by the shaded ovals in Figure 5.1, and it includes SSID-1 and SSID-3. SSID-2 is excluded because although it is currently visible to the driver, SSID-2 has not been preauthorized by the user and thus is not part of the CCSL. SSID-4 and SSID-5 are excluded from the intersection because, while configured as acceptable, they are not currently visible to the driver. The client applies a client-specific heuristic to determine with which of the SSIDs in the intersection to associate. The criteria for this decision are varied, but they may include a requirement that the SSID supports some minimal level of security in order to be considered a candidate. Some clients may allow the CCSL to be set to the universe, thus permitting association with any SSID in the CCSL.

- If 802.11 security measures requiring a preshared key are required by the selected APs, the client will provide the key information to the driver.

- If 802.1X security is required on the interface, the client will ensure that the 802.1X port-authentication protocol is run prior to allowing any user traffic to be transmitted or received by user applications. The actual filtering of the user traffic is performed by the driver, under the control of 802.1X.

- The client may provide human-readable statistics on both the relative signal strength and the security measures of the APs on the scan list reported to the client by the device driver.

As of this writing, there are three general-purpose 802.11 client applications in wide use. Many chipset-specific clients are available from the 802.11 hardware vendors, but these clients will only work with the hardware for which they are sold. Such chipset-specific clients may offer some specialized feature for that chipset, which the generalized clients cannot offer. The most popular general-purpose clients are the Microsoft *Wireless Zero Config* (*WZC*) client shown in Figure 5.2, Funk Software's Odyssey client depicted in

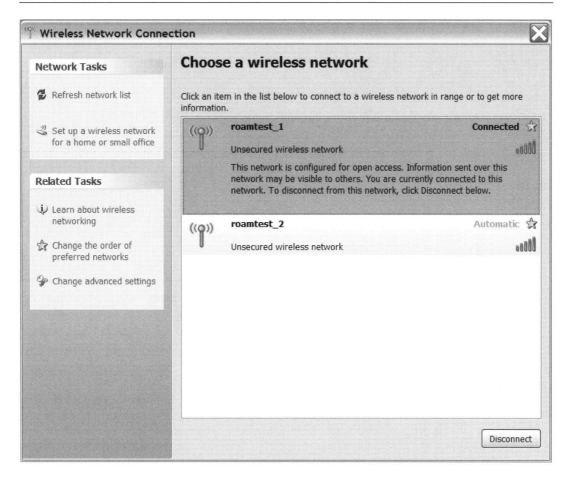

Figure 5.2: Microsoft's Wireless Client Which is Called Wireless Zero Config.

Figure 5.3, and Meetinghouse's Aegis client shown in Figure 5.4. These three clients will be used in the roaming tests presented later in this chapter and in chapter 7.

5.2.4 Scanning Considerations

Broadly defined, 802.11 scanning means looking for available APs. In this section, we discuss the many nuances and trade-offs related to how and when to scan. We first point out that one can look at scanning from two different perspectives. If we imagine a casual user who wants to find any available AP, this angle implies a "bottom-up" view of scanning. That is, any AP that the card can find on any channel is of interest to this user. The net that we cast to find APs

Chapter 5

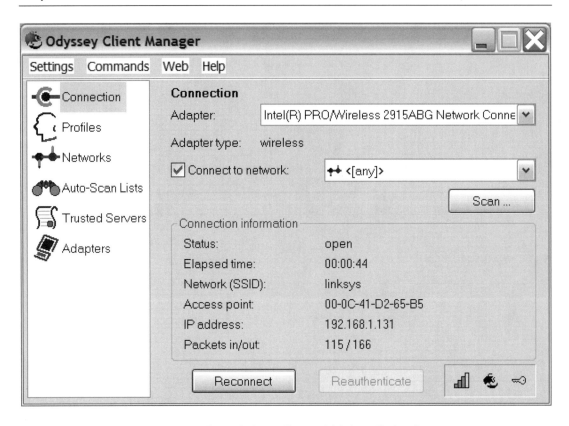

Figure 5.3: Funk's Wireless Client Which is Called Odyssey.

should be as wide as possible from this perspective. A different perspective is one where the user already knows the desired APs. This situation may be the case of a corporate user who is restricted to using certain SSIDs with specific security parameters that are approved for use with company wireless devices. In this case, it may be inefficient to fish with the biggest possible net.

By convention, the 802.11 drivers do not posses the ability to exclude certain SSIDs from the current driver scan list. As we have noted earlier, the ability to decide that only certain APs are candidates for association is very much a part of the client's architecture, but that matter is outside the scope of scanning. While the 802.11 paradigm does not prevent RF-accessible APs from finding their way onto the driver's scan list, it is possible to increase the likelihood that a certain SSID will be found before others. This prioritization is done by performing SSID-specific probes (also called *directed* probes) while doing *active* scans. However, if we want to cast our net wide, we can either use broadcast probes (also called *undirected* probes) or restrict ourselves to *passive* scans. Extending the time spent listening during those scans also extends the reach of our net.

Practical Aspects of Basic 802.11 Roaming

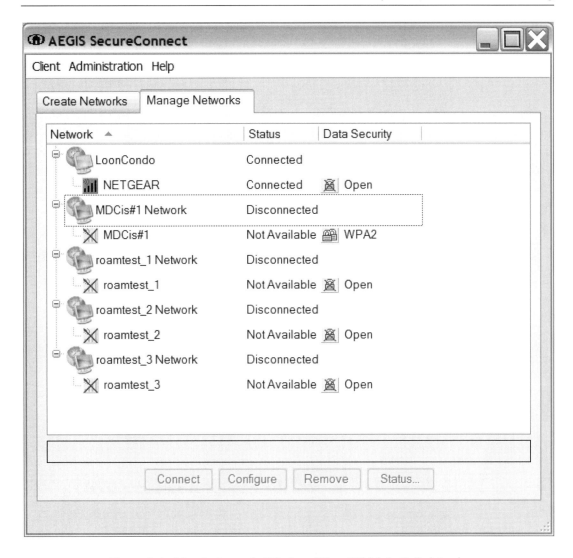

Figure 5.4: Meetinghouse's Wireless Client Which is Called Aegis.

We should point out that the primary use of directed probes is to ascertain the existence of preconfigured, nonbeaconing APs. Some APs will be configured to turn off beacons as a rudimentary security measure. More commonly, as of this writing, an AP will have multiple SSIDs, and only one of these can be advertised via beacons. The rationale for multiple SSIDs for a single radio is that the different virtual APs represented by these SSIDs have different characteristics. For example, one virtual AP using that radio might be configured to have open security, providing access to the public Internet via a dedicated VLAN, whereas there may be another with WPA2 enterprise security configured that provides access to a private corporate network via another VLAN. Because 802.11 requires that a single radio only beacon a single SSID, one of these two SSIDs will be *hidden*. A client wishing to associate with a hidden SSID must have *a priori* knowledge of this SSID through configuration information, and must discover its presence via directed probes since the second SSID cannot also send beacons.

In the roaming context, as truly fast roaming is only realistic within the same SSID family, an implementation may want to maintain information about neighboring APs with the same SSID as current as possible, yet relax that urgency with regard to APs with different SSIDs. Thus, to drift back one last time to our fishing metaphor, we may want to cast a small net frequently for the active SSID and a larger net less frequently for other SSIDs. If the STA is not associated, this fact diminishes the requirement that scan list information be frequently refreshed as there can be no fast roam from a nonexistent association. In the next section, we consider the concept of background scanning.

5.2.5 Background Scanning

In Section 5.2.2, we described a standard NDIS-driver OID that permits the 802.11 client application to request that a scan occurs. This interface (OID-802-11-BSSID-LIST-SCAN) takes no parameters. The facts about which channels should be scanned and whether active, passive, or a combination of both scanning techniques should be used are determined by driver parameters configured independently of the call to OID-802-11-BSSID-LIST-SCAN.

As explained earlier, active scanning entails the transmission of broadcast or SSID-specific 802.11 probe frames and then waiting for a probe response. *The standard NDIS-driver interfaces do not allow programming of the duration of the waiting period for a probe response.* Passive scanning does not involve transmitting any frames, only listening for 802.11 beacons.

If it is desired that nonbeaconing APs be added to the scan list reported by the driver, the client must do at least one active scan in order for the driver to learn of the AP's existence, and enter the AP into its scan list. This situation is due to the fact that the client-driver protocol does not permit a client to force the driver to associate to an AP which is not already in the driver's scan list. Such nonbeaconing APs are also known as *hidden* APs.

Scans may be either *explicit* or *implicit*. An explicit scan is one that is directly initiated by the client. In Windows, this scan is the result of the client calling OID-802-11-BSSID-LIST-SCAN. An implicit scan is one that is initiated by the driver without an immediately preceding call to OID-802-11-BSSID-LIST-SCAN. As the implicit scan is controlled by the driver that has immediate access to the virtual-carrier sensing logic on the active channel, the driver can initiate implicit scans with a timing and a duration that is sensitive to the expected level of activity on the active channel.

Practical Aspects of Basic 802.11 Roaming

We define *background scans* as any scan, explicit or implicit, that is performed by the STA on nonactive channels when associated on an active channel. Figure 5.5 provides a snapshot of the beaconing and probing activity between an AP beaconing on channel 1 and an STA that is currently associated and sending application traffic, with another AP on channel 6. Both APs are configured with SSID *roamtest_3*. This STA is a Dell laptop with a built-in Intel Centrino 802.11 chipset. The client is the Aegis client. We observe explicit background scanning in the form of the probe request-response exchanges.

These probes requests are the indirect result of the 802.11 client calling OID-802-11-BSSID-LIST-SCAN. It is interesting to note the high number of retransmissions that are necessary for the probe responses. We see in frames 627 through 633 in Figure 5.5 that the AP retransmits the probe response six times without any acknowledgment from the STA. Indeed, it is not until a new probe request is sent in frame 634 that we see a corresponding probe response and acknowledgment in frames 635 and 636, respectively. The failure of the STA to acknowledge the probe responses earlier is most likely attributed to its switching back to the active channel to deal with the ongoing application traffic.

Figures 5.6 and 5.7 provide exploded views of the probe requests in frames 634 and 637, respectively. We see that the probe request in frame 634 is a directed probe to the active

Figure 5.5: Listing of Background Implicit Scanning Information in Active Mode.

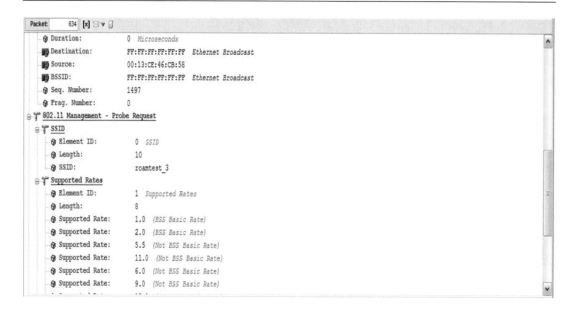

Figure 5.6: Listing of Background Scanning Details for a Specific Probe.

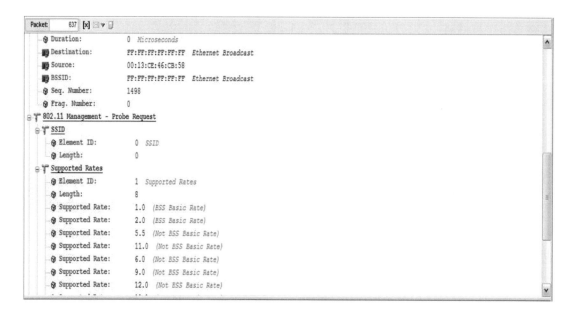

Figure 5.7: Listing of Background Scanning Details for a Wildcard Probe.

SSID roamtest_3, whereas frame 637 shows an undirected probe. This reflects the driver's priority in maintaining up-to-date information about the currently active SSID, while still casting a wide net for any other SSIDs that may be present via the single undirected probe. It should be noted that even the directed probe is being sent the broadcast Ethernet address, as the STA is interested in any AP with SSID roamtest_3 on that channel. It should also be noted in frame 626 and in frames 640 through 646 that the AP continually broadcasts its beacons on a 100-millisecond schedule interleaved with the probe request-response exchanges.

The scan is finally completed with the acknowledgment of the probe response in frame 639, at which point we see only beacons. As this example uses the Aegis client, and assuming the active channel continues to operate undisturbed, the subsequent background explicit scan would occur in 63 seconds. We remind the reader that the timing of such background explicit scans is client specific.

Figure 5.8 shows a similar traffic scenario but on channel 6. The STA is actively sending traffic via channel 11 in this case. The most striking difference in the figures is the fact that we have filtered beacons from the sniffer display. This filtering is often done when analyzing 802.11 traffic as the beacon frames clutter the output and usually do not affect the traffic analysis that is being performed. Both Figures 5.5 and 5.8 are instances of active scanning.

Packet	Source	Destination	BSSID	Channel	Signal	Size	Relative Time	Protocol	Summary
11	00:13:10:07:48:AC	00:13:CE:46:CB:58	00:13:10:07:48:AC	6	55%	85	15.416896	802.11 Probe Rsp	FC=........,SN=3762,F
12	00:13:CE:46:CB:58	00:13:10:07:48:AC		6	100%	14	15.442872	802.11 Ack	FC=........
13	00:13:CE:46:CB:58	Ethernet Broad...	FF:FF:FF:FF:FF:FF	6	100%	56	15.466598	802.11 Probe Req	FC=........,SN=1461,F
14	00:13:CE:46:CB:58	Ethernet Broad...	FF:FF:FF:FF:FF:FF	6	100%	46	15.467270	802.11 Probe Req	FC=........,SN=1462,F
15	00:13:10:07:48:AC	00:13:CE:46:CB:58	00:13:10:07:48:AC	6	51%	85	15.468169	802.11 Probe Rsp	FC=........,SN=3764,F
16	00:13:CE:46:CB:58	00:13:10:07:48:AC		6	97%	14	15.468473	802.11 Ack	FC=........
17	00:13:10:07:48:AC	00:13:CE:46:CB:58	00:13:10:07:48:AC	6	52%	85	15.520191	802.11 Probe Rsp	FC=........,SN=3766,F
18	00:13:10:07:48:AC	00:13:CE:46:CB:58	00:13:10:07:48:AC	6	50%	85	15.545997	802.11 Probe Rsp	FC=........,SN=3767,F
19	00:13:10:07:48:AC	00:13:CE:46:CB:58	00:13:10:07:48:AC	6	50%	85	15.547349	802.11 Probe Rsp	FC=...R....,SN=3767,F
20	00:13:10:07:48:AC	00:13:CE:46:CB:58	00:13:10:07:48:AC	6	52%	85	15.571569	802.11 Probe Rsp	FC=........,SN=3768,F
21	00:13:10:07:48:AC	00:13:CE:46:CB:58	00:13:10:07:48:AC	6	55%	85	15.573608	802.11 Probe Rsp	FC=...R....,SN=3768,F
22	00:13:10:07:48:AC	00:13:CE:46:CB:58	00:13:10:07:48:AC	6	50%	85	15.575311	802.11 Probe Rsp	FC=...R....,SN=3768,F
23	00:13:10:07:48:AC	00:13:CE:46:CB:58	00:13:10:07:48:AC	6	55%	85	15.577425	802.11 Probe Rsp	FC=...R....,SN=3768,F
24	00:13:10:07:48:AC	00:13:CE:46:CB:58	00:13:10:07:48:AC	6	55%	85	15.579954	802.11 Probe Rsp	FC=...R....,SN=3768,F
25	00:13:10:07:48:AC	00:13:CE:46:CB:58	00:13:10:07:48:AC	6	50%	85	15.583880	802.11 Probe Rsp	FC=...R....,SN=3768,F
26	00:13:10:07:48:AC	00:13:CE:46:CB:58	00:13:10:07:48:AC	6	54%	85	15.592719	802.11 Probe Rsp	FC=...R....,SN=3768,F
27	00:13:10:07:48:AC	00:13:CE:46:CB:58	00:13:10:07:48:AC	6	51%	85	16.406142	802.11 Probe Rsp	FC=........,SN=3777,F
28	00:13:10:07:48:AC	00:13:CE:46:CB:58	00:13:10:07:48:AC	6	51%	85	16.431706	802.11 Probe Rsp	FC=........,SN=3778,F
29	00:13:10:07:48:AC	00:13:CE:46:CB:58	00:13:10:07:48:AC	6	55%	85	16.457306	802.11 Probe Rsp	FC=........,SN=3780,F
30	00:13:CE:46:CB:58	00:13:10:07:48:AC		6	100%	14	16.457625	802.11 Ack	FC=........
31	00:13:CE:46:CB:58	Ethernet Broad...	FF:FF:FF:FF:FF:FF	6	97%	56	16.481337	802.11 Probe Req	FC=........,SN=1501,F
32	00:13:CE:46:CB:58	Ethernet Broad...	FF:FF:FF:FF:FF:FF	6	98%	46	16.482008	802.11 Probe Req	FC=........,SN=1502,F
33	00:13:10:07:48:AC	00:13:CE:46:CB:58	00:13:10:07:48:AC	6	50%	85	16.482909	802.11 Probe Rsp	FC=........,SN=3781,F
34	00:13:CE:46:CB:58	00:13:10:07:48:AC		6	98%	14	16.483217	802.11 Ack	FC=........
35	00:13:10:07:48:AC	00:13:CE:46:CB:58	00:13:10:07:48:AC	6	51%	85	16.508529	802.11 Probe Rsp	FC=........,SN=3782,F

Figure 5.8: Listing of Background-Implicit Scanning Details in Active Mode with Beacons Suppressed.

A sniffer example of passive-only scanning would be of little interest to the reader, as such an example would only show beacons emanating from the AP; there is no external manifestation of the STA switching its radio to nonactive channels, unless the STA transmits something when it does so. As there is no such transmission in passive mode, such a trace is not interesting.

There are a number of decisions that are left to the driver's discretion when OID-802-11-BSSID-LIST-SCAN is called. For example, the driver may implement background scanning independently of the calls to this interface so that up-to-date scan list information is available shortly after the call. Decisions about the order in which channels are scanned are also left to the interpretation of the driver implementer. The description of this OID imposes the vague requirement that the implementation of this scan request should not result in the STA disassociating from the current AP.

Some device drivers offer finer-grained control over these background scanning decisions than the generic NDIS interface described previously. One 802.11 chipset manufacturer allows the client the following additional control over the background scanning process:

- The number of channels to be scanned per each scan instance may be specified. This value effectively limits the amount of time that the radio spends away from the active channel, thus minimizing the likelihood of missed frames.

- The time in milliseconds between each scan instance can be specified.

- If active scan is chosen, the ability to specify how many broadcast probes are sent per scan instance is provided. Also, the driver provides the ability to select the SSIDs that should receive specific probes and the number of probes that should be sent per scan instance.

- It is also possible to exercise further control over which APs in the scan list are reported back to the client. The rationale is that there may be a signal-strength threshold, which is considered so weak that if an AP that shows weaker signal strength than that threshold is found, it is preferable that this information is not reported to the client. This pruned scan list can simplify the client's decision regarding viable candidate APs.

An obvious drawback of such vendor-specific extensions is that a card-neutral client will not be able to exploit the richer feature set exposed by such extensions. Most likely, only a client provided by the card manufacturer will utilize that manufacturer's extensions. As 802.11 clients have evolved far more complex functions than those originally thought possible, vendor-specific extensions leave users in the uncomfortable position of having to choose between a general-purpose client that may be feature rich and the manufacturer's own client that is feature poor but can fully exploit the uniqueness of their driver. Also, having a uniform client interface across all corporate wireless clients is of great importance to IT managers. As it is highly unlikely that all 802.11 devices in an enterprise contain the same manufacturer's 802.11 chipset, this situation is forcing IT managers to move toward generic 802.11 clients.

The additional granularity of control provided by these vendor-specific interfaces reflects both the importance and the risks associated with background scanning. The conundrum of background scanning is that it is not possible to optimize simultaneously for maximum connectivity with the current AP and for preparedness for the subsequent roam. This situation is attributed to the fact that to be fully prepared for the subsequent roam requires a priori knowledge of the best candidate AP. If this information is not available when the decision to leave the current AP is being made, valuable time will be wasted in scanning during the roam, thereby introducing some delay in the resumption of connectivity.

In order to maintain the freshness of the CDSL, the driver must leave the active channel and listen on other channels for candidate APs. During the time spent listening on the nonactive channel, the active channel may receive frames destined for the STA that does the scanning. In particular, there is the danger that an important 802.11 management frame might be received during this period, resulting in the STA becoming out of sync with the AP. While retry mechanisms should mitigate this risk, the risk exists, and we provide a real-life example to illustrate this problem in Section 5.2.7.

As of this writing, research in background scanning heuristics is very active in the 802.11 community. One particularly interesting heuristic, called *SyncScan*, will be discussed in Section 5.2.8. In the next couple of sections, we discuss scanning timers and some pitfalls with scanning.

5.2.6 Scanning Timers

In the case of active scanning, after the driver sends a probe request, the driver waits for a period of time for the probe response before giving up. We call this probe-response timer the active-scan timeout. This timer is not provided in the standard NDIS-driver interface. A typical active-scan timeout value would be in the range of 25–40 milliseconds. It should be noted that this means that in the case of the eleven channels of 802.11b, a full active scan could take on the order of a total of 400 ms, although this time is not necessarily spent contiguously.

Timeout periods related to passive scanning are more nuanced. The most common default value for a beacon interval is 100 milliseconds, although simply assuming that all APs are using this convention could result in gross miscalculations of the optimal passive-scan timer value. As the beacon interval is included in the beacon itself, receipt of a single beacon from each AP means that the STA knows the beacon interval of all APs beaconing on that channel. Once the STA has this information and the largest of these intervals is determined, we know that the passive-scan timer value need not be longer than the longest of these intervals in order to capture every beacon in a single scan.

Using this technique to determine actual passive-scan timers would be quite crude and likely to result in too much time away from the active channel. Also, this approach does not contemplate the situation of a new AP appearing on that channel, whose beacon interval is, by definition, unknown. Utilizing too short of a passive-scan timer value could result in missing that new arrival's beacons for an arbitrary length of time. This situation provides yet another illustration of why the research on background scan heuristics remains active. For example, it may be reasonable to adjust the time spent scanning a channel as a function of the relative activity sensed on the channel; a scan of a totally silent channel is less likely to detect a beacon of an AP than a channel busy with data traffic.

The timing of explicit scans is under the control of the client application. Some clients may only initiate explicit scans when directly prompted by the user (for example, WZC's "Refresh Network List" command). Other clients may do so both when prompted by the user and also on some regular interval. The WZC client and the Aegis client 4.0 both initiate an explicit scan approximately once per minute. Implicit scan timing is controlled by the driver, so this timing will vary from one 802.11 chipset vendor to another. We conducted a simple black-box test to examine the implicit-scan timing differences between two popular 802.11 interfaces: the Intel 2200 miniPCI and Atheros AR5004G. The Intel-driver version was 9.0.4.3 and the Atheros-driver version was 3.1.2.12. We describe the tests and the results in the following text.

The first test category measured implicit scan timing when there was no active association. In order to be certain that the scans were exclusively implicit, a special-test client which never initiates explicit scans was written. This test client requests the scan list from the driver at 500-ms intervals in order to obtain a reasonably fine-grained picture of when the scan list changes in the driver. The time was measured relative to the moment that the AP was powered on. As the same APs were used with both manufacturer's cards, any boot-up delay in the AP was assumed to be constant across all tests.

The measurements were taken five times for each interface, and the results were averaged. The Intel driver reported the addition of the AP after 15 seconds; the Atheros driver reported the AP after 20 seconds. The AP was then powered down, and the time for the AP to be removed from the scan list was measured. For the Intel interface, 140 seconds elapsed before the AP was removed from the scan list. Only 126 seconds passed before the AP was removed from the scan list in the case of the Atheros card. These test results are summarized in Table 5.7.

Table 5.7: Implicit Scanning Timing Test Results.

Manufacturer	Implicit Scan Add	Implicit Scan Remove	Background Implicit Scan Add	Background Implicit Scan Remove
Intel	15 seconds	140 seconds	15 seconds	140 seconds
Atheros	20 seconds	126 seconds	34 seconds	120 seconds

The second test category measured implicit scan timing on the same channel as the first test, but in this case the card was associated with another AP on another channel throughout the test. Thus, this test is a measure of *background implicit scanning*. The same measures were taken relative to the powering on and off of the test AP. In the case of the Intel driver, the results for background scanning were the same as those for scanning while disconnected. In the case of the Atheros driver, it took 34 seconds to appear on the scan list and 120 seconds to disappear from the scan list.

The relatively small variations between the two cards in the same test category reflects differences in the manufacturers' implementations. The more significant difference between the times reported for the two classes of tests is due to the practical consideration that receipt of a beacon is a sure indication that the corresponding AP is present at that moment. The *absence* of a beacon does not necessarily imply that the AP is no longer available; it is very possible that the beacon was lost due to a collision or some transient RF problem. Thus, any reasonable driver implementation should include an aging mechanism that will reset if a beacon is seen before the maximum age is reached. The longer times that we observed in the second and fourth columns of Table 5.7 are the likely consequences of such an aging mechanism.

5.2.7 Pitfalls with Scanning: Real-Life Example

Engineers are acutely aware that practical experience often differs from what theory predicts. The case described here shows that background scanning worsened overall system performance. This implementation was on an embedded Linux operating system running on a consumer-electronics device. As expected, while the scan was underway, the 802.11 STA would not successfully receive any frames. What was not expected was that the scan resulted in the 802.11 interface disregarding the active channel for an uninterrupted period of 2–3 seconds. During this time, the AP transmitted 802.1X frames to the STA and there was no response because the STA was scanning on other channels. The AP then transmitted DEAUTH frames to the STA to disassociate. None of these frames were received, and the AP gave up, yet the STA believed that it was still associated when it returned to the channel after the scan. The 802.11 procedures allow for the AP and STA to resynchronize and to re-establish communications, but the fact remains that a significant and unnecessary interruption in communications between the STA and the AP resulted from a well-intentioned initiation of a scan.

This example refers to an explicit scan. Thus, the scan was initiated by the client, not the driver. As the client is more aware of the current higher layer protocol context, it was possible to modify the client code to be more intelligent and not initiate a scan at that critical juncture. However, in the case of implicit background scans, the decisions are made by the device driver, which does not have the full protocol context that the client does, and thus the device driver can easily retune its radio to another channel at a particularly inopportune moment.

We mentioned previously that the driver had the advantage of having direct access to low-level information like virtual carrier sensing information. From the perspective of our example, it is the client that has access to contextual information that the driver does not. We pointed out earlier that an NDIS-compliant driver must not spend so much time executing a client-initiated scan request so as to lose an association. Clearly, the Linux driver used in this case did not obey this rule. Even without the higher layer contextual information, a well-written driver should have been able to avoid this pitfall.

5.2.8 SyncScan: Enhancement to Background Scanning

In this section, we describe the proposed enhancement to simple background scanning called *SyncScan*. We stated previously that there is an inherent conflict when trying to optimize for both roaming and throughput: The more time spent scanning other channels, the less time spent attending to the active channel and the more chance of missed frames. Thus, any heuristic that can increase the amount of usable information obtained per unit time spent away from the active channel would benefit the overall roaming throughput performance. This is precisely what SyncScan attempts to do. SyncScan is described in [4].

An important premise of the SyncScan work is that beacons are emitted by most APs at regular intervals. If the STA receives a beacon from an AP and knows the periodicity of that AP's beacon transmission, it follows that the STA can accurately predict when future beacons will be sent. As the periodicity, or *beacon interval* of an AP is included as data within the beacon itself, the future beacon schedule for an AP can be calculated once the first beacon from that AP has been received. The STA can organize its own background scanning schedule for an AP as a function of its predicted beacon schedule. By performing this operation on the set of APs that represent viable roaming candidates, the STA can gradually reduce the amount of time spent away from the active channel and still have a very high probability of receiving a beacon from one of those roaming candidates. This strategy allows the SyncScan-enabled STA to maintain truly up-to-date information about signal strength from the roaming candidates without unduly affecting the active channel.

Practical Aspects of Basic 802.11 Roaming

It is important to recognize that the algorithm that we just sketched only pertains to those cases where the AP actively transmits beacons. For those cases where the AP does not beacon, only active scanning will be useful, thereby rendering SyncScan irrelevant. For this algorithm to work well in an environment with a large number of access points, it is important that some other criterion such as minimum security level or QoS capabilities be used to help winnow the list of APs from all APs that are beaconing to a smaller list of APs that are truly viable roaming candidates. SyncScan represents an interesting innovation that combines the STA-controlled timing of active scanning with the efficiency of the one-way beacon transmission of passive scanning.

5.3 Detailed Analyses of Real-Life Roams

5.3.1 Introduction

In this section, we describe the tools used for conducting our roaming tests and also present the network topology that is used in our tests. We then discuss the Wi-Fi Alliance's WPA test-bed's simple roaming test. In Sections 5.4 and 5.5, we present our results and a discussion of these results in relation to global and local roaming, respectively.

The results shown in this chapter reflect actual measurements made with the test beds described. These test beds are composed of SOHO-class equipment, and the results are useful to study the steps involved in basic 802.11 roaming. We remind the reader that significantly faster roaming performance can be achieved on enterprise-class 802.11 platforms that have been optimized for real-time applications.

5.3.2 Tools Used in Roaming Study

The traffic analysis tool used to capture the 802.11 frames in our roaming tests is AiroPeek for Windows Version 2.0 from WildPackets, Inc. For the global roaming tests, the 802.11 client we used is Microsoft's Wireless Zero Config. The testing STA was a Dell laptop with a built-in Intel Centrino 802.11 chipset. The test bed is depicted in Figure 5.9, where we show a wired Ethernet network that connects two APs to the Internet.

In our tests, the application traffic stream is generated by repeatedly issuing IP *ICMP Echo Requests* (pings) to the Internet gateway 209.100.0.66. Assuming that the *ICMP Echo Replies* (ping responses) are promptly received, the ping requests are issued at one-second intervals. When no ping response is received, a longer delay will occur before the next ping is attempted. As we see in Figure 5.10, the pings fail during the roam itself and resume once full connectivity is established via the new AP. For this discussion, it is crucial to note the importance of how we define the completion of the roam. For these tests, we do not consider

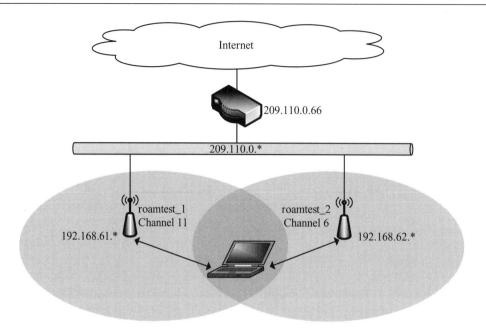

Figure 5.9: Topology for the Basic Global Roaming Tests.

the roam complete once the STA has associated with the new AP but only when the application traffic that was working prior to the roam is able to resume operation. This crude, but concrete measure of roaming time is the one used as part of the WPA certification by the Wi-Fi Alliance.

5.3.3 Wi-Fi Alliance's WPA Simple Roaming Test

The Wi-Fi Alliance WPA test bed defines a very simple roaming test. The test is simple in that it trivializes the roaming subtleties regarding gradual changes in signal strength that normally play a large role in the roaming decision. The test includes two APs configured with the same SSID and a single STA. In order to develop a simple repeatable test, the STA is forced to associate with a fixed AP due to the fact that the other AP is powered off. The second AP is powered on, resulting in a viable roaming candidate. The two APs are located such that signal strength is strong from both. At this point, by powering off the currently associated AP, its signal strength immediately goes to zero, and as only one other AP is available, there is no subjectivity in the roaming decision.

Also, the use of the ping application facilitates determination of pass or fail based on it receiving a response from the stipulated destination within the defined timeout period. In the case of the WPA test, the pings must resume within 90 seconds after having powered

Practical Aspects of Basic 802.11 Roaming

```
Pinging 209.110.0.66 with 32 bytes of data:

Reply from 209.110.0.66: bytes=32 time=3ms TTL=63
Reply from 209.110.0.66: bytes=32 time=4ms TTL=63
Reply from 209.110.0.66: bytes=32 time=2ms TTL=63
Reply from 209.110.0.66: bytes=32 time=2ms TTL=63
Reply from 209.110.0.66: bytes=32 time=2ms TTL=63
Reply from 209.110.0.66: bytes=32 time=2ms TTL=63
Reply from 209.110.0.66: bytes=32 time=2ms TTL=63
Reply from 209.110.0.66: bytes=32 time=2ms TTL=63
Request timed out.
Request timed out.
Request timed out.
Destination host unreachable.
Destination host unreachable.
Destination host unreachable.
Reply from 209.110.0.66: bytes=32 time=3ms TTL=63
Reply from 209.110.0.66: bytes=32 time=2ms TTL=63
Reply from 209.110.0.66: bytes=32 time=2ms TTL=63
Reply from 209.110.0.66: bytes=32 time=2ms TTL=63
Reply from 209.110.0.66: bytes=32 time=2ms TTL=63
Reply from 209.110.0.66: bytes=32 time=2ms TTL=63
Reply from 209.110.0.66: bytes=32 time=2ms TTL=63
Reply from 209.110.0.66: bytes=32 time=2ms TTL=63
```

Figure 5.10: Crude Roaming Test Using Ping.

off the first AP in order to receive a passing grade on the roaming test. A typical time for the pinging to resume is around five seconds. The WPA roaming test steps are outlined below:

1. Associate and authenticate with AP1.

2. Start ping to the host across AP1.

3. Turn on AP2.

4. Turn off the radio on AP1, and ping should continue.

As we noted, the ping must resume within 90 seconds in order to pass the test. No one would claim that a 90-second roaming delay is acceptable for most applications, but this value is a uniform benchmark against which WPA-certified devices are measured. WMM certification tests may involve much more stringent performance criteria. We use this WPA test methodology unchanged for local roaming tests in Section 5.5. For the global roaming tests in

Section 5.4, we will alter the test description slightly in that the two APs will have different SSIDs. Now we turn our attention to the global roam.

5.4 Dissection of a Global Roam

5.4.1 Test-Bed Description for Global Roam

We begin by explaining the framework of our test bed. In Figure 5.9, we showed the simple topology that we use for our first global roaming tests, wherein the wired Ethernet infrastructure connects the Internet gateway router and the two access points. For our global roaming tests, the APs we use are the Linksys 802.11G wireless router model WRT54G. The two APs advertise SSIDs roamtest_1 and roamtest_2, respectively. The roamtest_1 AP operates on channel 11 and roamtest_2 operates on channel 6. These APs were located approximately 30 meters apart in two different rooms of the same wooden structure. Received signal strength from both APs was excellent at most locations in the building. The 802.11 client that we used was WZC.

Referring back to Figure 5.2, we see that WZC has been configured to use either roamtest_1 or roamtest_2 and is currently connected to roamtest_1. This WZC display reflects the fact that at the time the network list was last refreshed, both roamtest_1 and roamtest_2 appeared on the current driver scan list passed from the driver to the client.

All the 802.11-traffic recordings shown in this section are of channel 6. Due to the crude nature of this roaming test, recordings of the starting channel, number 11, would be of little interest, which is attributed to the fact that channel 11 is carrying a steady state of application ping traffic and then is suddenly powered off to produce an absolute and instantaneous stimulus for the STA to roam. From a traffic point of view, there is no relevant dialog on channel 11, as the roamtest_1 AP simply becomes immediately silent. We will revisit this topic of what happens with the traffic with the now-idle AP in Section 5.5.3. At present, we take a closer look at what really happens during a global roam by exploring some test results.

5.4.2 Test Results for Global Roam

The traffic recorded in Figures 5.11 and 5.12 represents two snapshots of the frames captured on channel 6 during the course of this experiment. Figure 5.11 captures the moment when the STA has realized that it needs to associate with another configured AP, which the STA has found on channel 6. This roam involves changing the SSID. The roam depends on the driver reporting this lack of connectivity to the client and the client informing the driver with which SSID to associate. As we described in Section 5.2.2, if the client had made a BSSID-specific association request to the driver (for example, OID-802-11-BSSID), the client would

Practical Aspects of Basic 802.11 Roaming

Packet	Source	Destination	BSSID	Channel	Signal	Size	Relative Time	Protocol	Summary
151	00:13:10:07:48:AC	00:13:CE:46:CB:58		6	50%	85	24.705087	802.11 Probe Rsp	FC=...R...,SN=3919,F
152	00:13:10:07:48:AC	00:13:CE:46:CB:58	00:13:10:07:48:AC	6	50%	85	24.708438	802.11 Probe Rsp	FC=...R...,SN=3919,F
153	00:13:CE:46:CB:58	Ethernet Broad...	FF:FF:FF:FF:FF:FF	6	100%	56	27.640345	802.11 Probe Req	FC=........,SN=1833,F
154	00:13:10:07:48:AC	00:13:CE:46:CB:58	00:13:10:07:48:AC	6	51%	85	27.641226	802.11 Probe Rsp	FC=........,SN=3949,F
155	00:13:CE:46:CB:58	00:13:10:07:48:AC		6	100%	14	27.641538	802.11 Ack	FC=........
156	00:13:CE:46:CB:58	00:13:10:07:48:AC	00:13:10:07:48:AC	6	100%	34	27.675558	802.11 Auth	FC=........,SN=1834,F
157	00:13:10:07:48:AC	00:13:CE:46:CB:58		6	50%	14	27.675844	802.11 Ack	FC=........
158	00:13:10:07:48:AC	00:13:CE:46:CB:58	00:13:10:07:48:AC	6	51%	42	27.676408	802.11 Auth	FC=........,SN=3950,F
159	00:13:CE:46:CB:58	00:13:10:07:48:AC		6	98%	14	27.676721	802.11 Ack	FC=........
160	00:13:CE:46:CB:58	00:13:10:07:48:AC	00:13:10:07:48:AC	6	100%	60	27.677621	802.11 Assoc Req	FC=........,SN=1835,F
161	00:13:10:07:48:AC	00:13:CE:46:CB:58		6	51%	14	27.677932	802.11 Ack	FC=........
162	00:13:10:07:48:AC	00:13:CE:46:CB:58	00:13:10:07:48:AC	6	45%	58	27.678745	802.11 Assoc Rsp	FC=........,SN=3951,F
163	00:13:CE:46:CB:58	00:13:10:07:48:AC		6	100%	14	27.679059	802.11 Ack	FC=........
164	IP-0.0.0.0	IP Broadcast	00:13:10:07:48:AC	6	100%	364	27.706141	DHCP	C DISCOVER 192.168.61
165	00:13:10:07:48:AC	00:13:CE:46:CB:58		6	54%	14	27.706445	802.11 Ack	FC=........
166	IP-0.0.0.0	IP Broadcast	00:13:10:07:48:AC	6	100%	364	27.732179	DHCP	C DISCOVER 192.168.61
167	00:13:10:07:48:AC	00:13:CE:46:CB:58		6	52%	14	27.732460	802.11 Ack	FC=........
168	IP-0.0.0.0	IP Broadcast	00:13:10:07:48:AC	6	51%	364	27.808604	DHCP	C DISCOVER 192.168.61
169	00:13:10:07:48:AA	Ethernet Broad...	00:13:10:07:48:AC	6	45%	82	27.809496	ARP Request	192.168.62.151 = ?
170	IP-0.0.0.0	IP Broadcast	00:13:10:07:48:AC	6	44%	364	27.812679	DHCP	C DISCOVER 192.168.61
171	IP-192.168.62.1	IP Broadcast	00:13:10:07:48:AC	6	54%	612	28.732116	DHCP	R OFFER 192.168.62.15
172	00:13:10:07:48:AA	Ethernet Broad...	00:13:10:07:48:AC	6	54%	82	28.733044	ARP Request	192.168.62.151 = ?
173	IP-192.168.62.1	IP Broadcast	00:13:10:07:48:AC	6	48%	612	29.756074	DHCP	R OFFER 192.168.62.15
174	IP-0.0.0.0	IP Broadcast	00:13:10:07:48:AC	6	100%	376	29.760756	DHCP	C REQUEST 192.168.62.
175	00:13:10:07:48:AC	00:13:CE:46:CB:58		6	50%	14	29.761043	802.11 Ack	FC=........

Figure 5.11: Transition from the Background Scan to the Association in the Global Roam for the WZC Case.

Packet	Source	Destination	BSSID	Channel	Signal	Size	Relative Time	Protocol	Summary
176	IP-0.0.0.0	IP Broadcast	00:13:10:07:48:AC	6	45%	376	29.856685	DHCP	C REQUEST 192.168.62.
177	IP-192.168.62.1	IP Broadcast	00:13:10:07:48:AC	6	52%	612	29.861896	DHCP	R ACK
178	00:13:CE:46:CB:58	Ethernet Broad...	00:13:10:07:48:AC	6	100%	64	29.867569	ARP Request	192.168.62.151 = ?
179	00:13:10:07:48:AC	00:13:CE:46:CB:58		6	52%	14	29.867878	802.11 Ack	FC=........
180	00:13:CE:46:CB:58	Ethernet Broad...	00:13:10:07:48:AC	6	100%	64	29.888243	ARP Request	192.168.62.151 = ?
181	00:13:10:07:48:AC	00:13:CE:46:CB:58		6	52%	14	29.888537	802.11 Ack	FC=........
182	00:13:CE:46:CB:58	Ethernet Broad...	00:13:10:07:48:AC	6	44%	64	29.956484	ARP Request	192.168.62.151 = ?
183	00:13:CE:46:CB:58	Ethernet Broad...	00:13:10:07:48:AC	6	44%	64	29.957332	ARP Request	192.168.62.151 = ?
184	00:13:CE:46:CB:58	Ethernet Broad...	00:13:10:07:48:AC	6	100%	64	30.888246	ARP Request	192.168.62.151 = ?
185	00:13:10:07:48:AC	00:13:CE:46:CB:58		6	47%	14	30.888539	802.11 Ack	FC=........
186	00:13:CE:46:CB:58	Ethernet Broad...	00:13:10:07:48:AC	6	47%	64	30.980523	ARP Request	192.168.62.151 = ?
187	IP-192.168.62.151	IP-192.168.62.255	00:13:10:07:48:AC	6	100%	132	32.061222	NB Name Svc	C REGISTER NAME=MDC-0
188	00:13:10:07:48:AC	00:13:CE:46:CB:58		6	50%	14	32.061513	802.11 Ack	FC=........
189	IP-192.168.62.151	IP-224.0.0.22	00:13:10:07:48:AC	6	100%	76	32.076300	IGMP	Version 3 Membership
190	00:13:10:07:48:AC	00:13:CE:46:CB:58		6	50%	14	32.076592	802.11 Ack	FC=........
191	IP-192.168.62.151	IP-192.168.62.255	00:13:10:07:48:AC	6	50%	132	32.107446	NB Name Svc	C REGISTER NAME=MDC-0
192	IP-192.168.62.151	IP-224.0.0.22	00:13:10:07:48:AC	6	45%	76	32.108354	IGMP	Version 3 Membership
193	00:13:CE:46:CB:58	Ethernet Broad...	00:13:10:07:48:AC	6	100%	64	32.388491	ARP Request	192.168.62.1 = ?
194	00:13:10:07:48:AC	00:13:CE:46:CB:58		6	52%	14	32.388779	802.11 Ack	FC=........
195	00:13:10:07:48:AA	00:13:CE:46:CB:58	00:13:10:07:48:AC	6	52%	64	32.389241	ARP Response	00:13:10:07:48:AA = 1
196	00:13:CE:46:CB:58	00:13:10:07:48:AC		6	100%	14	32.389266	802.11 Ack	FC=........
197	IP-192.168.62.151	IP-209.110.0.66	00:13:10:07:48:AC	6	100%	96	32.390387	PING Req	Echo: 209.110.0.66
198	00:13:10:07:48:AC	00:13:CE:46:CB:58		6	52%	14	32.390643	802.11 Ack	FC=........
199	IP-209.110.0.66	IP-192.168.62.151	00:13:10:07:48:AC	6	51%	96	32.392710	PING Reply	Echo Reply: 192.168.6
200	00:13:CE:46:CB:58	00:13:10:07:48:AC		6	100%	14	32.392735	802.11 Ack	FC=........

Figure 5.12: Acquisition of a New IP Address and the Resumption of the Ping Application for the WZC Global Roam.

need to be involved in roams even where the SSID does not change (that is, local roams). The general-purpose clients examined in this chapter do not make BSSID-specific association requests.

As our example pertains to one of a global roam, there are purposely no other APs belonging to the roamtest_1 ESS. If there were other APs, the driver software would automatically request that the driver associate with another AP member of that ESS, thereby accomplishing a local roam. In frame 156 of Figure 5.11, we see the transition from pure background scanning mode on channel 6 to an active attempt to associate. The frame sequence 156–159 reflects the 802.11 open-authentication request from the STA to the AP and its acknowledgment, followed by the successful open-authentication response from the AP to the STA and its corresponding acknowledgment.

The next four exchanges in frames 160–163 show the STA associating with the roamtest_2 AP. These eight frames are precisely what one would expect in a straightforward 802.11 roam. The reader should note that this point in the frame exchange occurs after the first printing of the "Request Timed Out" message in Figure 5.10. The time between that message and the resumption of successful pings was approximately 20 seconds. We now look more closely at what transpires in this exchange.

The first thing that we observe in frame 164 in Figure 5.11 is the STA making a DHCP request soliciting the same IP address 192.168.61.151 that it had been using with roamtest_1. If this request was granted, it would minimize the disruption to the IP-based application. Since we are dealing with a global roam here, the STA has moved to the new subnet 192.168.62.*, and it is not possible to renew the DHCP lease with the same address used earlier. Frame 169 heralds the AP preparing to grant a new DHCP lease to the STA. First it ARPs to ascertain that the candidate address that it wishes to lease, 192.168.62.151, is not in use. Frame 170 is the fourth repeat of the STA's first attempt to renew 192.168.61.151. The fact that there was no response to the ARP frees the AP to offer the new address 192.168.62.151 to the STA, which we observe in frame 173. The DHCP exchange concludes in frames 176 and 177, as seen in Figure 5.12 with the actual granting of the lease for the new address. However, this lengthy exchange following the association seen in frames 164–177 is still not sufficient for the application to resume following the roam.

Frames 176 through 197 shown in Figure 5.12 reveal that there is additional overhead related to the multicast-group membership (*Internet Group Management Protocol* (*IGMP*)) protocol and name services that is incurred before the resumption of the ping traffic in frame 197. The number of non-802.11 specification frames exchanged to complete the roam is much larger than the number of frames to switch the PHY- and MAC-layer connections to the new access point. This intertwined relationship has a great impact on what expectations we should have about the reduction of roaming times due to fine tuning of only the 802.11 specifications. We remind the reader of the multiple phases of roaming that we described in Section 3.3.3. What we describe here is an example of what can occur in the *Infrastructure Routing Delay* phase.

Another interesting observation is that the total interruption to the application shown in Figure 5.10 was measured as approximately 20 seconds. In Figures 5.11 and 5.12, the total time from the association to the first application ping is approximately 5 seconds. Thus, of the 20 seconds of application interruption, about 15 seconds are spent detecting and reporting the loss of roamtest_2 and selecting roamtest_1, versus a mere 5 seconds connecting to the new AP. The fact that application traffic does not resume for 20 seconds reflects the degree to which our widely used 802.11 implementations actually impact the user experience during a global roam. We will look more closely at the individual components of this delay in Section 5.5.3.

Figure 5.13 shows the WZC client when it appears during the same time period as frames 164–197. We should point out that the details of the frame exchange between the association and the resumption of the application traffic will vary from one 802.11 client to another.

5.5 Dissection of a Local Roam

5.5.1 Test Description for Local Roam

The tests in this section were performed using the same network topology and test methodology as was used for global roaming, except as noted here. As this test was for local roaming, the two access points advertise the same SSID, unlike the global roaming tests. The topology used for these tests is depicted in Figure 5.14. As we can see from the figure, both the APs advertise the SSID roamtest_3. The APs are Buffalo Wireless Switches Model Air Station WLA2-644L. The channels used in this test are 1 and 6, where the initial connection is via channel 6 and the roam is to the AP working on channel 1. The ping is to the address 209.110.0.101 in lieu of the Internet gateway 209.110.0.66, as the client use an internally generated ping to the gateway to determine whether or not the subnet has changed. In order to distinguish clearly this client-generated ping from our test application's pings, we change the address of our test application's pings to 209.110.0.101. In these tests, we observed the results for the Odyssey clients.

5.5.2 Test Results for Local Roam

In these tests, we measure and analyze the interval between the association on the new channel and the resumption of the application pings. The interval of interest is captured in the frame trace shown in Figure 5.15.

Chapter 5

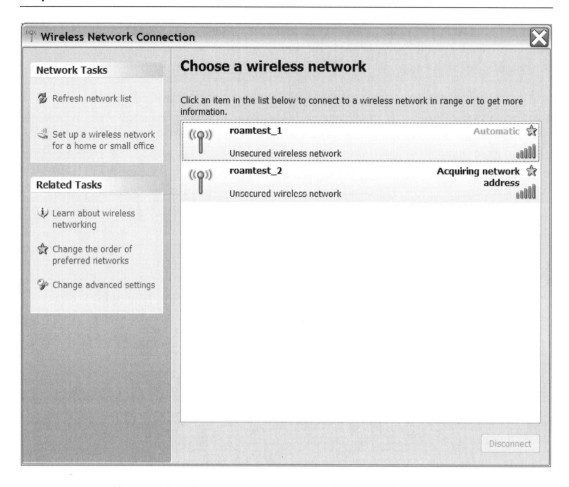

Figure 5.13: Microsoft's Wireless Client During a Global Roam.

The 802.11 open authentication that starts the association appears in frame 65 at 01:52:50. The association proceeds normally at that point. The next noteworthy frame is number 74 at 01:56:37—an ARP request to the gateway address 209.110.0.66. This ARP reflect the STA trying to determine if the subnet has changed as a result of the roam. This method conforms to the recommendation found in [1]. This Internet Draft is not specific to 802.11 roaming. It can apply to any situation where an IP-protocol stack has just been informed of a LINK DOWN to LINK UP transition and where the attached subnet may be different from that before the transition. This switch happens regularly in 802.3 networks when changing VLANs, but it is also relevant to the case of roaming in 802.11 networks. This ARP is actually the indirect result of the client flushing the STA's ARP table and then transmitting a ping to the gateway address. This ping can be seen in frame 78 in Figure 5.15. We will refer to this ARP as the *subnet-detection* ARP. The application pings to 209.110.0.101 finally resume in frame 87 at

Practical Aspects of Basic 802.11 Roaming

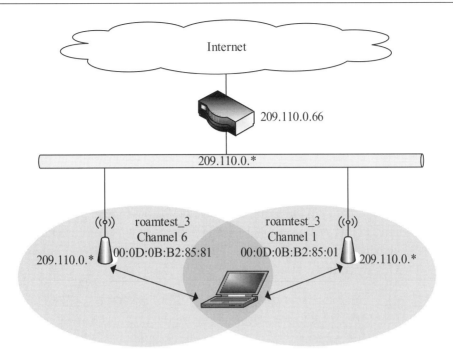

Figure 5.14: Topology for Basic Local Roaming Tests.

Packet	Source	Destination	BSSID	Channel	Size	Relative Time	Protocol	Summary
62	00:13:CE:46:CB:58	Ethernet Broadcast	FF:FF:FF:FF:FF:FF	1	56	0:01:52.470510	802.11 Probe Req	FC=........,SN=3538,FN
63	00:0D:0B:B2:85:01	00:13:CE:46:CB:58	00:0D:0B:B2:85:01	1	85	0:01:52.471410	802.11 Probe Rsp	FC=........,SN=1153,FN
64	00:13:CE:46:CB:58	00:0D:0B:B2:85:01		1	14	0:01:52.471704	802.11 Ack	FC=........
65	00:13:CE:46:CB:58	00:0D:0B:B2:85:01	00:0D:0B:B2:85:01	1	34	0:01:52.505745	802.11 Auth	FC=........,SN=3539,FN
66	00:0D:0B:B2:85:01	00:13:CE:46:CB:58		1	14	0:01:52.506047	802.11 Ack	FC=........
67	00:0D:0B:B2:85:01	00:13:CE:46:CB:58	00:0D:0B:B2:85:01	1	42	0:01:52.506744	802.11 Auth	FC=........,SN=1154,FN
68	00:13:CE:46:CB:58	00:0D:0B:B2:85:01		1	14	0:01:52.507046	802.11 Ack	FC=........
69	00:13:CE:46:CB:58	00:0D:0B:B2:85:01	00:0D:0B:B2:85:01	1	60	0:01:52.507945	802.11 Assoc Req	FC=........,SN=3540,FN
70	00:0D:0B:B2:85:01	00:13:CE:46:CB:58		1	14	0:01:52.508246	802.11 Ack	FC=........
71	00:0D:0B:B2:85:01	00:13:CE:46:CB:58	00:0D:0B:B2:85:01	1	58	0:01:52.509152	802.11 Assoc Rsp	FC=........,SN=1155,FN
72	00:13:CE:46:CB:58	00:0D:0B:B2:85:01		1	14	0:01:52.509455	802.11 Ack	FC=........
73		Xerox:00:00:01		1	14	0:01:55.007578	802.11 Ack	
74	00:13:CE:46:CB:58	Ethernet Broadcast	00:0D:0B:B2:85:01	1	64	0:01:56.377734	ARP Request	209.110.0.66 = ?
75	00:0D:0B:B2:85:01	00:13:CE:46:CB:58		1	14	0:01:56.378027	802.11 Ack	FC=........
76	Runtop:D3:92:77	00:13:CE:46:CB:58	00:0D:0B:B2:85:01	1	82	0:01:56.378765	ARP Response	Runtop:D3:92:77 = 209.
77	00:13:CE:46:CB:58	00:0D:0B:B2:85:01		1	14	0:01:56.378794	802.11 Ack	FC=........
78	IP-209.110.0.106	IP-209.110.0.66	00:0D:0B:B2:85:01	1	69	0:01:56.380243	PING Req	Echo: 209.110.0.66
79	00:0D:0B:B2:85:01	00:13:CE:46:CB:58		1	14	0:01:56.380551	802.11 Ack	FC=........
80	IP-209.110.0.66	IP-209.110.0.106	00:0D:0B:B2:85:01	1	82	0:01:56.381297	PING Reply	Echo Reply: 209.110.0.
81	00:13:CE:46:CB:58	00:0D:0B:B2:85:01		1	14	0:01:56.381323	802.11 Ack	FC=........
82	00:13:CE:46:CB:58	Ethernet Broadcast	00:0D:0B:B2:85:01	1	64	0:01:56.429332	ARP Request	209.110.0.66 = ?
83	00:13:CE:46:CB:58	Ethernet Broadcast	00:0D:0B:B2:85:01	1	64	0:01:57.377350	ARP Request	209.110.0.101 = ?
84	00:0D:0B:B2:85:01	00:13:CE:46:CB:58		1	14	0:01:57.377643	802.11 Ack	FC=........
85	Asustek Comp:4...	00:13:CE:46:CB:58	00:0D:0B:B2:85:01	1	82	0:01:57.378332	ARP Response	Asustek Comp:47:D4:29
86	00:13:CE:46:CB:58	00:0D:0B:B2:85:01		1	14	0:01:57.378359	802.11 Ack	FC=........
87	IP-209.110.0.106	IP-209.110.0.101	00:0D:0B:B2:85:01	1	96	0:01:57.380008	PING Req	Echo: 209.110.0.101

Figure 5.15: Transition to a New AP in the Local Roam with the Odyssey Client.

Chapter 5

01:57:38. Thus, in this test, the total time between the start of the association and the application resumption is 4.93 seconds.

5.5.3 A Closer Look at Roaming Delay

We have mentioned during our discussion of global roaming that the delay observed before the ping application resumed was approximately 20 seconds. In the case of local roaming, the sniffer trace showed the interval between the association with the new AP and the resumption of application pings to be approximately 5 seconds. Thus far, our analysis has focused on roaming delay once the association to the new AP occurs. We have explained earlier that we would concentrate on examining the frame exchange on the new active channel rather than the one before the roam. In this section, we will briefly look at what transpires on the formerly active channel once that AP is powered down. This interval is captured in Figures 5.16 and 5.17 for the Odyssey client.

In Figure 5.16, we see that the last frame transmitted from the AP was frame 702—a ping reply from 209.110.0.101. This frame was transmitted at 02:07:74. We will consider this time as the one when the AP was powered off. As the STA is not yet aware that the AP is

Packet	Source	Destination	BSSID	Channel	Size	Relative Time	Protocol	Summary
697	IP-209.110.0.101	IP-209.110.0.106	00:0D:0B:B2:85:81	6	96	0:02:06.749812	PING Reply	Echo Reply: 209.11(
698	00:13:CE:46:CB:58	00:0D:0B:B2:85:81		6	14	0:02:06.749843	802.11 Ack	FC=........
699	IP-209.110.0.100	IP-209.110.0.255	00:0D:0B:B2:85:81	6	253	0:02:07.738692	SMB	C Browser Request I
700	IP-209.110.0.106	IP-209.110.0.101	00:0D:0B:B2:85:81	6	96	0:02:07.749031	PING Req	Echo: 209.110.0.10:
701	00:0D:0B:B2:85:81	00:13:CE:46:CB:58		6	14	0:02:07.749064	802.11 Ack	FC=........
702	IP-209.110.0.101	IP-209.110.0.106	00:0D:0B:B2:85:81	6	96	0:02:07.749761	PING Reply	
703		00:0D:0B:B2:85:81		6	14	0:02:07.749780	802.11 Ack	FC=........
704	IP-209.110.0.106	IP-209.110.0.101	00:0D:0B:B2:85:81	6	96	0:02:08.749026	PING Req	Echo: 209.110.0.10:
705	IP-209.110.0.106	IP-209.110.0.101	00:0D:0B:B2:85:81	6	96	0:02:08.749343	PING Req	Echo: 209.110.0.10:
706	IP-209.110.0.106	IP-209.110.0.101	00:0D:0B:B2:85:81	6	96	0:02:08.749905	PING Req	Echo: 209.110.0.10:
707	IP-209.110.0.106	IP-209.110.0.101	00:0D:0B:B2:85:81	6	96	0:02:08.750872	PING Req	Echo: 209.110.0.10:
708	IP-209.110.0.106	IP-209.110.0.101	00:0D:0B:B2:85:81	6	96	0:02:08.753050	PING Req	Echo: 209.110.0.10:
709	IP-209.110.0.106	IP-209.110.0.101	00:0D:0B:B2:85:81	6	96	0:02:08.755264	PING Req	Echo: 209.110.0.10:
710	IP-209.110.0.106	IP-209.110.0.101	00:0D:0B:B2:85:81	6	96	0:02:08.764974	PING Req	Echo: 209.110.0.10:
711	IP-209.110.0.106	IP-209.110.0.101	00:0D:0B:B2:85:81	6	96	0:02:08.778613	PING Req	Echo: 209.110.0.10:
712	IP-209.110.0.106	IP-209.110.0.101	00:0D:0B:B2:85:81	6	96	0:02:08.786947	PING Req	Echo: 209.110.0.10:
713	IP-209.110.0.106	IP-209.110.0.101	00:0D:0B:B2:85:81	6	96	0:02:08.805418	PING Req	Echo: 209.110.0.10:
714	IP-209.110.0.106	IP-209.110.0.101	00:0D:0B:B2:85:81	6	96	0:02:08.812746	PING Req	Echo: 209.110.0.10:
715	IP-209.110.0.106	IP-209.110.0.101	00:0D:0B:B2:85:81	6	96	0:02:08.827918	PING Req	Echo: 209.110.0.10:
716	IP-209.110.0.106	IP-209.110.0.101	00:0D:0B:B2:85:81	6	96	0:02:08.833273	PING Req	Echo: 209.110.0.10:
717	IP-209.110.0.106	IP-209.110.0.101	00:0D:0B:B2:85:81	6	96	0:02:08.844303	PING Req	Echo: 209.110.0.10:
718	IP-209.110.0.106	IP-209.110.0.101	00:0D:0B:B2:85:81	6	96	0:02:08.845716	PING Req	Echo: 209.110.0.10:
719	00:13:CE:46:CB:58	00:0D:0B:B2:85:81	00:0D:0B:B2:85:81	6	28	0:02:09.968099	802.11 Data	FC=T...P...,SN= 14
720	00:13:CE:46:CB:58	00:0D:0B:B2:85:81	00:0D:0B:B2:85:81	6	28	0:02:09.968201	802.11 Data	FC=T..RP...,SN= 14
721	00:13:CE:46:CB:58	00:0D:0B:B2:85:81	00:0D:0B:B2:85:81	6	28	0:02:09.968423	802.11 Data	FC=T..RP...,SN= 14

Figure 5.16: View of the Active Channel as the AP is Powered Down in the Local Roam with the Odyssey Client.

Practical Aspects of Basic 802.11 Roaming

Packet	Source	Destination	BSSID	Channel	Size	Relative Time	Protocol	Summary
803	00:13:CE:46:CB:58	00:0D:0B:B2:85:81	00:0D:0B:B2:85:81	6	28	0:02:10.403686	802.11 Data	FC=T..RP...,SN= 17!
804	00:13:CE:46:CB:58	00:0D:0B:B2:85:81	00:0D:0B:B2:85:81	6	28	0:02:10.403957	802.11 Data	FC=T..RP...,SN= 17!
805	00:13:CE:46:CB:58	00:0D:0B:B2:85:81	00:0D:0B:B2:85:81	6	28	0:02:10.404616	802.11 Data	FC=T..RP...,SN= 17!
806	00:13:CE:46:CB:58	00:0D:0B:B2:85:81	00:0D:0B:B2:85:81	6	28	0:02:10.405144	802.11 Data	FC=T..RP...,SN= 17!
807	00:13:CE:46:CB:58	00:0D:0B:B2:85:81	00:0D:0B:B2:85:81	6	28	0:02:10.405919	802.11 Data	FC=T..RP...,SN= 17!
808	00:13:CE:46:CB:58	00:0D:0B:B2:85:81	00:0D:0B:B2:85:81	6	28	0:02:10.406452	802.11 Data	FC=T..RP...,SN= 17!
809	00:13:CE:46:CB:58	00:0D:0B:B2:85:81	00:0D:0B:B2:85:81	6	28	0:02:10.407587	802.11 Data	FC=T........,SN= 17(
810	00:13:CE:46:CB:58	00:0D:0B:B2:85:81	00:0D:0B:B2:85:81	6	28	0:02:10.407729	802.11 Data	FC=T..R.....,SN= 17(
811	00:13:CE:46:CB:58	00:0D:0B:B2:85:81	00:0D:0B:B2:85:81	6	28	0:02:10.407858	802.11 Data	FC=T..R.....,SN= 17(
812	00:13:CE:46:CB:58	00:0D:0B:B2:85:81	00:0D:0B:B2:85:81	6	28	0:02:10.408050	802.11 Data	FC=T..R.....,SN= 17(
813	00:13:CE:46:CB:58	00:0D:0B:B2:85:81	00:0D:0B:B2:85:81	6	28	0:02:10.408785	802.11 Data	FC=T..R.....,SN= 17(
814	00:13:CE:46:CB:58	00:0D:0B:B2:85:81	00:0D:0B:B2:85:81	6	28	0:02:10.409334	802.11 Data	FC=T..R.....,SN= 17(
815	00:13:CE:46:CB:58	00:0D:0B:B2:85:81	00:0D:0B:B2:85:81	6	28	0:02:10.410204	802.11 Data	FC=T..R.....,SN= 17(
816	00:13:CE:46:CB:58	00:0D:0B:B2:85:81	00:0D:0B:B2:85:81	6	28	0:02:10.410870	802.11 Data	FC=T..R.....,SN= 17(
817	00:13:CE:46:CB:58	00:0D:0B:B2:85:81	00:0D:0B:B2:85:81	6	30	0:02:10.412376	802.11 Disassoc	FC=.........,SN= 17
818	00:13:CE:46:CB:58	00:0D:0B:B2:85:81	00:0D:0B:B2:85:81	6	30	0:02:10.413167	802.11 Disassoc	FC=...R....,SN= 17
819	00:13:CE:46:CB:58	00:0D:0B:B2:85:81	00:0D:0B:B2:85:81	6	30	0:02:10.413942	802.11 Disassoc	FC=...R....,SN= 17
820	00:13:CE:46:CB:58	00:0D:0B:B2:85:81	00:0D:0B:B2:85:81	6	30	0:02:10.414630	802.11 Disassoc	FC=...R....,SN= 17
821	00:13:CE:46:CB:58	00:0D:0B:B2:85:81	00:0D:0B:B2:85:81	6	30	0:02:10.415526	802.11 Disassoc	FC=...R....,SN= 17
822	00:13:CE:46:CB:58	00:0D:0B:B2:85:81	00:0D:0B:B2:85:81	6	30	0:02:10.416795	802.11 Disassoc	FC=...R....,SN= 17
823	00:13:CE:46:CB:58	00:0D:0B:B2:85:81	00:0D:0B:B2:85:81	6	30	0:02:10.417584	802.11 Disassoc	FC=...R....,SN= 17
824	80:4B:C7:06:EF:58	Ethernet Broadcast	AF:37:F7:9F:75:D8	6	46	0:02:30.223704	802.11 Probe Req	
825	00:13:CE:46:CB:58	Ethernet Broadcast	FF:FF:FF:FF:FF:FF	6	56	0:02:30.248645	802.11 Probe Req	FC=.........,SN= 21.
826	00:13:CE:46:CB:58	Ethernet Broadcast	FF:FF:FF:FF:FF:FF	6	46	0:02:30.249323	802.11 Probe Req	FC=.........,SN= 21.

Figure 5.17: View of the Active Channel as the STA is about to Change the Channel in the Local Roam with the Odyssey Client.

unavailable, in frame 704 the STA transmits another ping request almost exactly one second following the previous request. We then observe a rapid series of 802.11 retransmissions of this ping request in frames 705–718. Beginning in frame 719 at time 02:09:96, we observe a long series of empty 802.11 data frames transmitted by the STA to the now-dead AP. This pattern continues until frame 817 at time 02:10:41, where we see the STA trying to disassociate from the AP. As the AP cannot respond to this request, the STA gives up the disassociation attempts after six retries. At this point, about 2.67 seconds after the AP was powered off, we can safely assume that the STA has given up on channel 6.

The monitoring equipment that we used does not allow us to synchronize measures taken on two different channels, so we cannot know precisely how much time passes between the disassociation and the association attempt that we denote as the start of the roam in our discussion in Section 5.5.2. However, one coarse inference can be made by comparing the application delay of 5 seconds for local roaming with the 20 seconds for global roaming. Regardless of whether the roaming is global or local, we measure a delay of 2.67 seconds from the powering down of the first AP to the disassociation and a separate delay of approximately 5 seconds between the association with the new AP and the resumption of the application traffic.

For this discussion, we will call the two delays just mentioned, D1 and D2. For clarity, D1 represents the time from the powering down of the first AP to the moment that the STA decides to abandon the AP, and D2 represents the time from the STA deciding to associate with the new AP and the resumption of application traffic. We approximate the sum D1 + D2 as 7 seconds. Allowing for the coarseness of the application delay measure performed, it is reasonable to assume that these 7 seconds correspond to the approximate 5-second delay that we observe in the application resumption in the local-roam case.

The timing of D1 and D2 are the same for the global roam case, so there must be a third component of the delay, D3, which is very small in the case of local roaming and approximately 13 seconds in the case of global roaming. As D1 commences with powering off the first AP and D2 ends with the resumption of application traffic, D3 occurs *between* D1 and D2. In a local roam, the driver takes full responsibility for selecting another BSSID within the same SSID, and no client involvement is required. However, in the case of global roaming the driver will not automatically select another SSID. The driver will notify the client of a LINK DOWN condition and wait for the client to request association with a new SSID. Client implementations will make this decision based on heuristics unique to that client.

The D3 delay can be attributed to the client processing involved in the global roam case; this client processing does not exist in the case of local roaming. It should be noted that in both the local and global roams, the driver sends a LINK UP event when it is associated with the new AP. This LINK UP event does not provoke any client action in the local roam case studied here because no 802.11 security is involved. When we study the details of *secure* roaming in Chapter 7, we will see that client processing is indeed involved even in the local roam case in order to perform the required security procedures.

Before leaving Figure 5.17, we call the reader's attention to the 20-second delay between the disassociation with the dead AP and the first active scan on channel 6 from the STA that is now busy sending application traffic on channel 1.

5.6 Access-Point Placement Methodologies

5.6.1 Introduction

Any discussion of the practical aspects of basic 802.11 roaming would be incomplete without the mention of the importance of the placement of the APs. Determining AP placement that provides good roaming support seems to correlate directly to a placement that provides good coverage, so the methods discussed in this section are not specific or tuned to the roaming problem, but for general 802.11 access. More background on large-scale 802.11 network design can be found in [3].

5.6.2 Access-Point Placement

There are many relevant questions that may be asked regarding the placement of access points. These include the following:

- How many APs should be placed? APs used to be so expensive that economic considerations were a major factor. Commodity APs are indeed so inexpensive that for those deployments using them, their cost is no longer a factor, as of this writing, whereas some enterprise class APs still cost many hundreds of dollars each, so the additional cost of unnecessary APs may still be important.

- Are there so few APs that there are dead spots?

- Are there so many APs that there is unnecessary channel overlap?

- Once the number of APs is fixed, have they been placed optimally?

The research community has been drawn to the problem of optimal AP placement because it is a timely problem that lends itself well to mathematical modeling and because a practical application could have real commercial value. Examples of some different modeling approaches can be found in [2] and [5].

Models for AP placement generally work by creating an abstraction of the physical space into which the APs are to be located. This space may be in two or three dimensions according to the problem definition. Some models will assume a fixed number of APs and try to place them optimally, whereas others will find the minimum number of APs for which a *complete coverage* placement can be found. Typically, *cost functions* are defined for which a minimum value is obtained in the execution of the model. In [2], for example, one cost function used was the reciprocal of the estimate of signal coverage, thereby yielding a lower cost for higher signal coverage. Thus, in this case by lowering cost we have better coverage. While such models are of great academic interest and have potential practical applications, hard commercial cash has instead been spent on empirically oriented *site surveys* rather than theoretical models. We now turn our discussion to site surveys.

5.6.3 Site Surveys for Access-Point Placement

A site survey is often conducted before a large commercial AP deployment. The process involved in the site survey sharply contrasts with the modeling techniques mentioned previously in that they are often almost entirely empirical. A common approach is to place AP(s) by intelligent guesswork and then walk around the physical plant recording received signal strength. This information is then fed into the site survey application. These pieces of software have varying degrees of sophistication. The application can suggest different positioning of the APs to obtain more complete coverage or a more even distribution of signal strength. These changes are tried; the measurements are taken again; the process is iterated upon until satisfactory results are obtained.

Beyond the evident brute-force nature of this approach, there are additional drawbacks. The RF signal observed during the site survey does not necessarily reflect the actual conditions when the network is operational. There are several reasons for this variation. First, there may be periodic atmospheric or other environmental fluctuations that affect the signal, but escaped the site survey. There may be shifts in the physical plant itself, such as changes in plumbing and/or electrical wiring, or even the construction and/or demolition of walls, which invalidate the initial observations and conclusions from the site survey. As the measurement device used in the site survey is often just an 802.11 card reporting received signal strength, the tool is not a good detector of other sources of RF interference, such as microwave ovens or portable telephones.

Because the RF equipment needed to detect this kind of interference is usually unavailable and requires a level of sophistication beyond most organizations that conduct site surveys, a more useful measure of interference might be to measure the *consequence* of the interference—reduced throughput—over varying periods of time. If we observe unexplained changes in the 802.11 throughput rates, it is very likely that these are the consequence of intermittent interference. The use of different observation periods provides a better opportunity to observe inherently transient interference from other devices when the interference actually occurs. Using lower throughput as an indicator of interference is valuable as unexplained low throughput is one of the most direct side effects of interference.

5.6.4 Self-Monitoring 802.11 Networks

The drawbacks of site surveys have led to a new branch in the 802.11 technology tree. This branch includes the self-monitoring, self-configuring, and even self-repairing of 802.11 networks. The genesis of this technology stems from the argument that if site surveys are deficient because they do not measure the network during its actual operation, why not ask the network to monitor itself? This tactic is particularly inviting since the very APs that we are already using as portals to our network infrastructure are by design agile RF sensors! Even the most simple AP reporting can reveal users clustering around one AP, overwhelming it, or delivering poor service throughout that cluster, whereas other nearby APs go underutilized. When not running at capacity on their active channel, an AP can listen for neighboring APs and can help draw an RF-aware channel coverage map of the network. It is even possible for the system to change the APs' signal strength and channel assignment automatically in reaction to these ongoing real-time observations.

There are now emerging custom 802.11 chip designs that incorporate sophisticated RF-spectrum-analysis capabilities. Such capabilities can greatly augment the self-monitoring capabilities of the network. Cognio is a current leader in this technology. Admittedly, such network-wide coordination between APs is possible with traditional AP architectures. Some providers of centralized wireless switches argue that their alternative architecture is better suited to self-monitoring, configuration, and repair. We will present the basic concepts of wireless switches in Chapter 8.

References

[1] Bernard Aboba, James Carlson, and Stuart Cheshire, Detecting Network Attachment in IPv4 (DNAv4), *Internet Draft—Proposed Standard*, Internet Engineering Task Force, 2005.

[2] Roberto Battiti, Mauro Brunato, and Andrea Delai, Optimal Wireless Access Point Placement for Location-Dependent Services, *Technical Report DIT-03-052*, University of Trento, 2003.

[3] Alex Hills, Large-Scale Wireless LAN Design, *IEEE Communications Magazine*, **39**(11): 98–107, 2001.

[4] Ishwar Ramani and Stefan Savage, SyncScan: Practical Fast Handoff for 802.11 Infrastructure Networks, *Proceedings of the IEEE Infocom*, March 2005.

[5] Hanif D. Sherali, Chandra M. Pendyala, and Ted S. Rappaport, Optimal Location of Transmitters for Micro-Cellular Radio Communication System Design, *IEEE Journal on Selected Areas in Communications*, **14**(4): 662–673, 1996.

[6] Windows Platform Design Notes, *IEEE 802.11 Network Adapter Design Guidelines for Windows XP*, Microsoft Corporation, 2002.

CHAPTER 6
Fundamentals of User Authentication in 802.11

6.1 Introduction

6.2 802.1X Port-Level Authentication

6.3 The AAA Server

6.4 The Extensible Authentication Protocol

6.5 Flexible and Strong Authentication in 802.11

6.6 Other 802.11 Authentication Methodologies

6.7 Network Access Control

6.8 Summary

References

6.1 Introduction

In this chapter, we review the standard IEEE 802.11 admission-control model. This model includes three components:

- A client-based 802.1X supplicant
- An access-point-based 802.1X authenticator
- A back-end AAA server

Chapter 6

These three items in combination provide a robust, efficient, and scalable admission-control system that works over many media types, including 802.11. Although it is not obvious that the authentication model is of direct importance to the roaming topic, the reader requires a basic understanding of this paradigm in order to absorb the material on fast secure roaming presented in Chapter 7. In addition, here we discuss some alternative non-IEEE standard authentication models that are in use, as of this writing.

At the end of this chapter, we present the nascent area of *Network Access Control* (*NAC*), which brings a new dimension to 802.11 user authentication. This process adds further to the complexity of secure roaming in 802.11. We now turn our attention to port-level authentication.

6.2 802.1X Port-Level Authentication

6.2.1 Introduction

While the initial development of *Extensible Authentication Protocol* (*EAP*) in AAA technology was for dial-up networks, the proliferation of LAN technology in the 1990s created the need for admission control in LANs. The IEEE decided as much as possible to base the LAN authentication model on existing technologies, notably EAP and AAA. The only new protocol defined by 802.1X was called *EAP Over LAN* (*EAPOL*), and this protocol is just EAP with two additional frame types that allow the protocol to start and stop layer-two point-to-point conversations.

EAPOL is used to transport EAP frames over the single layer-two hop between the user machine and the *Network Access Server* (*NAS*). The NAS is the port-of-entry onto the network and is most commonly embodied in an Ethernet-switch port for 802.3 networks or the access point in 802.11 networks. The client end of the EAPOL protocol is called the *supplicant* and the NAS end of the EAPOL protocol is called the *authenticator*. We show the relationship between these components in an 802.11 setting in Figure 6.1.

The 802.1X standard was ratified by the IEEE 802.1 workgroup in 2001 and was updated in 2004. Adherence to this standard ensures a high degree of interoperability between

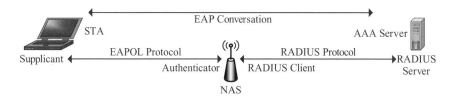

Figure 6.1: The 802.1X Paradigm.

Fundamentals of User Authentication in 802.11

various vendors' 802.1X supplicants, authenticators, and AAA servers. In addition to providing this interoperability, the 802.1X standard has provided a foundation for the Network Admission Control initiatives that we will introduce in Section 6.7.1. Although 802.1X is an IEEE standard, it is flexible enough to handle the different choices vendors and enterprises make in deploying 802.1X technology. We examine the features of 802.1X in the next subsection.

6.2.2 Flexibility of 802.1X

The 802.1X standard exhibits flexibility in the following areas:

- *Multiple EAP methods:* The 802.1X standard can be deployed using a myriad of credential types by simply modifying the EAP method that is used. While EAP itself is now an IETF standard, the specifications for most of the popular EAP methods are at most IETF informational RFCs, and many of these will never become IETF standards. However, this fact has not impeded their adoption by industry.

- *Multiple authenticated attributes:* By using some of the published EAP types, the network may authenticate the physical machine, the user working on the machine, and the state of the machine. These three items provide a robust system of verification for network access.

- *Protocol flexibility:* The 802.1X standard provides *Remote Access Dial-In User Server* (*RADIUS*) as an example for the communications between the authenticator and the authentication server, but the standard makes it clear that other protocols such as DIAMETER and *Terminal Access Controller Access Control System* (*TACACS*) may be used.

- *Uncontrolled port usage:* The 802.1X standard permits selected types of network traffic to be allowed into the network before and during the authentication process. For example, it is feasible to implement systems that utilize login scripts that allow access to file and print services prior to entering the authenticated state.

- *Extensibility:* The combination of multiple EAP types, protocol flexibility, and uncontrolled port usage can be combined with other mechanisms to leverage the 802.1X system as the foundation for a number of additional features and functions such as policy-based networks and Network Admission Control. For example, an enterprise may have a policy that executives be assigned to VLANs with bandwidth guarantees higher than those of their staff. When these users are admitted, the 802.1X components that we have described here can easily be used to implement such a policy.

The issues that we have just covered indicate a high degree of flexibility in the 802.1X standard.

6.2.3 Evolution of 802.1X

Within the original 802.1X framework, port access refers to access to a *user port*. The ability to scale for very large deployments at low cost, like most IEEE network endeavors, was a main design consideration by the developers of 802.1X. This scalability was achieved by weaving together pre-existing and highly effective network infrastructure components, such as EAP, RADIUS, *Lightweight Directory Access Protocol* (*LDAP*), and *Active Directory*. The 802.1X standard itself is not an authentication method but a layer-two transport protocol and a port-access control standard that uses the other components for the authentication task. EAP is the protocol that carries on authentication dialogs between the end user and the AAA server in the 802.1X paradigm; the 802.1X standard defines how EAP may be used on LANs for port access control.

In order to meet the changing market needs, the IEEE ratified 802.1XRev in 2004. The 802.1XRev standard supersedes the original 802.1X and the revision features the following notable changes:

1. The introduction of a controlled port on the supplicant so that the supplicant may protect itself from the network. While the focus of the original 802.1X was to protect the network from the supplicant, it is now widely accepted that a mobile device (supplicant) needs to protect itself from a potentially malicious network, including a rogue access point. The controlled port allows the supplicant to limit network access to the client device until 802.1X has ascertained the validity of the network to which it is attempting to connect.

2. The 802.1XRev standard additionally improves interoperability and security by clarifying and delineating the roles of network transport (EAPOL), port access, EAP, and RADIUS. In addition, it defines the EAPOL-Key frame that has been used by 802.11 implementations to pass session-key information between the NAS and the STA after authentication is complete. The actual description of how to use the EAPOL-Key frames was intentionally left unspecified by the authors of 802.1XRev. They consciously left the responsibility for this definition to the 802.11i workgroup. The original 802.1X specification left excessive room for interpretation by developers, which in turn led to interoperability problems and security holes, notably weak key exchanges in 802.11 wireless access. Closing those security holes with respect to the use of EAPOL-Key frames is the responsibility of the 802.11i standard.

Next, we shift our focus to the AAA server.

6.3 The AAA Server

6.3.1 Introduction

We introduced the concepts of Authentication, Authorization, and Accounting in Chapter 1. Figure 6.2 illustrates a generalized view of a traditional dial-up AAA environment, including the remote access server and AAA server. The generic term for the remote access server in an AAA architecture is the *Network Access Server* that we introduced earlier in the context of 802.1X. The role of the AAA server is to make authentication and authorization decisions for the network. The server processes the credentials transmitted from the client requesting admission and the server grants or denies access based on those credentials and the access being requested. When access is permitted, the granting may be accompanied by the authorization of different kinds of access. The AAA server may also start and stop the recording of accounting information related to that user's session.

The concept of a server dedicated to providing AAA service for a large number of network access ports is not new, predating 802.11 by many years. Of the many AAA technologies that have been conceived, three technologies, namely, TACACS, RADIUS, and DIAMETER, stand out.

These names refer to formal definitions of client–server communication protocols rather than the implementations of the client or server themselves. Kerberos, named after the three-headed guard dog of Hades, was developed in the late 1980s at MIT to protect network services associated with that university. We mention Kerberos because it has evolved into a general-purpose client–server authentication system that has enjoyed widespread deployment both in

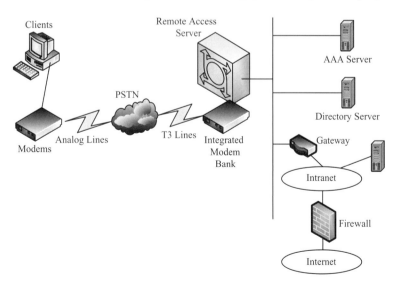

Figure 6.2: The Traditional Dial-Up AAA Configuration.

academia and in industry. However, Kerberos is not an AAA server because it only addresses authentication, and not authorization or accounting; so this protocol will not be discussed further in this chapter.

6.3.2 Remarks about TACACS and DIAMETER

TACACS [3] and its replacement *TACACS-plus* were relatively early entrants onto the AAA stage. The original TACACS was developed in the mid-1980s, when the embryonic Internet was embodied by the ARPANET and the *Military Network* (*MILNET*), and dial-up access to that network was via terminal servers called *Terminal Access Controllers* (*TACs*). The TACACS protocol allowed user access to the TAC to be granted based on credentials that were provided by the user, and passed through the TAC to an authentication server somewhere else in the network, where the correct credentials were held. The credentials were not held in the TAC itself, but when the TAC received the response from the authentication server, it would grant or deny access, as appropriate. While TACACS-plus is considered to be more CPU and memory intensive than RADIUS, TACACS-plus is considered to be more versatile, and thus, as of this writing, TACACS-plus is still in use in some major networking products.

Much like IPv6 that was developed to address all the shortcomings of IPv4, DIAMETER was conceived to be the replacement for RADIUS. The purported advantages of DIAMETER with respect to RADIUS include improved support for roaming users, the ability to share a user's accounting information between ISPs, and the ability to solicit additional information at login time beyond just basic authentication information. DIAMETER's parallel to IPv6 holds with regard to adoption in enterprise networks; neither IPv6 nor DIAMETER have significantly displaced their progenitors, at least as of this writing. Despite DIAMETER's lack of traction in the enterprise market, the 3GPP and IMS networks that we will discuss in Chapter 10 do use DIAMETER.

Much as the address-space limitation of the 32-bit IPv4 address has been mitigated by the use of address-translation technologies like *Network Address Translation* (*NAT*). Extensions to RADIUS have kept this technology vibrant. We have saved the discussion of RADIUS AAA technology for last, since as of this writing, RADIUS has more deployments and commercial impact than any of the others. We provide a dedicated treatment for RADIUS in the following section.

6.3.3 The RADIUS Protocol

The RADIUS protocol was first developed by Livingston Enterprises, Inc. Livingston was subsequently acquired by Lucent Technologies. A number of IETF RFCs have been published documenting the basic protocol and extensions to it. These include RFC 2865 [5], RFC 2866 [4], RFC 2868 [8], and others. While RADIUS has been widely applied to a number of access technologies, including the 802.11 networks that are the focus of this book, the original application was for dial-in users, indicated by the middle two letters of its name.

RADIUS remains the authentication technology of choice for the increasingly obsolete but still ubiquitous dial-up *Point of Presence* (*POP*) *Internet Service Providers* (*ISPs*) around the globe. A typical installation such as the ones mentioned previously would include a modem pool attached to a remote access server, a LAN including a RADIUS server behind that server, and a public Internet gateway also attached to the LAN. We provided an example of such a topology in Figure 6.2.

In the early to mid-1990s, it was common for companies to offer dial-up access to their private networks for their employees who wished to access the company network from home. This application involved a similar topology to that shown in Figure 6.2 with the primary difference being that the gateway led to the private corporate Intranet rather than the public Internet. The rapid evolution of VPN technology has reduced the need for such direct dial-in access to private networks and most private organizations now rely on remote dial-in users accessing the network through ISP POPs and using VPN to gain protected access to the private network. An AAA technology like RADIUS will still be used to verify the user's credentials when establishing the VPN connection.

RADIUS is a client/server protocol. The complexity of the protocol is asymmetrical in that a RADIUS client implementation is typically much simpler and more compact than the server implementations, which are often complex. The client is implemented in the NAS, which, being on the user data path, has a primary responsibility of forwarding user traffic, whereas AAA functions are secondary. This situation is in contrast to the RADIUS server, which most often runs on a machine that is not on the data path and is dedicated to the AAA function. The asymmetry between the client and the server was intentional in that it shunts the most complex operations to a back-end server such that this server's processing power could be shared by many simple NASs.

The traditional RADIUS dial-up authentication and authorization processes include the following steps:

1. The physical-layer connection is established when the user dials the phone number of a local ISP POP.

2. The user starts a *Point to Point Protocol* (*PPP*) session with the dial-up remote access server. (Note that this setup is equivalent to the EAPOL layer-two conversation in 802.11.)

3. The authentication method that will be used is negotiated over the PPP session. The chosen method must be agreed upon by both the client and the remote access server.

Some of the earlier methods were *Password Authentication Protocol* (*PAP*), *Challenge Handshake Authentication Protocol* (*CHAP*), and *Microsoft Password Authentication Protocol* (*MS-CHAP*). EAP was introduced later; it could be carried by PPP or carried by other transport protocols and so allowed greater flexibility than the earlier methods.

4. The authentication method, which was selected, is used to pass the user's credentials.

5. The RADIUS client on the remote access server forms a set of n-tuples, sometimes called a *checklist*, and with that information builds a RADIUS *authentication request*. The client sends this request to the RADIUS server. It should be noted that RADIUS messages are carried as the payload of UDP packets. As UDP is used, the RADIUS server may be located many layer-three network hops away from the access server. The only restriction on the location of the RADIUS server would be that as the number of network hops increases or the speed of those hops decreases, increased delays in authentication are likely.

6. When the authentication request packet is received and processed by the RADIUS server, the user's credentials are checked against a database of user information. In early implementations, this information was a local database, but modern RADIUS implementations more often have the RADIUS server querying an external user database such as LDAP or Active Directory. This method has the advantages of increased scalability and the ability to reuse a database, which may already exist for other reasons within the entity that manages the network.

7. If the user's credentials are acceptable, the RADIUS server will transmit a RADIUS *Access Accept*. If they are rejected, a RADIUS *Access Reject* is sent. The contents of these messages will again be an n-tuple containing information that may be used for further processing beyond the simple *accepted/rejected* paradigm of early RADIUS implementations. Such additional processing could include different levels of authorization, such as restricting the user to specific parts of the network.

Once this authentication and authorization process is complete, the RADIUS protocol may then be used to gather accounting information during the course of the user session. As we mentioned earlier, the RADIUS client is normally located in the NAS. Since the NAS is on the user data path, it is possible for the RADIUS client to gather statistics such as packet count and elapsed connect time for that user session. The RADIUS client will periodically emit RADIUS accounting messages to the RADIUS server for subsequent processing.

6.3.4 The RADIUS Protocol in Use

We noted previously that RADIUS was commonly used in the early 1990s as part of a private dial-in service to corporate networks. The popularity of this type of roaming access to corporate Intranets has exploded in the last decade, both due to an increasingly mobile workforce and also due to the burgeoning 802.3 and 802.11 access available at public facilities around the world. RADIUS remains a key component for a traveling businessperson accessing a home network via dial-up access as well as for hotel-based 802.3 wired networks and a variety of public 802.11 facilities. The authentication request that emanates from such globally roaming users will often first reach a RADIUS server that has no knowledge about the authenticating individual. However, such an authentication request may be accompanied by information that identifies an organization that does know that particular user and is willing to accept any charges associated with that user's connection. The use of RADIUS *proxying* can be relied on for addressing such situations.

When RADIUS proxying is in use, the RADIUS server acts as a proxy RADIUS client to other RADIUS servers or, in some cases, other types of AAA servers. This feature has allowed RADIUS to support roaming users in a way beyond what was conceived of in the early days of the technology. Basically, RADIUS proxying allows the server receiving the initial authentication and authorization request to forward this request to the server identified by the user's domain. This feature, and the infrastructure deployments it enabled, is one of the reasons why RADIUS has not been easily displaced by its intended successor, DIAMETER. While DIAMETER can provide the same function, this new protocol is not sufficiently superior to RADIUS to make it worthwhile for most organizations to deploy DIAMETER.

In the application of RADIUS to 802.11 networks, the basic process described previously still applies with some modifications and expansion. The biggest differences are that the authentication method used is always EAP and that the carrier protocol between the user and the NAS is EAPOL rather than PPP. The extensible nature of EAP has allowed this authentication paradigm to expand to areas of authentication far removed from its humble origins. We introduce some of these new areas of authentication in Section 6.7.1.

6.4 The Extensible Authentication Protocol

6.4.1 Introduction

The early applications of RADIUS technology relayed user credentials to the AAA server via protocols such as CHAP and PAP. Over time, it was necessary to modify and to extend these and similar protocols in order to accommodate new aspects of authentication. These modifications and extensions forced changes in different network components, some of which were not directly involved in the details of authentication. This natural evolution was the genesis of the EAP standard RFC 2284 [1], which was intended to allow different and extensible authentication methodologies to be encapsulated within the EAP transport protocol. This encapsulation permits the authentication methods to change and to mature over time without forcing unnecessary modifications to different network components. In its simplest form,

Chapter 6

EAP-encapsulated user credentials may be sent to the AAA server for validation much as they are for CHAP and PAP.

6.4.2 The EAP State Machine

Figure 6.3 shows the relationship between the EAP protocol itself and the individual EAP methods that may be transported by EAP. While the original EAP specification RFC 2284 [1] was first published in 1998, the burgeoning applications of this technology outpaced the full formalization of the details of the specification. Much of this pressure for new and expanded uses of EAP arose from 802.11 security issues. Many implementations, in the absence of clear published guidelines, deviated from the original premises. RFC 4137 [7] was written in part to clarify some of this confusion.

Part of this clarification is that a basic notion of EAP is that conversations occur between EAP-transport endpoints. These EAP-transport endpoints are called EAP switches in Figure 6.3. Peer sessions form between equivalent EAP methods behind those endpoints. The EAP endpoints are called EAP switches since they handle method-agnostic EAP messages and route EAP method-specific messages to method-specific functions. Figure 6.3 shows M1↔M1,

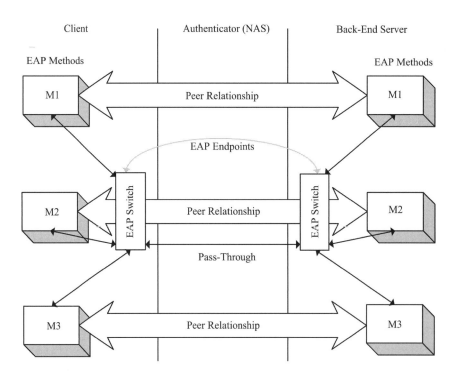

Figure 6.3: The EAP Protocol Hierarchy.

M2↔M2, and M3↔M3 as examples of peer sessions between equivalent EAP methods. While EAP is a protocol transported by other carrier protocols such as PPP, EAPOL, and RADIUS, EAP itself is a transport for the EAP methods. The EAP protocol should be communicated transparently by any intermediate network entities (for example, the NAS) between the peers at the EAP endpoints.

In order to help avoid confusion about the boundary between EAP and the EAP methods, the IETF has attempted to define a finite-state machine that clearly describes the operation of the EAP endpoints as well as their interaction with EAP methods. The corresponding definitions are presented in RFC 4137 [7].

Among other things, RFC 4137 [7] provides guidelines for how multiple EAP methods may be *chained* (executed in succession) to provide more flexible and expandable authentication. Such chaining is only permitted if the channel over which they are carried is a protected channel. In the EAP context, *protected channel* refers to processing performed within a *tunneled* EAP method. In the following sections we explain the concept of EAP tunneling and clarify which of the EAP types are tunneled methods.

6.4.3 Prominent EAP Methods

As of this writing, user authentication in 802.11 encompasses a range of EAP authentication methods, including the following:

- *Message Digest 5* (*MD5*)

- *Transport Layer Security* (*TLS*)

- *Tunneled Transport Layer Security* (*TTLS*)

- *Light Extensible Authentication Protocol* (*LEAP*)

- *Protected Extensible Authentication Protocol* (*PEAP*)

- *Flexible Authentication via Secure Tunneling* (*FAST*)

- *Subscriber Identity Module* (*SIM*)

- *Authentication and Key Agreement* (*AKA*)

Each of these EAP methods has its own advantages and disadvantages depending on the environment. Other EAP methods will likely be developed to address new problems. The following inventory of EAP types should convey to the reader that a wide variety of EAP

Chapter 6

methods are available, each with tradeoffs, including ease of deployment and level of security provided.

In Table 6.1, we list the most prominent EAP types that are in use, as of this writing. Some of the EAP methods presented offer *mutual authentication*, where the client is authenticated to the network and the network is authenticated to the client. Without mutual authentication, a valid user may connect to a *rogue AP* which is pretending to be a legitimate network. The rogue AP can then steal the user's credentials by mounting a *man-in-the-middle* (*MITM*) attack.

As man-in-the-middle attacks are common, it is worth a brief digression here to explain these types of attacks. To mount an MITM attack, the attacker first poses as a valid AP. The client then associates with the rogue AP. Next, the rogue AP poses as the valid supplicant to the network and passes all the supplicant's responses off as its own. When the network indicates a successful authentication, the rogue AP takes control of the connection and sends a "disassociate" to the valid supplicant.

In the following text we explain that the tunneled authentication methods TTLS, PEAP, and FAST represent major advancements in WLAN authentication. Despite these advances, a completely unbreakable security system covering every eventuality will likely remain an elusive target. Now we examine EAP's numerous authentication methods in more detail.

Message Digest 5 (MD5):   Message Digest 5 is the EAP equivalent to CHAP, where a one-way hash algorithm is used in combination with a shared secret and a challenge to verify that the supplicant knows the shared secret [6]. As with any method that uses a random challenge combined with a password and hash algorithm, this method is open to a *dictionary attack*. If the attackers can obtain the challenge and the hashed response, they can then run a program off-line with the same algorithm as the supplicant, plugging in words from a

Table 6.1: Prominent EAP Methods and Their Characteristics.

EAP Type	Dynamic Session Keys	Mutual Authentication	Requires Client Certificates	Tunneled
AKA	Yes	Yes	No	No
FAST	Yes	Yes	No	Yes
LEAP	Yes	Yes	No	No
MD5	No	No	No	No
PEAP	Yes	Yes	No	Yes
SIM	Yes	Yes	No	No
TLS	Yes	Yes	Yes	No
TTLS	Yes	Yes	No	Yes

dictionary until their hashed response matches the supplicant's. Therefore, the hackers get to know of the supplicant's password and can steal the supplicant's identity to gain access to the network. This process is made much easier in wireless LANs, where the challenge and response are passed through air. IT managers admonish their users to choose passwords that are not based on dictionary words, in order to render the dictionary attack more difficult. The term *dictionary* is used liberally within this context. In reality, combinations of true dictionary words, proper nouns, and numbers are automatically generated and tried during such an attack. MD5 is considered a base-level authentication method and is not appropriate where strong security is required. In addition, MD5 does not provide keying information, which renders it unusable for 802.11i.

Transport Layer Security (TLS): Transport Layer Security offers a very secure authentication process that replaces simple passwords with client- and server-side certificates through the use of a *Public Key Infrastructure* (*PKI*). A certificate is a record of information about an entity (for example, a person, a corporation, and so on) that is verified by an *asymmetric* mathematical algorithm. By "asymmetric," we mean a process that is not easily reversed. Mutual authentication is supported by TLS, as are dynamic session keys.

TLS is a good choice where strong authentication and data privacy are required, and PKI technology is already deployed. TLS enjoys the distinction of being an IETF standard. As of this writing, all the remaining methods that we present in the subsequent sections of this chapter are at most IETF informational RFCs. However, the use of PKI, where every client has a personal certificate, is expensive when compared with password-based systems. The expense is attributed to the software tools that are required as well as the training and education required to make the system effective.

Tunneled Transport Layer Security (TTLS): Tunneled Transport Layer Security is an extension to TLS and was developed to overcome the need created by TLS for client-side certificates. Server-side certificates are required with TTLS. However, as they only need to be deployed to the relatively few servers in the organization, this issue is not considered onerous. Similar to other tunneling authentication methods, TTLS is a two-step authentication method. In the first step, an asymmetrical algorithm based on server keys is used to verify the server's identity and to set up a symmetric encryption tunnel. The second step involves verifying the client's identity by using a second authentication method passed through the symmetric encryption tunnel for the actual authentication negotiation.

With TTLS, the second authentication method used within the tunnel may be an EAP type (often MD5) or a legacy method such as PAP, CHAP, MS CHAP, or MS CHAP V2. The symmetric encryption tunnel of TTLS is used only for protecting the client authentication method.Once authentication is complete, the encryption tunnel is collapsed. In 802.11, data privacy is maintained by hardware-based symmetrical encryption using dynamic keys derived by the TTLS process. This approach offers strong security during authentication while accommodating existing end-user working methods (user ID/password), thus prolonging the useful life of legacy non-EAP AAA servers.

Protected Extensible Authentication Protocol (PEAP): Protected Extensible Authentication Protocol is similar to TTLS. Similar to TTLS, PEAP is a tunneled EAP method. As with TTLS, a symmetric encryption tunnel is established using a server certificate over which the client authentication process is securely conducted. PEAP supports EAP methods through the tunnel, but, unlike TTLS, legacy methods are not supported for the client authentication negotiation step.

Light Extensible Authentication Protocol (LEAP): Light Extensible Authentication Protocol was developed by Cisco for use on WLANs that use Cisco 802.11 access points. LEAP features mutual authentication, secure session-key derivation, and dynamic per-user per-session WEP keys. However, LEAP is vulnerable to dictionary attacks. Cisco elected to maintain close control over the LEAP technology, and while a number of STA implementations support LEAP, only access points from Cisco support this EAP method. The 802.1X standard is quite clear that the NAS should not take actions based on the EAP packets that pass through it, but LEAP was implemented by Cisco prior to the ratification of the 802.1X standard. The function of NAS according to the standard, is to act as an EAP relay between the supplicant and the authentication server. This stipulation allows new EAP methods to be created and implemented without any modifications to the NAS devices. LEAP violates this premise. Similar to MD5, LEAP's use of unencrypted challenges and responses does leave it open to potential off-line dictionary attacks. Still, when LEAP is combined with a rigorous user-password policy, LEAP can offer reasonable authentication security without the use of certificates. Due to its technical strengths, ease of deployment, and early availability in the market, LEAP has been widely deployed.

EAP Subscriber Identity Module (SIM): As of this writing, Subscriber Identity Module is the standard authentication method used by the world's dominant cellular phone technology, GSM. SIM is a smartcard-like authentication method. *Smartcards* themselves are credit card-sized cards with microprocessors and other electronics that can store small programs (for example, encryption programs) and information about the history of a device, its use, and its owner. In their application in the cellular world, these smartcards have been even further downsized. Clients use these small cards, which contain a chip holding their credentials, to authenticate to the network.

EAP-SIM enables an EAP-based authentication solution to utilize the existing GSM-roaming infrastructure for authentication. EAP-SIM provides for key derivation as well as for mutual authentication. There is some concern that the keys used for authenticating the server are derived keys, and this fact makes the keys more vulnerable to *spoofing* (someone else pretending to be the valid user) than permanent keys.

EAP Authentication and Key Agreement (AKA): As of this writing, Authentication and Key Agreement is a new standard being developed by some cellular service providers. AKA is not yet a standard, but, like SIM, the draft is to the point of being a relatively mature version. EAP-AKA is similar to EAP-SIM except that EAP-AKA utilizes the *User Service Identity Module* (*USIM*) device and the AKA-authentication algorithms contained in that device rather than SIM cards and GSM-authentication algorithms. The USIM device is also smartcard-like. The USIM is defined in the Universal Mobile Telecommunications System (UMTS). While EAP-AKA is similar to EAP-SIM, it is generally acknowledged that EAP-AKA is more secure than EAP-SIM because of the permanent keys it uses for mutual authentication.

EAP-FAST: Cisco developed EAP-FAST to provide an EAP method with the security strengths of EAP-PEAP, yet without the complications of PKI technology. As of this writing, EAP-FAST is an Internet Draft, and Cisco intends to publish this paper as an Informational RFC. Cisco had enjoyed great success with LEAP, but its security vulnerabilities generated the need for a successor. Unlike LEAP, Cisco eventually intends to make the protocol available in the public domain. The protocol is documented in [2]. The fact that Cisco's NAC program is based in large part on the EAP-FAST method (see Section 6.7.2) indicates that this method will be widely adopted.

With EAP-FAST, Cisco has introduced the *Protected Access Credential* (*PAC*). Cisco's claim is that the PAC provides a secure user credential with many of the advantages of a PKI-based digital certificate and few of the drawbacks. Many enterprises hesitate to adopt digital certificates because of the complex processing involved with PKI technology and the complexity of deploying digital certificates on thousands of corporate machines. Indeed, the amount of memory required for PKI software may be prohibitive in certain memory-constrained handheld mobile devices like cellular phones. EAP-FAST has neither of these drawbacks. The credentials are protected with simpler software and may be automatically provisioned on the clients via an optional phase zero of EAP-FAST authentication.

Phase one and two of EAP-FAST resemble those of EAP-PEAP. In phase one, mutual authentication is performed between the client and the AAA server, and this authentication is followed by the establishment of a secure tunnel between them. In phase two, the actual client authentication is performed through the tunnel. EAP-FAST passes the credentials using *Type-Length-Value* (*TLV*) triplets. The NAC technology presented in Section 6.7.2 uses these TLVs to check the state of the machine and for VLAN assignments that occur as a function of those checks.

Summary of EAP Methods: We have covered eight different EAP methods in this section. Each method has several advantages and disadvantages. New methods continue to be developed and old methods are still being updated and enhanced. We expect this evolution to continue into the foreseeable future. The reader is referred to Table 6.1 for a summary of the characteristics of the methods that we have presented.

6.5 Flexible and Strong Authentication in 802.11

6.5.1 Introduction

In this section we illustrate through the use of diagrams and examples as to how 802.1X, EAP, and the AAA server collaborate to provide the flexible and strong authentication demanded in 802.11 networks. Particular emphasis is placed on explaining how the apparently simple task of user authentication has evolved into a complicated multiple round-trip packet exchange involving considerable communications and processing overhead. This information is relevant to the discussion of secure roaming in subsequent chapters because of the latency introduced by these processes and the extensions being made to them.

6.5.2 Basic Authentication Process in 802.11

As explained earlier, in the 802.1X model, EAP packets are encapsulated in EAPOL frames. As shown in Figure 6.4, EAPOL communication occurs between the end-user station (supplicant) and the wireless access point (authenticator). Figure 6.4 shows the RADIUS

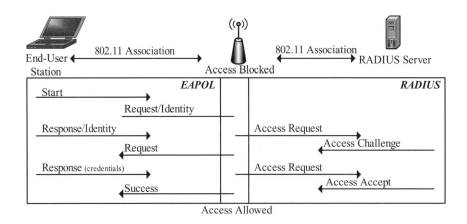

Figure 6.4: Authentication in 802.11 Using Basic 802.1X.

protocol used for communication between the authenticator and the AAA server, although other AAA technologies could be used. The figure shows the mapping between the basic EAPOL frames and the RADIUS messages that transport the EAP contents of those frames to the RADIUS server. We remind the reader that EAPOL and RADIUS are collaboratively used to permit an end-to-end EAP conversation between the client and the AAA server. The basic authentication process described in the following steps begins immediately after the 802.11 association has occurred between the STA and the AP.

1. Upon detecting the 802.11 association, either the client or the authenticator may send an EAPOL-Start message.

2. The authenticator opens the uncontrolled port for the 802.1X authentication session, leaving all non-802.1X-traffic blocked at the controlled port.

3. The access point sends an EAP-Request/Identity message.

4. The client's EAP-Response packet with client's identity is passed to the authentication server in the first RADIUS Access-Request message shown in Figure 6.4.

5. The authentication server challenges the client to prove itself, and the server may send its credentials to prove itself to the client (if the client requires mutual authentication). This information is encoded in an EAP message and sent to the AP in the payload of a RADIUS Access-Challenge message and forwarded to the client by the AP in an EAPOL message.

6. The client sends its credentials to the server in order to allow the server to verify the client's identity. The client's credentials are embedded in an EAP message and transported to the AP in the second EAP Response shown in Figure 6.4 and relayed by the AP to the AAA server in the second RADIUS Access-Request message shown in the figure.

7. Based on the credentials received, the authentication server accepts or rejects the client's request for connection. In the example shown in the figure, the request is accepted via an EAP-Success message encapsulated in the RADIUS Access-Accept message and relayed via EAPOL to the STA.

8. If the AAA server accepted the request, the authenticator changes the user's controlled port on the AP to an authorized state enabling non-802.1X traffic.

9. When EAP-Logoff is received, the user's controlled port on the AP is changed back to the unauthorized state. The authenticator should change the state to unauthorized, if the STA disassociates from that AP in the event that no EAP-Logoff is received.

Chapter 6

10. There may be session timers determined by the AAA server that will force reauthentication after the expiration of the timer. In that case, all the steps just outlined will be repeated periodically.

6.5.3 Discussion of Tunneling in EAP Methods

As we pointed out in Section 6.4.3, the requirements of mutual authentication and protection against man-in-the-middle attacks have created a strong demand for tunneled EAP methods such as EAP-TTLS, EAP-PEAP, and EAP-FAST. The use of tunneled EAP methods results in the addition of a number of steps to the process described in Subsection 6.5.2. We use Figure 6.5 in conjunction with Figure 6.4 to illustrate these additional steps.

When a tunneled EAP method is in use, the initial RADIUS Access-Request and Access-Accept messages are not used to request and to verify the actual user credentials. The first function accomplished by any tunneled EAP method is to establish the TLS tunnel itself. To this end, the TLS-tunnel-establishment protocol, beginning with a TLS *ClientHello* message, is transported as the payload of the inner method's EAP messages until the TLS-based tunnel is established. We should emphasize here that this highly secure tunnel may be established without the use of client-side certificates, which was one of the principal drawbacks of EAP-TLS. One of the most important side effects of establishing this secure tunnel is that it permits the client to verify the identity of the network—a key step in accomplishing mutual authentication.

With the establishment of the tunnel, we focus on Figure 6.5. The dialog shown in Figure 6.5 occurs between the RADIUS Access-Request and Access-Accept messages in Figure 6.4. We call the reader's attention to the *tunnel server* shown in Figure 6.5. This server is the endpoint of the TLS tunnel over which the user credentials will be passed. While the tunnel server is a

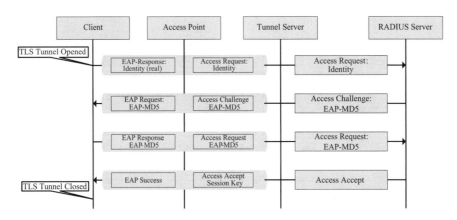

Figure 6.5: 802.11 Authentication Using EAP Tunneling.

protocol entity that is distinct from the RADIUS server, in many implementations the tunnel server is only a software component running on the same hardware platform as the RADIUS server. The details of the tunneled authentication process are shown in Figure 6.5. The figure shows EAP-MD5 as the inner method, although other EAP methods are often used. The RADIUS dialog that emerges from the tunnel server is normally the same as that which would emerge from an access point in a nontunneled EAP application.

6.6 Other 802.11 Authentication Methodologies

6.6.1 Introduction

The growing commercial acceptance of 802.1X-based authentication for 802.11 networks as well as the adoption of this standard by the IEEE and the Wi-Fi Alliance have led us to emphasize this authentication paradigm. Nonetheless, there are other authentication methodologies in use in 802.11 networks. In this section we provide a brief treatment of the best known of these methodologies—*MAC-based* and *Web-based* authentication.

6.6.2 MAC-Based Authentication

MAC-based authentication requires that the MAC address of authorized devices be configured into the access point or RADIUS server that provides service to those devices. This method had some popularity in the early days of 802.11 when deployments were experimental in scale. One severe drawback of this technique is the great difficulty in managing the database of all the MAC addresses of the machines that might access the network. When the total number of potential users is in the two-digit range, this accounting might be feasible, but quickly becomes unmanageable in public- or large-enterprise environments.

The MAC-based authentication method is really not suitable for home environments because the MAC address of a device can be programmed such that it resembles that of an authorized device, and thus one can easily gain illicit access to the network. This faking of an address is called *MAC spoofing*. Finally, this form of authentication is not coupled with any form of dynamic key generation to provide reliable data privacy, unlike the 802.11-sponsored security mechanisms.

Despite the drawbacks that we have mentioned, MAC-based authentication is still widely used for *Voice Over Wireless Internet Protocol* (*VoWIP*) applications where fast roaming times are of paramount importance. The Wi-Fi Alliance maintains a strong position that this authentication is not adequate and that WMM certification will not be granted for devices only capable of MAC-based authentications.

6.6.3 Web-Based Authentication

The fact that a standard Web browser is the only software required for Web-based authentication has made this authentication method very popular at public hotspots. The normal procedure for Web-based authentication, otherwise known as a *walled garden*, is for the user to connect to what appears as an open (security-free) access point. Once the 802.11 association is complete and an IP address has been obtained through DHCP, the user launches a Web browser.

In the scenario just described, the user often runs the browser while thinking that this association with an open AP will be a free-of-charge service. Regardless of the URL that the user inputs into the browser, the connection is redirected to an authentication Web server. This Web server will request that the user be identified as a subscriber of the hotspot provider, or, if the user is not a subscriber, offer the ability to become a subscriber or to pay for a period of connection time with a credit card. While most providers of such services will ensure that the login transaction itself is protected via a secure Web session (`https`), any subsequent user traffic will not enjoy the layer-two encryption that is built into all new 802.11 hardware. A user who relies on an encrypted layer-three VPN connection for all private traffic may find this acceptable.

Another drawback of the Web-based authentication is that other than the need to start the entire manual login process all over again, there is no standard approach for the reauthentication of the user if the user roams to another AP. While such a standardization approach could be developed, no such effort appears to be underway at the time of this writing. Finally, when access to the hotspot is via a mobile handheld device, it is likely that the user will expect to authenticate via some smartcard or other mechanism that minimizes user interaction during the authentication process.

In concluding this section, we should at least mention open 802.11 authentication merely to avoid confusion on the part of the reader. This very rudimentary form of authentication, which we discussed briefly in Chapter 3, is too weak and too ill-defined to be of any practical use for real 802.11 user authentication.

6.7 Network Access Control

6.7.1 Introduction

Access control in 802.11 networks was initially based on verifying the identity of the user against the user database accessed by the AAA server. The unfortunate proliferation of worms, viruses, and spyware creates a common situation where a valid user attempts network access on a compromised computer. Such a user's machine should be denied access lest the computer contaminate the network. This eventuality has resulted in the expansion of the concept of admission control to include the checking of the state of the machine in addition to the user's credentials. The state of the machine may include the latest update of the antivirus software, the last time that a full virus scan was performed, or the patch level of the operating system, to name but a few possibilities.

Three well-publicized efforts are underway to standardize the way in which the 802.11 admission-control paradigm should perform the checking of machine state as well as the checking of user's credentials. These paradigms are as follows:

- *Cisco's Network Admission Control (NAC)*
- *Trusted Computing Group: Trusted Network Connect (TNC)*
- *Microsoft's Network Access Protection (NAP)*

This section will include a brief description of the basic technical models on which all three of these public initiatives are based. The first of these three, Cisco's Network Admission Control, already has commercial deployments. While Cisco's NAC enjoys this advantage over the other two, all three of these technologies are very much in a state of flux.

6.7.2 Cisco Network Admission Control

Network Admission Control is a set of technologies and solutions built on an industry initiative led by Cisco Systems. It uses the network infrastructure to enforce security-policy compliance on all devices seeking to access network computing resources, thereby limiting exposure and damage related to emerging security threats such as viruses, worms, and spyware. NAC will provision network access only to compliant and to trusted endpoint devices (for example, PCs, laptops, servers, and PDAs) and can restrict the access of noncompliant devices, thereby maintaining the security posture of the entire network infrastructure.

Cisco's NAC initiative represents an extension of the 802.1X standard to deliver an unified model for the authentication of the machine, the person, and the state of the device being used to access the network. With greater than 60 technology partners, the Cisco NAC initiative represents a significant leap forward in the adoption of 802.1X as the authentication mechanism of choice.

NAC works by extending the 802.1X standard to support the verification of the state of the machine in terms of the presence and status of software components on the machine. Following the authentication process for verifying the username/credentials of the person or the machine accessing the network, NAC also uses EAP to communicate the attributes of the machine to Cisco's *Access Control Server* (*ACS*) server. These attributes are collected by the

Chapter 6

Cisco Trust Agent (*CTA*), which also resides on the user's machine. The CTA, in turn, gathers this information from third party *posture plugins*. ACS then refers the attributes of the machine to the appropriate *policy servers* such as the antivirus, anti-SPAM, or *Operating System* (*OS*) patch servers.

These systems interact with ACS to determine whether or not the machines are compliant with the company's requirements for machines connecting to the network. If a machine is found to be out of compliance, it can be quarantined to a secured portion of the LAN, where the machine can be patched or updated. Once the machine has achieved compliance, it is returned to the production network. An architectural view of the components involved in a NAC implementation are provided in Figure 6.6. The sample posture plugins and policy servers in Figure 6.6 are McAfee, Symantec, and Trend Micro.

6.7.3 Trusted Network Connect

The *Trusted Computing Group* (*TCG*) is a not-for-profit organization formed to develop, define, and promote open standards for hardware-enabled trusted computing and security technologies, including hardware building blocks and software interfaces across multiple platforms, peripherals, and devices. The *Trusted Network Connect* (*TNC*) is an effort by the TCG to define and promote an open solution architecture that will enable network administrators to enforce security policies for endpoint-host connections to their multivendor networks. The TNC specification assists in protecting networks from viruses, worms, denial of service attacks, and host-software vulnerabilities by allowing users to enforce

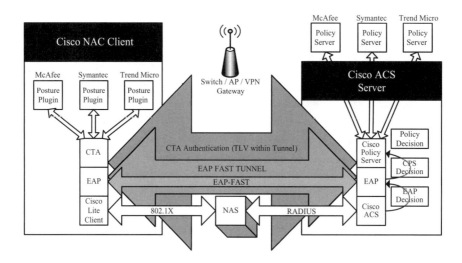

Figure 6.6: Cisco's Network Admission Control Architecture.

security policies to prevent vulnerable or untrusted systems from connecting to the network.

Figure 6.7 shows the architectural components of TNC. A quick comparison of Figures 6.6 and 6.7 shows the strong similarity between the two architectures. In the TNC architecture, client software on the *Network Access Requester* (*NAR*) begins the network access attempt. In a TNC environment, the NAR may be an 802.1X supplicant, VPN client, or Web browser initiating a *Secure Sockets Layer* (*SSL*) connection. The *Policy Enforcement Point* (*PEP*) forwards information about the NAR and its network connection attempt to a *Policy Decision Point* (*PDP*), where a *Network Access Authority* (*NAA*) determines whether the endpoint should be admitted to the network.

In an 802.11 context, the PEP would likely be the NAS, and the enforcement would be accomplished by the 802.1X authenticator on that NAS, enabling and disabling the controlled port for that user. TNC defines a client-side component called an *Integrity Measurement Collector* (*IMC*) whose role resembles that of NAC's posture plugin. Similarly, the NAC policy server is analogous to TNC's *Integrity Measurement Verifier* (*IMV*). Figure 6.7 shows sample IMCs and IMVs from IPass, Intel/TNC, and InfoExpress.

6.7.4 Microsoft's Network Access Protection

As we mentioned earlier, Microsoft's answer to Cisco's NAC is called Network Access Protection. While TNC is the only truly open NAC technology of the three technologies presented in this section, both Cisco and Microsoft present their technologies as open

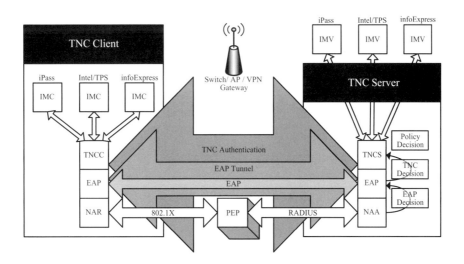

Figure 6.7: The Trusted Network Connect Architecture.

standards and available to the industry. From a high-level view, NAP resembles the NAC and TNC technologies presented in the previous two subsections. *Health Policy Validation*, *Health Policy Compliance*, and *Network Isolation* are the three key features of NAP.

NAP defines a *System Health Agent* (*SHA*) that runs on the endpoint device attempting to access the network. The SHA will gather information about the current security health of the device in order to perform *Health Policy Validation*. As is the case with NAC and TNC, the examples of OS patch level and virus-detection updates are most often used as measures of the device's security health. The SHA is analogous to the CTA in NAC in that it may rely on third-party applications to perform these checks. The SHA will communicate this snapshot of the current security health to the *System Health Validator* (*SHV*) that resides in the network infrastructure. The SHV is the entity where the IT administrator has configured policies regarding acceptable health criteria for admission and actions to be taken in the event of noncompliance.

Health Policy Compliance can be obtained via the *Systems Management Server* (*SMS*) component of NAP. If the SHV determines that the system is noncompliant, the SMS may be used to remedy the cause of the noncompliance by providing, for example, the missing OS patches or the missing antivirus updates. Until the device can be made compliant, *network isolation* may be performed through the use of DHCP, VPN, *Internet Protocol Security* (*IPSec*), or 802.1X VLAN-based quarantine methods.

6.8 Summary

This chapter has focused on the authentication building blocks that are relevant to 802.11 security. The most fundamental of these building blocks to the 802.11 standard are 802.1X, EAP, and AAA. In Chapter 7, we will examine the changes in 802.11 roaming when these technologies are applied to provide secure roaming. The examples that we provide in Chapter 7 will consequently be more complex than those covered in Chapter 5, where the roaming examples were in unsecured networks. In Chapter 7, we will see that securing the 802.11 network involves more than only authentication. Securing the 802.11 network also means providing data privacy through strong encryption. Each additional security piece adds complexity to secure roaming.

References

[1] Larry Blunk and John Vollbrecht, PPP Extensible Authentication Protocol (EAP), RFC 2284, Internet Engineering Task Force, 1998.

[2] Nancy Cam-Winget, Drew McGrew, Joseph Saloway, and Hao Zhou, The Flexible Authentication via Secure Tunneling Extensible Authentication Protocol Method (EAP-FAST), 2005.

[3] Craig Finseth, An Access Control Protocol, Sometimes Called TACACS, RFC 1492, Internet Engineering Task Force, 1993.

[4] Carl Rigney, RADIUS Accounting, RFC 2866, Internet Engineering Task Force, 2000.

[5] Carl Rigney, Steve Willens, Allan Rubens, and William Simpson. Remote Authentication Dial In User Service (RADIUS), RFC 2865, Internet Engineering Task Force, 2000.

[6] Ronald Rivest, The MD5 Message-Digest Algorithm, RFC 1321, Internet Engineering Task Force, 1992.

[7] John Vollbrecht, Pasi Eronen, Nick Petroni, and Yoshihiro Ohba, State Machines for Extensible Authentication Protocol (EAP) Peer and Authenticator, RFC 4137, Internet Engineering Task Force, 2005.

[8] Glen Zorn, Dory Leifer, Allan Rubens, John Shriver, Matt Holdrege, and Ignacio Goyret, RADIUS Attributes for Tunnel Protocol Support, RFC 2868, Internet Engineering Task Force, 2000.

CHAPTER 7

Roaming Securely in 802.11

7.1 Introduction

7.2 The 802.11 Security Staircase

7.3 Preauthentication in 802.11i

7.4 Detailed Analysis of Real-Life Secured Roams

7.5 Dissection of a WPA-PSK Protected Roam

7.6 Dissection of a WPA2 Enterprise Roam

7.7 Dissection of an 802.11i Preauthentication

7.8 Summary

7.1 Introduction

Security concerns in 802.11 were a complete afterthought; the design team did no upfront planning. Therefore, it is rather amazing that security in 802.11 turns out to be such a critical component of what has become one of the most important data communications standards of the first decade of the 21st century. The designers of 802.11 focused so much attention on providing similar layer-two access as was provided in 802.3 that they discounted the consequences of the fact that radio waves are trivially eavesdropped upon and attacked. Ultimately, although, first users and then the standards body itself recognized how important it was to plug some of these gaping security holes. This effort resulted in a great deal of friction in the 802.11 debates and put pressure on industry to implement changes before the 802.11i group could agree upon a standard to ratify. In this chapter, we examine the evolution of 802.11-specific security and discuss many key issues related to secure roaming.

Chapter 7

7.2 The 802.11 Security Staircase

7.2.1 Introduction

Addressing security concerns by patching up holes in the 802.11 standard has led to a very interesting set of continuously evolving developments. Figure 7.1 shows the evolution of standards-based 802.11-security technologies over recent years. The diagram is split into two halves along the vertical axis. The top half refers to the encryption technologies introduced during a particular time period. The bottom half lists EAP methods introduced at that time. We use the term "introduced" here to refer to the availability of commercial implementations rather than the mere availability of a specification. WEP technology was introduced in 1997. We begin the detailed chronology in Figure 7.1 in 2001 since this year brought a widespread industry consensus that WEP provided little deterrent to the determined hacker.

Numerous articles regarding security flaws in WEP surfaced in trade journals during 2001, and freely available WEP-cracking software was published on the World Wide Web. Thus, the year 2001 represents a watershed one in 802.11 security. A torrent of incremental improvements in technologies, intertwined with the discovery of new attacks, resulted in an incremental plugging of security holes and an evolution of 802.11 security standards—each one attaining some degree of commercialization. While each incremental technology represented an improvement over its predecessors, not all achieved the same level of success and commercial adoption.

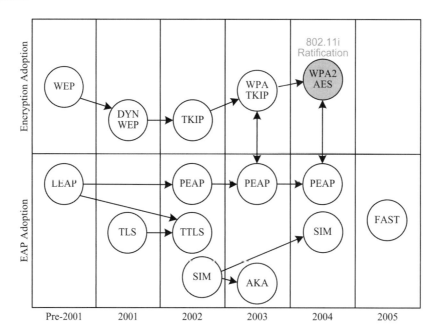

Figure 7.1: The 802.11 Security Staircase.

We illustrate the *degree of adoption* of the technologies in Figure 7.1 by the relative position of the elements on their *y*-axis. That is, the closer to the bottom of its segment a bubble appears, the lower was its adoption rate. For example, in 2002, both TTLS and PEAP were introduced as new EAP technologies. As we explained in Chapter 6, these technologies share many fundamental characteristics. The fact that PEAP achieved and has maintained a higher market penetration is indicated by its higher *y*-axis position relative to TTLS. Keeping this convention in mind, we will now explain the remaining parts of Figure 7.1.

7.2.2 Evolution of Security Technologies

As the debates raged on in 2001 about the inadequacy of WEP, discussions began in the IEEE and the Wi-Fi Alliance about how to address this problem. The wireless community recognized that these bodies would need years to come to a consensus, which could be ratified as a standard, and a number of companies realized that some type of stop-gap solution needed to be brought to market quickly. In order to develop a solution rapidly, two existing technologies were leveraged. First, there was already an encryption technology partially built into 802.11 hardware and, second, 802.1X was an already ratified IEEE standard for port authentication.

Discussions involving encryption built into 802.11 hardware and the 802.1X standard resulted in an association of these two technologies to create a quick-to-market solution to address WEP's weaknesses. The idea was to extend 802.1X to derive a dynamic key at the end of the authentication phase. This dynamically derived key was to be programmed into the RC4-based WEP encryption engines in the 802.11 chips. The fact that the key changed from session to session was a radical improvement in security, as compared with static WEP. This approach became known as *dynamic WEP keying* and is shown in Figure 7.1 as *DYN WEP*. Cisco had already been shipping its LEAP technology for 802.11 security since 2000. As Cisco's LEAP model was mostly based on the then open technology, it is not surprising that the initial tentative steps toward standards-based solutions resembled LEAP.

Two notable drawbacks of LEAP were that it was a Cisco-controlled technology and that nonstandard intervention in the EAP stream is performed by the access points in a LEAP-enabled system. The EAP method that was promoted by the first commercial implementations of dynamic WEP keying was EAP-TLS. We discussed this method in detail in Chapter 6. Most notably, EAP-TLS was available in Microsoft's initial support of dynamic WEP keying, released with the initial version of Windows XP in the fall of 2001. It should be noted that in terms of actual deployments of the technology, LEAP saw far more deployments than the other EAP types, as reflected in its higher position in the figure.

Dynamic WEP keying laid the foundation for the entire family of standards-based 802.11 security solutions which succeeded it, as shown in Figure 7.1. This solution was promulgated in an atmosphere so focused on expediency and security that it left us with a legacy of roaming latency, which continues to plague the technology, even as of this writing. The 802.11 security staircase actually led to some of the greatest challenges that face the 802.11r roaming workgroup, as they struggle to provide fast roaming times, while carrying forward the security legacy reflected in Figure 7.1.

The year 2002 witnessed the introduction of the TKIP- and AES-encryption ciphers, although essentially no commercial adoption of them occurred this early. Tunneled authentication was introduced first via TTLS and then via PEAP. EAP-SIM was also introduced at that time but saw no commercial implementations. In 2003, PEAP emerged as the dominant EAP type for 802.11i, and the two new encryption ciphers began to gain commercial traction. In 2004, the 802.11i standard was ratified and, with PEAP and EAP-TLS as the predominant EAP types in 802.11 deployments, commercial deployment of this technology was underway, buoyed in large part by the commercial certification and branding programs of WPA and WPA2. While WPA and WPA2 were by 2004 enjoying growing commercial success, LEAP, the incumbent in enterprize networks, still remained extremely popular.

The introduction of the NAC technologies from Cisco and TNC in 2005 added a new twist to what is becoming a security spiral staircase; the fact that EAP is being used to transport much more information than user credentials and keying information has introduced an emerging need for new EAP types such as EAP-FAST, which we saw in Figure 7.1 was introduced in 2005.

7.3 Preauthentication in 802.11i

7.3.1 Introduction

The finally ratified 802.11i standard reflected the workgroup's best effort to allow fast secure roaming within the existing constraints of 802.11. The key component of 802.11i that relates to fast roaming is called *preauthentication*. While the actual handoff in 802.11 is always a strictly hard handoff, the security related to that association is actually handled in a manner similar to the soft handoffs, which we presented in Chapter 1. In the remainder of this chapter, we will describe the complex process of preauthentication with examples and diagrams, and then illustrate how this method can reduce secure-roaming times. Despite the decreased latency, even preauthentication really falls short of being able to provide sufficiently low latency to permit acceptable roaming for voice connections. This shortfall has led to some of the techniques for roaming, which we discuss in Chapter 8, and ultimately to the *Fast BSS Transition* work of 802.11r, which we cover in Chapter 9.

7.3.2 Steps Involved in 802.11i Preauthentication

Figure 7.2 shows the basic steps involved in 802.11i preauthentication. We remind the reader that support for preauthentication is not mandatory for 802.11i compliance. The client

Roaming Securely in 802.11

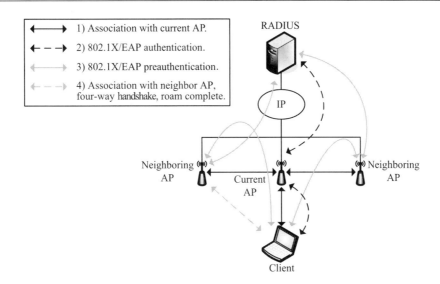

Figure 7.2: The 802.11i Preauthentication.

STA first associates with the *current AP* and then performs normal 802.11i authentication procedures. At this point, preauthentication diverges from the mandatory 802.11i steps. First, the STA becomes aware of neighboring candidate APs through the scanning procedures that we described in Chapter 5. These APs are placed on the current-driver scan list. These candidates will normally be other APs in the same ESS with sufficiently strong *Received Signal Strength* (*RSS*) to be considered candidates. The STA then initiates a normal EAPOL conversation with each of the selected candidates *via the current access point*.

Figure 7.2 shows the preauthentication EAPOL exchanges in the form of the arrows passing through the current AP between the STA and the two neighboring APs. This step is the key one in preauthentication in that it solves the problem of not being able to communicate with the candidate APs without breaking its association with the current AP. (Recall that 802.11 infrastructure mode requires that an STA be associated with an AP in order to send data frames to it and that an STA can only associate with one AP at a time.) By conversing with the candidates through the current AP and hence through the wired network, the STA can carry out the normal EAPOL conversation that is necessary to authenticate with that AP as per the 802.11i specification.

As we discussed in Chapter 6, the EAP conversation carried in the EAPOL frames must be relayed to the AAA server, but this portion can be done as it normally is, using RADIUS packets as the transport over the wired network. We see these conversations reflected in the arrows between the two neighboring APs and the RADIUS server in Figure 7.2.

Each of these preauthentications will derive a unique PMK, one for each pairing of a STA with an access point. The STA must keep track of which PMK belongs to each of the access points with which it has preauthenticated. Both the STA and AP maintain an identifier called the *PMK Identity* (*PMKID*) that permits the two to confirm that they are using the correct PMK when they begin their four-way exchange.

While the EAPOL authentication exchange between the client and the current access point closely resembles the EAPOL preauthentication exchange with the neighboring access points, their Ethernet types differ. As defined by the 802.1X standard, the Ethernet type of the normal EAPOL frames is 0x888E, whereas the preauthentication frames use the Ethernet type 0x88C7. It is important that the access point be able to distinguish between these classes of frames, as some access point implementations may "swallow up" all normal EAPOL frames, yet must forward the preauthentication EAPOL frames onto the wired network.

7.4 Detailed Analysis of Real-Life Secured Roams

7.4.1 Introduction

The Wi-Fi Alliance's WPA and WPA2 certification programs have led to widespread adoption of 802.11i technology. The WPA-PSK mode is easily configured and implemented in SOHO environments, and the more complex WPA Enterprise mode provides the higher security required in more demanding enterprise environments. In Chapter 5, we provided a close look at detailed frame traces of unsecured roams. In the following sections, we will provide a similar detailed view of a local roam for three security modes:

- WPA-PSK
- WPA Enterprise
- Preauthentication-enabled WPA Enterprise

The detailed analysis that we provide here builds on the understanding developed in Chapter 5.

All three secure-roaming technologies studied in this chapter represent increased complexity, as compared with the unsecured roams of Chapter 5. This complexity ranges from four extra frames in the case of WPA-PSK to the scores of additional frames and vastly greater complexity of preauthentication-enabled WPA Enterprise mode. The level of detail presented in the following sections will allow the reader to see the exact amount of frame overhead incurred when different modes are used for performing secure roams. While the timing in empirical tests such as those presented in the following sections are influenced by external factors such as the RF environment and the presence of unrelated network traffic, the frame-count overhead represents an absolute floor for the hand-off time. As the frame count grows, this floor approaches a level that represents unacceptable roaming times for technologies like *Voice over Wireless Fidelity* (*VoWi-Fi*). As VoWi-Fi hand-off time constraints normally assume local and not global roaming, we restrict our analysis in this chapter to local roams.

7.5 Dissection of a WPA-PSK Protected Roam

7.5.1 Test Description for WPA-PSK Roam

Figure 7.3 shows the simple topology that we use for the WPA-PSK roaming test. The topology, test tools, and test methodology are the same as those for the local-roaming tests presented in Chapter 5. The important difference is that the two access points are now configured for WPA-PSK security, which, as we will see, adds a small amount of protocol overhead to the roam.

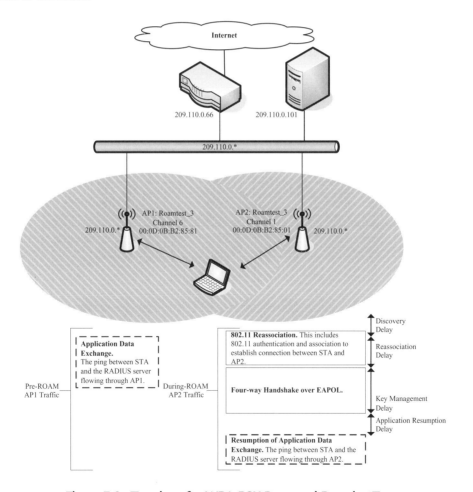

Figure 7.3: Topology for WPA-PSK Protected Roaming Tests.

In the tests, passphrase *bondgarden1* has been preconfigured in the two access points and in the 802.11 client on the STA for the Roamtest_3 SSID. The 802.11 client used for this test was Microsoft WZC. The tests we describe will illustrate that the delay components at play in WPA-PSK roaming are *Discovery Delay*, *Reassociation Delay*, *Key Management Delay*, and *Application Resumption Delay*. *Authentication Delay* is absent in the case of WPA-PSK.

7.5.2 Test Results for WPA-PSK Roam

Tests results for our WPA-PSK experiments are shown in Figure 7.4. Frames 48 through 54 of the figure record the STA associating with AP2 following the roam. These frames correspond to the Reassociation Delay shown in Figure 7.3. Next, in Figure 7.4, we observe the Key Management Delay in the four-way handshake as four EAPOL-Key frames are exchanged between frames 58 and 64. These are the frames that immediately follow the association and comprise the only protocol-overhead penalty incurred for using WPA-PSK. As we described in Section 4.6, this four-way exchange is used to derive the PTK from the PMK. This EAP dialog is very brief compared with the WPA-Enterprise EAP dialogs, which will be discussed later in this chapter. This dialog is limited because in the case of WPA-PSK, the PMK is already known by both the access point and the STA prior to the association. Indeed,

Packet	Source	Destination	BSSID	Channel	Size	Relative Time	Protocol	Summary
45	00:13:CE:46:CB:58	Ethernet Broadcast	FF:FF:FF:FF:FF:FF	1	56	0:01:40.815600	802.11 Probe Req	FC=........,SN=3005,F
46	00:0D:0B:B2:85:01	00:13:CE:46:CB:58	00:0D:0B:B2:85:01	1	109	0:01:40.816696	802.11 Probe Rsp	FC=........,SN= 853,F
47	00:13:CE:46:CB:58	00:0D:0B:B2:85:01		1	14	0:01:40.816995	802.11 Ack	FC=........
48	00:13:CE:46:CB:58	00:0D:0B:B2:85:01	00:0D:0B:B2:85:01	1	34	0:01:40.850425	802.11 Auth	FC=........,SN=3006,F
49	00:0D:0B:B2:85:01	00:13:CE:46:CB:58		1	14	0:01:40.850717	802.11 Ack	FC=........
50	00:0D:0B:B2:85:01	00:13:CE:46:CB:58	00:0D:0B:B2:85:01	1	42	0:01:40.851310	802.11 Auth	FC=........,SN= 854,F
51	00:13:CE:46:CB:58	00:0D:0B:B2:85:01		1	14	0:01:40.851609	802.11 Ack	FC=........
52	00:13:CE:46:CB:58	00:0D:0B:B2:85:01	00:0D:0B:B2:85:01	1	86	0:01:40.852736	802.11 Assoc Req	FC=........,SN=3007,F
53	00:0D:0B:B2:85:01	00:13:CE:46:CB:58		1	14	0:01:40.853035	802.11 Ack	FC=........
54	00:0D:0B:B2:85:01	00:13:CE:46:CB:58	00:0D:0B:B2:85:01	1	58	0:01:40.853965	802.11 Assoc Rsp	FC=........,SN= 855,F
55	00:13:CE:46:CB:58	00:0D:0B:B2:85:01		1	14	0:01:40.854275	802.11 Ack	FC=........
56	00:13:CE:46:CB:58	00:0D:0B:B2:85:01	00:0D:0B:B2:85:01	1	41	0:01:40.872588	EAPOL-Start	FC=T.......,SN=1484,F
57	00:0D:0B:B2:85:01	00:13:CE:46:CB:58		1	14	0:01:40.872881	802.11 Ack	FC=........
58	00:0D:0B:B2:85:01	00:13:CE:46:CB:58	00:0D:0B:B2:85:01	1	135	0:01:40.958153	EAPOL-Key	FC=.F......,SN= 857,F
59	00:13:CE:46:CB:58	00:0D:0B:B2:85:01		1	14	0:01:40.958190	802.11 Ack	FC=........
60	00:13:CE:46:CB:58	00:0D:0B:B2:85:01	00:0D:0B:B2:85:01	1	161	0:01:40.962013	EAPOL-Key	FC=T.......,SN=1485,F
61	00:0D:0B:B2:85:01	00:13:CE:46:CB:58		1	14	0:01:40.962315	802.11 Ack	FC=........
62	00:0D:0B:B2:85:01	00:13:CE:46:CB:58	00:0D:0B:B2:85:01	1	159	0:01:40.964273	EAPOL-Key	FC=.F......,SN= 858,F
63	00:13:CE:46:CB:58	00:0D:0B:B2:85:01		1	14	0:01:40.964300	802.11 Ack	FC=........
64	00:13:CE:46:CB:58	00:0D:0B:B2:85:01	00:0D:0B:B2:85:01	1	135	0:01:40.966265	EAPOL-Key	FC=T.......,SN=1486,F
65	00:0D:0B:B2:85:01	00:13:CE:46:CB:58		1	14	0:01:40.966573	802.11 Ack	FC=........
66	00:13:CE:46:CB:58	Runtop:D3:92:77	00:0D:0B:B2:85:01	1	89	0:01:40.967756	802.11 WEP Data	FC=T.....W.,SN=1487,F
67	00:0D:0B:B2:85:01	00:13:CE:46:CB:58		1	14	0:01:40.968068	802.11 Ack	FC-........
68	00:0D:0B:B2:85:01	00:13:CE:46:CB:58	00:0D:0B:B2:85:01	1	187	0:01:40.969504	802.11 WEP Data	FC=.F....W.,SN= 859,F
69	00:13:CE:46:CB:58	00:0D:0B:B2:85:01		1	14	0:01:40.969524	802.11 Ack	FC=........
70	00:13:CE:46:CB:58	00:0D:0B:B2:85:01	00:0D:0B:B2:85:01	1	155	0:01:40.971805	802.11 WEP Data	FC=T.....W.,SN=1488,F

Figure 7.4: Postroam Association and Four-Way Handshake for WPA-PSK.

the PMK is derivable from the static WPA-PSK passphrase at any time because both the STA and the access point are statically configured with that passphrase.

One of the advantages of WPA-PSK mode is that after frame 64 is received, the process of securing the wireless link for unicast traffic is complete. As one can see in Figure 7.4, all data frames following frame 64 appear as "WEP data" in the sniffer traces because the sniffer does not know how to decrypt them. Since our example is for WPA-PSK, the GTK two-way exchange actually follows the four-way exchange and is thus part of the "WEP data." In WPA2, the GTK exchange is piggybacked onto the third and fourth frames of the four-way exchange. The GTK is needed so that multicast and broadcast traffic are also encrypted, as well as the unicast traffic. By examining the sending and receiving MAC addresses, the frames sizes, and the traffic patterns, we can deduce that the application pings actually resume in frame 84 (see Figure 7.5), at which point the user perceives the roam to be complete.

7.6 Dissection of a WPA2 Enterprise Roam

7.6.1 Introduction

WPA Enterprise mode involves much more complex and robust security measures than WPA-PSK mode. The complexity of the authentication-protocol flow increases dramatically

Packet	Source	Destination	BSSID	Channel	Size	Relative Time	Protocol	Summary
71	00:0D:0B:B2:85:01	00:13:CE:46:CB:58		1	14	0:01:40.972112	802.11 Ack	FC=........
72	00:13:CE:46:CB:58	Runtop:D3:92:77	00:0D:0B:B2:85:01	1	98	0:01:41.718526	802.11 WEP Data	FC=T.....W.,SN=1489,F
73	00:0D:0B:B2:85:01	00:13:CE:46:CB:58		1	14	0:01:41.718814	802.11 Ack	FC=........
74	Runtop:D3:92:77	00:13:CE:46:CB:58	00:0D:0B:B2:85:01	1	102	0:01:42.072384	802.11 WEP Data	FC=.F....W.,SN= 871,F
75	00:13:CE:46:CB:58	00:0D:0B:B2:85:01		1	14	0:01:42.072415	802.11 Ack	FC=........
76	Runtop:D3:92:77	00:13:CE:46:CB:58	00:0D:0B:B2:85:01	1	102	0:01:42.262855	802.11 WEP Data	FC=.F....W.,SN= 874,F
77	00:13:CE:46:CB:58	00:0D:0B:B2:85:01		1	14	0:01:42.262884	802.11 Ack	FC=........
78	00:13:CE:46:CB:58	Runtop:D3:92:77	00:0D:0B:B2:85:01	1	89	0:01:42.328149	802.11 WEP Data	FC=T.....W.,SN=1490,F
79	00:0D:0B:B2:85:01	00:13:CE:46:CB:58		1	14	0:01:42.328438	802.11 Ack	FC=........
80	Runtop:D3:92:77	00:13:CE:46:CB:58	00:0D:0B:B2:85:01	1	102	0:01:42.329281	802.11 WEP Data	FC=.F....W.,SN= 875,F
81	00:13:CE:46:CB:58	00:0D:0B:B2:85:01		1	14	0:01:42.329311	802.11 Ack	FC=........
82	00:13:CE:46:CB:58	Runtop:D3:92:77	00:0D:0B:B2:85:01	1	96	0:01:42.374699	802.11 WEP Data	FC=T.....W.,SN=1491,F
83	00:0D:0B:B2:85:01	00:13:CE:46:CB:58		1	14	0:01:42.374979	802.11 Ack	FC=........
84	00:13:CE:46:CB:58	Asustek Comp:47...	00:0D:0B:B2:85:01	1	116	0:01:43.328169	802.11 WEP Data	FC=T.....W.,SN=1492,F
85	00:0D:0B:B2:85:01	00:13:CE:46:CB:58		1	14	0:01:43.328465	802.11 Ack	FC=........
86	Asustek Comp:4...	00:13:CE:46:CB:58	00:0D:0B:B2:85:01	1	116	0:01:43.329072	802.11 WEP Data	FC=.F....W.,SN= 886,F
87	00:13:CE:46:CB:58	00:0D:0B:B2:85:01		1	14	0:01:43.329096	802.11 Ack	FC=........
88	00:13:CE:46:CB:58	Ethernet Broadcast	00:0D:0B:B2:85:01	1	84	0:01:44.328033	802.11 WEP Data	FC=T.....W.,SN=1493,F
89	00:0D:0B:B2:85:01	00:13:CE:46:CB:58		1	14	0:01:44.328265	802.11 Ack	FC=........
90	00:13:CE:46:CB:58	Ethernet Broadcast	00:0D:0B:B2:85:01	1	84	0:01:44.329052	802.11 WEP Data	FC=T.....W.,SN=1494,F
91	00:0D:0B:B2:85:01	00:13:CE:46:CB:58		1	14	0:01:44.329297	802.11 Ack	FC=........
92	Asustek Comp:4...	00:13:CE:46:CB:58	00:0D:0B:B2:85:01	1	102	0:01:44.329505	802.11 WEP Data	FC=.F....W.,SN= 897,F
93	00:13:CE:46:CB:58	00:0D:0B:B2:85:01		1	14	0:01:44.329526	802.11 Ack	FC=........
94	00:13:CE:46:CB:58	Asustek Comp:47...	00:0D:0B:B2:85:01	1	116	0:01:44.330633	802.11 WEP Data	FC=T.....W.,SN=1495,F
95	00:0D:0B:B2:85:01	00:13:CE:46:CB:58		1	14	0:01:44.330869	802.11 Ack	FC=........
96	Runtop:D3:92:77	00:13:CE:46:CB:58	00:0D:0B:B2:85:01	1	102	0:01:44.330954	802.11 WEP Data	FC=.F....W.,SN= 898,F

Figure 7.5: Resumption of User Traffic Following a WPA-PSK Roam.

Chapter 7

with respect to the simple four-way exchange discussed in Section 7.5. The tests that follow will illustrate that the delay components at play in WPA2 Enterprise roaming are Discovery Delay, Reassociation Delay, Authentication Delay, Key Management Delay, and Application Resumption Delay.

Figure 7.6 presents the authentication-protocol flow for normal WPA enterprise mode in a manner that will facilitate the understanding of the detailed traces provided in the next section. The most striking difference in the protocol flow is that there is now an AAA server intimately involved in the authentication. There is a sharp contrast between the simplicity of the security steps that we just described in Section 7.5, and the complexity revealed by Figure 7.6.

7.6.2 Test Description for WPA2 Enterprise Roam

Figure 7.7 shows the topology that we use for the WPA Enterprise roaming test. The test tools and methodology resemble those in the WPA-PSK roaming tests described in Section 7.5.1

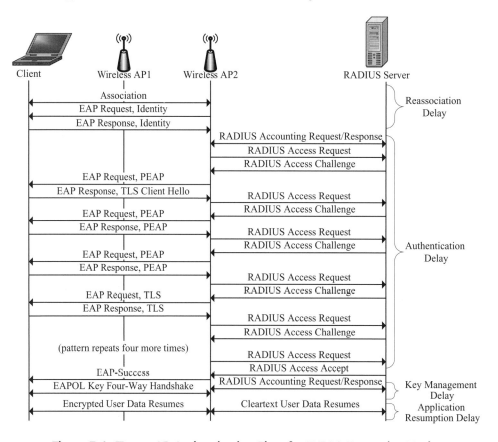

Figure 7.6: Target AP Authentication Flow for WPA2 Enterprise Mode.

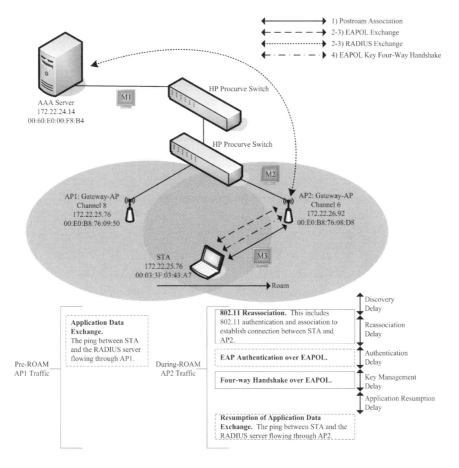

Figure 7.7: Topology for WPA2 Enterprise Protected Roaming Tests.

with a number of modifications. The topology has become considerably more complex. We now require an AAA server to provide the more robust authentication performed in WPA Enterprise mode. This AAA server runs on an Intel Pentium-3 machine that runs the Aegis 1.1.6 RADIUS server application with IP address 172.22.24.14. The two APs are Gateway AP model 7001 80211A+G, and both are configured for WPA Enterprise security with SSID Gateway-AP. AP1 operates on channel 8 and AP2 on channel 6. The roam is from AP1 to AP2.

The two wired-Ethernet switches connecting the various network components are HP Procurve model 1600M switches. The STA is a Dell laptop with an Atheros 5004 chipset on a PCMCIA card as the wireless interface. The 802.11 client used for this test is the Aegis SecureConnect, which has been configured to use EAP-PEAP when authenticating with the Gateway-AP

SSID. The application traffic for the test is generated by the ping application that runs on the STA and pinging 172.22.24.14, which is the IP address of the RADIUS-server machine.

As expected in an 802.11i-enabled network, our test application cannot begin to send user traffic until the enterprise-class authentication has been performed with the AAA server shown in Figure 7.7. We place an 802.11 monitor on channel 6 to record the 802.11 portion of the postroam association. In addition, we place an 802.3 monitor on the 802.3 wired network to record the RADIUS packets that pass the authentication requests from the wireless environment to the AAA RADIUS server.

In roaming tests described earlier in this book, we have forced the roam to occur by powering off the initially associated AP. Here and in the remaining tests in this chapter, the roams are produced "naturally" by moving away from the first access point until the signal strength drops below a threshold that results in the driver choosing to roam to the now-nearby second access point. As we force the roam to occur in this manner, we record the reassociation and new authentication on channel 6 and the corresponding wire-line RADIUS traffic across the entire roaming transition. As the monitoring equipment is separate for these two media and not time-correlated, we will depict the overall frame exchange on both channel 6 and the 802.3 network shown in Figure 7.6 and correlate the frames graphically depicted in that figure with the actual recordings of those frames in the corresponding sniffer trace. The fact that we have recorded the ping-application traffic on the wired Ethernet hop directly connected to the RADIUS server allows us to obtain precise measurements of the relative timing of the cessation of application traffic over the first AP, the association and authentication processes related to the secure roam, and the resumption of the application traffic.

7.6.3 Test Results for WPA2 Enterprise Roam

Figure 7.8 captures the traffic on the wired Ethernet hop that connects the RADIUS server to the network. The traffic was captured by an Ethereal sniffer denoted as M1 in Figure 7.7. As our application pings are sent to 172.22.24.14, the machine on which the RADIUS server runs, we are able to see three key components of this test time-synchronized. These are the ping-application traffic prior to the roam, the RADIUS authentications that take place after the roam as the STA authenticates with the second AP and then the resumption of the application pings. In order to avoid clutter on the sniffer traces, we have filtered out all the traffic that has not been explicitly addressed to 172.22.24.14. Figures 7.9, 7.10, and 7.11 show the frames sent to or from the STA from the second AP.

If we follow the chronological steps of the roam, we first note the application pings that are visible in frames 68 to 74 in Figure 7.8. These pings correspond to the preroam Application Data Exchange shown in Figure 7.7. The cessation of these pings is due to the fact that the

Figure 7.8: WPA2-Enterprise Mode Monitor M1 Frame Trace.

Figure 7.9: WPA2-Enterprise Mode Monitor M3 Frames 2524–2549.

Chapter 7

Packet	Source	Destination	BSSID	Channel	Size	Absolute Time	Protocol	Summary
2550	00:03:7F:03:43:A7	00:E0:B8:76:08:D8	00:E0:B8:76:08:D8	6	82	8:35:24.519321	EAP Response	FC=T......,SN= 65,F
2551	00:E0:B8:76:08:D8	00:03:7F:03:43:A7		6	14	8:35:24.519632	802.11 Ack	FC=........
2552	00:E0:B8:76:08:D8	00:03:7F:03:43:A7	00:E0:B8:76:08:D8	6	1372	8:35:24.535465	EAP Request	FC=.F......,SN=2714,F
2553	00:03:7F:03:43:A7	00:E0:B8:76:08:D8	00:E0:B8:76:08:D8	6	248	8:35:24.671828	EAP Response	FC=T......,SN= 66,F
2554	00:E0:B8:76:08:D8	00:03:7F:03:43:A7		6	14	8:35:24.672119	802.11 Ack	FC=........
2555	00:E0:B8:76:08:D8	00:03:7F:03:43:A7	00:E0:B8:76:08:D8	6	109	8:35:24.690367	EAP Request	FC=.F......,SN=2718,F
2556	00:03:7F:03:43:A7	00:E0:B8:76:08:D8	00:E0:B8:76:08:D8	6	82	8:35:24.706749	EAP Response	FC=T......,SN= 67,F
2557	00:E0:B8:76:08:D8	00:03:7F:03:43:A7		6	14	8:35:24.707060	802.11 Ack	FC=........
2558	00:E0:B8:76:08:D8	00:03:7F:03:43:A7	00:E0:B8:76:08:D8	6	124	8:35:24.713788	EAP Request	FC=.F......,SN=2719,F
2559	00:03:7F:03:43:A7	00:E0:B8:76:08:D8	00:E0:B8:76:08:D8	6	120	8:35:24.783997	EAP Response	FC=T......,SN= 68,F
2560	00:E0:B8:76:08:D8	00:03:7F:03:43:A7		6	14	8:35:24.784313	802.11 Ack	FC=........
2561	00:E0:B8:76:08:D8	00:03:7F:03:43:A7	00:E0:B8:76:08:D8	6	156	8:35:24.843159	EAP Request	FC=.F......,SN=2727,F
2562	00:03:7F:03:43:A7	00:E0:B8:76:08:D8	00:E0:B8:76:08:D8	6	184	8:35:24.929727	EAP Response	FC=T......,SN= 69,F
2563	00:E0:B8:76:08:D8	00:03:7F:03:43:A7		6	14	8:35:24.930024	802.11 Ack	FC=........
2564	00:E0:B8:76:08:D8	00:03:7F:03:43:A7	00:E0:B8:76:08:D8	6	172	8:35:24.936124	EAP Request	FC=.F......,SN=2730,F
2565	00:03:7F:03:43:A7	00:E0:B8:76:08:D8	00:E0:B8:76:08:D8	6	120	8:35:24.954757	EAP Response	FC=T......,SN= 70,F
2566	00:E0:B8:76:08:D8	00:03:7F:03:43:A7		6	14	8:35:24.955063	802.11 Ack	FC=........
2567	00:E0:B8:76:08:D8	00:03:7F:03:43:A7	00:E0:B8:76:08:D8	6	124	8:35:24.960127	EAP Request	FC=.F......,SN=2731,F
2568	00:03:7F:03:43:A7	00:E0:B8:76:08:D8	00:E0:B8:76:08:D8	6	82	8:35:24.974080	EAP Response	FC=T......,SN= 71,F
2569	00:E0:B8:76:08:D8	00:03:7F:03:43:A7		6	14	8:35:24.974393	802.11 Ack	FC=........
2570	00:E0:B8:76:08:D8	00:03:7F:03:43:A7	00:E0:B8:76:08:D8	6	44	8:35:24.981532	EAP Success	FC=.F......,SN=2732,F
2571	00:E0:B8:76:08:D8	00:03:7F:03:43:A7	00:E0:B8:76:08:D8	6	135	8:35:24.983129	EAPOL-Key	FC=.F......,SN=2733,F
2572	00:03:7F:03:43:A7	00:E0:B8:76:08:D8	00:E0:B8:76:08:D8	6	175	8:35:25.011725	EAPOL-Key	FC=T......,SN= 72,F
2573	00:E0:B8:76:08:D8	00:03:7F:03:43:A7		6	14	8:35:25.012030	802.11 Ack	FC=........
2574	00:E0:B8:76:08:D8	00:03:7F:03:43:A7	00:E0:B8:76:08:D8	6	239	8:35:25.016703	EAPOL-Key	FC=.F......,SN=2734,F
2575	00:03:7F:03:43:A7	00:E0:B8:76:08:D8	00:E0:B8:76:08:D8	6	135	8:35:25.022601	EAPOL-Key	FC=T......,SN= 73,F

Figure 7.10: WPA2-Enterprise Mode Monitor M3 Frames 2550–2575.

STA has moved sufficiently far from the first AP that the driver has determined that it must roam, so there is a brief lack of connectivity corresponding to the Discovery Delay in Figure 7.7.

The Reassociation Delay shown in Figure 7.7 starts in frame 2529, as shown Figure 7.9. We attempted to synchronize the clocks of the different sniffers used in these tests using TIMEP at the start of the tests, so there should be some correspondence in the timestamps across sniffers. Nevertheless, the time granularity that we can obtain from this kind of multiclock synchronization is still coarse, so we have to compare carefully the time stamps from the different sniffers. The timestamp of frame 2529 is 8:35:24:030368, whereas the timestamp of the last application ping was 08:34:27.558069.

Referring back to Figure 7.8, the next frame that is shown following the last application ping is the first RADIUS packet related to the authentication of the STA on the new AP. This frame, number 75 in Figure 7.8, was sent at 08:35:26:305884. As we stated previously, the last application ping before the roam arrived at 8:34:27:558069. All the timestamps from M1 are correlated, and the RADIUS authentication would immediately follow the association with AP2. Thus, the 58.8 second difference between these timestamps is a close measure of the interval between the end of effective communications with AP1 and the

Roaming Securely in 802.11

Packet	Source	Destination	BSSID	Channel	Size	Absolute Time	Protocol	Summary
2576	00:E0:B8:76:08:D8	00:03:7F:03:43:A7		6	14	8:35:25.022906	802.11 Ack	FC=........
2577	00:03:7F:03:43:A7	FF:FF:FF:FF:FF:FF	00:E0:B8:76:08:D8	6	80	8:35:25.112560	802.11 WEP Data	FC=T.....W.,SN= 74,F
2578	00:E0:B8:76:08:D8	00:03:7F:03:43:A7		6	14	8:35:25.112586	802.11 Ack	FC=........
2579	00:03:7F:03:43:A7	FF:FF:FF:FF:FF:FF	00:E0:B8:76:08:D8	6	84	8:35:25.113675	802.11 WEP Data	FC=.F....W.,SN=2737,F
2580	11:08:83:0C:88:00	00:03:7F:03:43:A7	00:E0:B8:B6:F4:76	6	98	8:35:25.114712	802.11 WEP Data	
2581	00:03:7F:03:43:A7	FF:FF:FF:FF:FF:FF	00:01:F4:96:1B:4A	6	102	8:35:25.117192	802.11 WEP Data	FC=.F....W.,SN=3927,F
2582	00:03:7F:03:43:A7	00:E0:B8:76:08:D8	00:E0:B8:76:08:D8	6	28	8:35:25.230607	802.11 Data	FC=T...P...,SN= 75,F
2583	00:E0:B8:76:08:D8	00:03:7F:03:43:A7		6	14	8:35:25.230624	802.11 Ack	FC=........
2584	00:03:7F:03:43:A7	00:60:E0:00:F8:B4	00:E0:B8:76:08:D8	6	112	8:35:26.774851	802.11 WEP Data	FC=T.....W.,SN= 76,F
2585	00:E0:B8:76:08:D8	00:03:7F:03:43:A7		6	14	8:35:26.774878	802.11 Ack	FC=........
2586	00:60:E0:00:F8:B4	00:03:7F:03:43:A7	00:E0:B8:76:08:D8	6	112	8:35:26.776099	802.11 WEP Data	FC=.F....W.,SN=2783,F
2587	00:03:7F:03:43:A7	00:E0:B8:76:08:D8	00:E0:B8:76:08:D8	6	28	8:35:26.939277	802.11 Data	FC=T...P...,SN= 77,F
2588	00:E0:B8:76:08:D8	00:03:7F:03:43:A7		6	14	8:35:26.939296	802.11 Ack	FC=........
2589	00:03:7F:03:43:A7	FF:FF:FF:FF:FF:FF	00:E0:B8:76:08:D8	6	80	8:35:27.028856	802.11 WEP Data	FC=T..R..W.,SN= 78,F
2590	00:E0:B8:76:08:D8	00:03:7F:03:43:A7		6	14	8:35:27.028884	802.11 Ack	FC=........
2591	00:03:7F:03:43:A7	FF:FF:FF:FF:FF:FF	00:E0:B8:76:08:D8	6	84	8:35:27.029970	802.11 WEP Data	FC=.F....W.,SN=2790,F
2592	00:08:83:0C:88:00	00:03:7F:03:43:A7	00:E0:B8:76:08:D8	6	98	8:35:27.037836	802.11 WEP Data	FC=.F....W.,SN=2791,F
2593	00:03:7F:03:43:A7	00:08:83:0C:88:00	00:E0:B8:76:08:D8	6	85	8:35:27.038186	802.11 WEP Data	FC=T.....W.,SN= 79,F
2594	00:E0:B8:76:08:D8	00:03:7F:03:43:A7		6	14	8:35:27.038211	802.11 Ack	FC=........
2595	00:08:83:0C:88:00	00:03:7F:03:43:A7	00:E0:B8:76:08:D8	6	98	8:35:27.038753	802.11 WEP Data	FC=.F.R..W.,SN=2791,F
2596	00:08:83:0C:88:00	00:03:7F:03:43:A7	00:E0:B8:76:08:D8	6	98	8:35:27.039504	802.11 WEP Data	FC=.F.R..W.,SN=2791,F
2597	00:08:83:0C:88:00	00:03:7F:03:43:A7	00:E0:B8:76:08:D8	6	85	8:35:27.040947	802.11 WEP Data	FC=.F....W.,SN=2792,F
2598	00:03:7F:03:43:A7	00:E0:B8:76:08:D8	00:E0:B8:76:08:D8	6	28	8:35:27.240958	802.11 Data	FC=T...P...,SN= 80,F
2599	00:E0:B8:76:08:D8	00:03:7F:03:43:A7		6	14	8:35:27.240985	802.11 Ack	FC=........
2600	00:03:7F:03:43:A7	FF:FF:FF:FF:FF:FF	00:E0:B8:76:08:D8	6	80	8:35:27.775647	802.11 WEP Data	FC=T.....W.,SN= 81,F
2601	00:E0:B8:76:08:D8	00:03:7F:03:43:A7		6	14	8:35:27.775679	802.11 Ack	FC=........

Figure 7.11: WPA2-Enterprise Mode Monitor M3 Frames 2576–2601.

association with AP2. This delay is much longer than what we have measured in earlier roaming tests when we disabled the radio to force the roam. This additional delay is due to the fact that the communications with AP1 had degraded to a point of being unusable long before the driver in the STA decided to roam to AP2.

The RADIUS packets that we see in frames 75 to 87 in Figure 7.8 correspond to the EAPOL exchange shown in frames 2535 through 2570 in Figures 7.9 and 7.10. These frames comprise the Authentication Delay shown in Figure 7.7. It should be noted that due to the filter applied in the sniffer, we can only see the RADIUS packets destined *to* the RADIUS server in Figure 7.8 and not *from* it. In the interest of conserving space, we leave it to the reader to deduce the RADIUS packets sent in response to those shown. As we explained in Chapter 6, the PEAP conversation between the STA and the RADIUS server is transported first by the EAPOL carrier protocol from the STA to the AP and then from the AP to the RADIUS server by the RADIUS carrier protocol. This authentication exchange is seen in the EAPOL exchange in frames 2535 through 2570 in Figures 7.9 and 7.10 and the corresponding RADIUS exchange in frames 76 through 85 in Figure 7.8.

Following the successful authentication, the Key Management Delay is show in the four-way handshake in frames 2571 through 2575 in Figure 7.10. We explained in Chapter 5 that the four-way handshake is used to develop the PTK from the PMK that was derived during the authentication process. At this point, the 802.11i TKIP-based encryption of the wireless link from the STA to the new AP starts. Beginning with frame 2577, the remaining sniffer trace in Figure 7.11 shows all user frames as "WEP data" because the sniffer does not know how to decrypt them. We can, however, see the end of the Application Resumption Delay with the resumption of the application pings in frame 88 in Figure 7.8 since the encryption is only of the wireless link, and when these frames emerge from the AP, they are transmitted in the clear. At this point, the roam is complete from the application's perspective.

7.7 Dissection of an 802.11i Preauthentication

7.7.1 Introduction

Preauthentication was introduced into the 802.11i standard as a means to abbreviate the number of frames needed to authenticate an 802.11i STA after it roams to a new AP in the same ESS. This efficiency is accomplished without sacrificing any of the security benefits of the full 802.1X security paradigm in the nonpreauthenticated 802.11i case. As discussed previously, achieving this degree of security requires significant EAPOL and RADIUS exchanges including the STA, the target AP, and the RADIUS server. Figure 7.12 presents the authentication-protocol flow for 802.11i preauthentication mode.

The dashed lines in Figure 7.12 represent frames sent between the access points over the distribution service, which in this topology is a wired Ethernet. Figure 7.12 should be contrasted to Figure 7.6, where the EAPOL frames are transmitted over the air, and the EAPOL and RADIUS exchanges for the roaming STA occur *after* the roam and before application traffic can resume. Figures 7.12 and 7.13 show that for preauthentication, the Authentication Delay occurs *in advance* of the roam—reducing the protocol complexity required during the roam itself.

7.7.2 Test Description for Preauthentication

The procedures in this test are identical to those described in Section 7.6.2, except that 802.11i preauthentication has been enabled in the STA and the APs. The wired and wireless sniffers were placed in the same locations. While the test topology is actually the same as that used in Section 7.6.2, we use Figure 7.13 to highlight the different paths that the authentication traffic takes in the case of preauthentication.

7.7.3 Test Results for Preauthentication

In Section 7.6.2, all the roaming-related frames that we studied emanated from AP2, except for the final-application traffic which was sent just before the roam commenced. This is

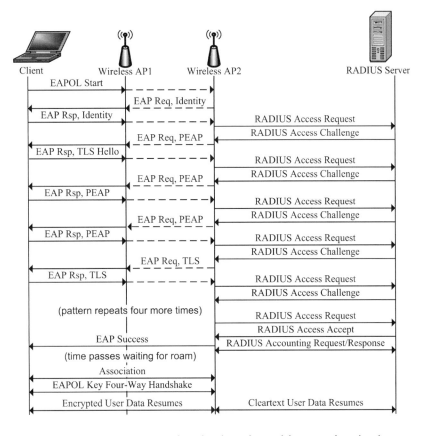

Figure 7.12: Target AP Authentication Flow with Preauthentication.

because all the relevant roaming-related protocol exchanges occur *after* the STA associates with AP2. This situation is not the case with preauthentication. We see evidence of this fact in Figure 7.14, where we observe RADIUS authentication requests coming from 172.22.26.92, AP2, although our STA still has not associated with it. Indeed, the process of roaming has not even begun, as we are well within the cell of AP1.

The RADIUS exchange between AP2 and the RADIUS server is captured in frames 162 through 184 of Figure 7.14. These frames comprise the Authentication Delay shown in Figure 7.13. In the 802.11i paradigm, this RADIUS exchange relays the EAPOL exchange between the STA and that AP with which it is attempting to authenticate. That EAPOL exchange is in fact occurring but is taking place over the wired network connecting the current and target APs, and is occurring in anticipation of a future association, which could happen in the event of a future roam.

Chapter 7

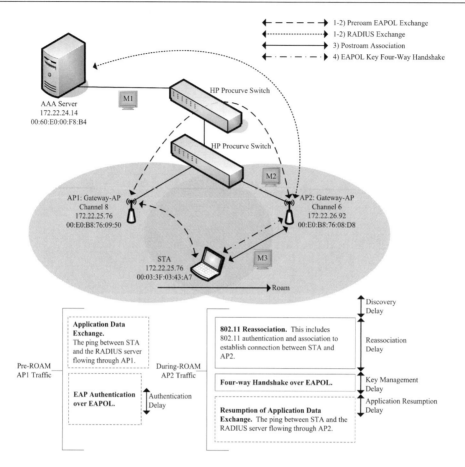

Figure 7.13: Topology for Preauthentication Roaming Tests.

We see this EAPOL exchange starting in frame 86 in Figure 7.15 and continuing through frame 128 in Figure 7.16. As the M2 sniffer sits on the wired Ethernet connecting AP2 to the network, we see not only the EAPOL exchange but also the interleaved RADIUS exchange, as that RADIUS traffic must travel over the same Ethernet link in the reverse direction to reach the RADIUS server. We encourage the reader to review Figure 7.13, which illustrates how the EAPOL preauthentication path and RADIUS paths converge on that Ethernet link. For the sake of clarity, we emphasize that the EAP conversation that occurs in frames 86 through 128 of Figure 7.15 and Figure 7.16 is *the same* EAP conversation shown in frames 162 through 184 of Figure 7.14, just observed closer to the STA.

The process just described all happens very shortly after the initial association of the STA and the first AP and before the application pings begin. We see the start of the preroam application pings in frame 193 of Figure 7.17. These frames correspond to the Application Data Exchange shown in Figure 7.13. They reflect a stable-state situation which lasts until we force a roam by

No.	Time	Source	Destination	Protocol	Info
162	08:44:02.831291	172.22.26.92	172.22.24.14	RADIUS	Access Request(1) (id=68, l=165)
163	08:44:02.832316	172.22.24.14	172.22.26.92	RADIUS	Access challenge(11) (id=68, l=65)
164	08:44:02.870698	172.22.26.92	172.22.24.14	RADIUS	Access Request(1) (id=69, l=276)
165	08:44:02.944681	172.22.24.16	172.22.27.255	NBNS	Name query NB MDC<1c>
166	08:44:02.945470	172.22.24.14	172.22.26.92	RADIUS	Access challenge(11) (id=69, l=1465)
167	08:44:02.959758	172.22.26.92	172.22.24.14	RADIUS	Access Request(1) (id=70, l=176)
168	08:44:02.960523	172.22.24.14	172.22.26.92	RADIUS	Access challenge(11) (id=70, l=1465)
169	08:44:02.972976	172.22.26.92	172.22.24.14	RADIUS	Access Request(1) (id=71, l=176)
170	08:44:02.973682	172.22.24.14	172.22.26.92	RADIUS	Access challenge(11) (id=71, l=1401)
171	08:44:03.124192	172.22.26.92	172.22.24.14	RADIUS	Access Request(1) (id=72, l=378)
172	08:44:03.162620	172.22.24.14	172.22.26.92	RADIUS	Access challenge(11) (id=72, l=128)
173	08:44:03.182699	172.22.26.92	172.22.24.14	RADIUS	Access Request(1) (id=73, l=176)
174	08:44:03.183876	172.22.24.14	172.22.26.92	RADIUS	Access challenge(11) (id=73, l=143)
175	08:44:03.215373	172.22.26.92	172.22.24.14	RADIUS	Access Request(1) (id=74, l=250)
176	08:44:03.222947	172.22.24.14	172.22.26.92	RADIUS	Access challenge(11) (id=74, l=175)
177	08:44:03.247298	172.22.26.92	172.22.24.14	RADIUS	Access Request(1) (id=75, l=314)
178	08:44:03.264053	172.22.24.14	172.22.26.92	RADIUS	Access challenge(11) (id=75, l=191)
179	08:44:03.278366	172.22.26.92	172.22.24.14	RADIUS	Access Request(1) (id=76, l=250)
180	08:44:03.337302	172.22.24.14	172.22.26.92	RADIUS	Access challenge(11) (id=76, l=143)
181	08:44:03.350495	172.22.26.92	172.22.24.14	RADIUS	Access Request(1) (id=77, l=176)
182	08:44:03.351851	172.22.24.14	172.22.26.92	RADIUS	Access Accept(2) (id=77, l=184)
183	08:44:03.360160	172.22.26.92	172.22.24.14	RADIUS	Accounting Request(4) (id=78, l=159)
184	08:44:03.361402	172.22.24.14	172.22.26.92	RADIUS	Accounting Response(5) (id=78, l=20)
185	08:44:04.172677	AironetW_5a:8f:11	01:40:96:ff:ff:00	LLC	U, func=UI; SNAP, OUI 0x004096 (Unknown), PID 0x0000
186	08:44:04.350426	0.0.0.0	255.255.255.255	DHCP	DHCP Discover - Transaction ID 0x5d7be5b4
187	08:44:04.351011	172.22.24.10	255.255.255.255	DHCP	DHCP Offer - Transaction ID 0x5d7be5b4
188	08:44:04.362503	0.0.0.0	255.255.255.255	DHCP	DHCP Request - Transaction ID 0x5d7be5b4

Figure 7.14: Preauthentication Monitor M3 Frames 162–187.

No.	Time	Source	Destination	Protocol	Info
86	08:44:10.606486	172.22.25.76	Gateway2_76:08:d8	EAPOL	Start
87	08:44:10.608300	Gateway2_76:08:d8	172.22.25.76	EAP	Request, Identity [RFC3748]
88	08:44:10.619592	172.22.25.76	Gateway2_76:08:d8	EAP	Response, Identity [RFC3748]
89	08:44:10.621343	172.22.26.92	172.22.24.14	RADIUS	Access Request(1) (id=68, l=165)
90	08:44:10.622610	172.22.24.14	172.22.26.92	RADIUS	Access challenge(11) (id=68, l=65)
91	08:44:10.623891	Gateway2_76:08:d8	172.22.25.76	EAP	Request, PEAP [Palekar]
92	08:44:10.659450	172.22.25.76	Gateway2_76:08:d8	TLS	Client Hello
93	08:44:10.660714	172.22.26.92	172.22.24.14	RADIUS	Access Request(1) (id=69, l=276)
94	08:44:10.734878	172.22.24.16	172.22.27.255	NBNS	Name query NB MDC<1c>
95	08:44:10.737201	172.22.24.14	172.22.26.92	RADIUS	Access challenge(11) (id=69, l=1465)
96	08:44:10.739928	Gateway2_76:08:d8	172.22.25.76	EAP	Request, PEAP [Palekar]
97	08:44:10.748405	172.22.25.76	Gateway2_76:08:d8	EAP	Response, PEAP [Palekar]
98	08:44:10.749811	172.22.26.92	172.22.24.14	RADIUS	Access Request(1) (id=70, l=176)
99	08:44:10.752244	172.22.24.14	172.22.26.92	RADIUS	Access challenge(11) (id=70, l=1465)
100	08:44:10.754793	Gateway2_76:08:d8	172.22.25.76	EAP	Request, PEAP [Palekar]
101	08:44:10.761784	172.22.25.76	Gateway2_76:08:d8	EAP	Response, PEAP [Palekar]
102	08:44:10.763021	172.22.26.92	172.22.24.14	RADIUS	Access Request(1) (id=71, l=176)
103	08:44:10.765339	172.22.24.14	172.22.26.92	RADIUS	Access challenge(11) (id=71, l=1401)
104	08:44:10.767680	Gateway2_76:08:d8	172.22.25.76	TLS	Server Hello, Certificate, Server Key Exchange, Serv
105	08:44:10.912645	172.22.25.76	Gateway2_76:08:d8	TLS	Client Key Exchange, Change Cipher Spec, Encrypted H
106	08:44:10.914141	172.22.26.92	172.22.24.14	RADIUS	Access Request(1) (id=72, l=378)
107	08:44:10.952979	172.22.24.14	172.22.26.92	RADIUS	Access challenge(11) (id=72, l=128)
108	08:44:10.954143	Gateway2_76:08:d8	172.22.25.76	TLS	Change Cipher Spec, Encrypted Handshake Message
109	08:44:10.971340	172.22.25.76	Gateway2_76:08:d8	EAP	Response, PEAP [Palekar]
110	08:44:10.972684	172.22.26.92	172.22.24.14	RADIUS	Access Request(1) (id=73, l=176)
111	08:44:10.974226	172.22.24.14	172.22.26.92	RADIUS	Access challenge(11) (id=73, l=143)

Figure 7.15: Preauthentication Monitor M2 Frames 86–111.

```
No..  Time                 Source              Destination         Protocol  Info
112   08:44:10.975340 Gateway2_76:08:d8        172.22.25.76        TLS       Application Data, Application Data
113   08:44:11.003907 172.22.25.76             Gateway2_76:08:d8   TLS       Application Data, Application Data
114   08:44:11.005345 172.22.26.92             172.22.24.14        RADIUS    Access Request(1) (id=74, l=250)
115   08:44:11.013328 172.22.24.14             172.22.26.92        RADIUS    Access challenge(11) (id=74, l=175)
116   08:44:11.014493 Gateway2_76:08:d8        172.22.25.76        TLS       Application Data, Application Data
117   08:44:11.035953 172.22.25.76             Gateway2_76:08:d8   TLS       Application Data, Application Data
118   08:44:11.037263 172.22.26.92             172.22.24.14        RADIUS    Access Request(1) (id=75, l=314)
119   08:44:11.054513 172.22.24.14             172.22.26.92        RADIUS    Access challenge(11) (id=75, l=191)
120   08:44:11.055780 Gateway2_76:08:d8        172.22.25.76        TLS       Application Data, Application Data
121   08:44:11.067110 172.22.25.76             Gateway2_76:08:d8   TLS       Application Data, Application Data
122   08:44:11.068359 172.22.26.92             172.22.24.14        RADIUS    Access Request(1) (id=76, l=250)
123   08:44:11.127684 172.22.24.14             172.22.26.92        RADIUS    Access challenge(11) (id=76, l=143)
124   08:44:11.128950 Gateway2_76:08:d8        172.22.25.76        TLS       Application Data, Application Data
125   08:44:11.139207 172.22.25.76             Gateway2_76:08:d8   EAP       Response, PEAP [Palekar]
126   08:44:11.140525 172.22.26.92             172.22.24.14        RADIUS    Access Request(1) (id=77, l=176)
127   08:44:11.142280 172.22.24.14             172.22.26.92        RADIUS    Access Accept(2) (id=77, l=184)
128   08:44:11.145501 Gateway2_76:08:d8        172.22.25.76        EAP       Success
129   08:44:11.150203 172.22.26.92             172.22.24.14        RADIUS    Accounting Request(4) (id=78, l=159)
130   08:44:11.151654 172.22.24.14             172.22.26.92        RADIUS    Accounting Response(5) (id=78, l=20)
131   08:44:11.962847 AironetW_5a:8f:11        01:40:96:ff:ff:00   LLC       U, func=UI; SNAP, OUI 0x004096 (Unknown), PID 0x0000
132   08:44:12.140798 0.0.0.0                  255.255.255.255     DHCP      DHCP Discover - Transaction ID 0x5d7be5b4
133   08:44:12.141415 172.22.24.14             255.255.255.255     DHCP      DHCP Offer    - Transaction ID 0x5d7be5b4
134   08:44:12.153014 0.0.0.0                  255.255.255.255     DHCP      DHCP Request  - Transaction ID 0x5d7be5b4
135   08:44:12.157243 172.22.24.10             255.255.255.255     DHCP      DHCP ACK      - Transaction ID 0x5d7be5b4
136   08:44:12.189182 172.22.25.76             Broadcast           ARP       Who has 172.22.25.76? Gratuitous ARP
137   08:44:12.272354 172.22.25.55             172.22.27.255       BROWSER   Domain/Workgroup Announcement MDC, NT Workstation, D
138   08:44:12.468411 172.22.25.76             Broadcast           ARP       Who has 172.22.24.14? Tell 172.22.25.76
```

Figure 7.16: Preauthentication Monitor M2 Frames 112–138.

```
No..  Time                 Source              Destination         Protocol  Info
187   08:44:04.331011 172.22.24.10             255.255.255.255     DHCP      DHCP Offer    - Transaction ID 0x5d7be5b4
188   08:44:04.362593 0.0.0.0                  255.255.255.255     DHCP      DHCP Request  - Transaction ID 0x5d7be5b4
189   08:44:04.366807 172.22.24.10             255.255.255.255     DHCP      DHCP ACK      - Transaction ID 0x5d7be5b4
190   08:44:04.368358 HP_88:da:fb              Spanning-tree-(for- STP       Conf. Root = 100/00:60:b0:88:da:80  Cost = 0  Port =
191   08:44:04.399642 172.22.25.76             Broadcast           ARP       Who has 172.22.25.76? Gratuitous ARP
192   08:44:04.482029 172.22.25.55             172.22.27.255       BROWSER   Domain/Workgroup Announcement MDC, NT Workstation, D
193   08:44:04.683398 172.22.25.76             172.22.24.14        ICMP      Echo (ping) request
194   08:44:04.683470 172.22.24.14             172.22.25.76        ICMP      Echo (ping) reply
195   08:44:05.352405 172.22.25.76             Broadcast           ARP       Who has 172.22.25.76? Gratuitous ARP
196   08:44:05.677770 172.22.25.76             172.22.24.14        ICMP      Echo (ping) request
197   08:44:05.677819 172.22.24.14             172.22.25.76        ICMP      Echo (ping) reply
198   08:44:05.884990 SymbolTe_4f:22:f0        01:a0:f8:f0:f0:02   0x8781    Ethernet II
199   08:44:06.349632 172.22.25.76             Broadcast           ARP       Who has 172.22.25.76? Gratuitous ARP
200   08:44:06.367246 HP_88:da:fb              Spanning-tree-(for- STP       Conf. Root = 100/00:60:b0:88:da:80  Cost = 0  Port =
201   08:44:06.646593 172.22.26.91             172.22.24.14        ARP       Who has 172.22.24.14? Tell 172.22.26.91
202   08:44:06.646617 172.22.24.14             172.22.26.91        ARP       172.22.24.14 is at 00:60:e0:00:f8:b4
203   08:44:06.651361 Cisco_8c:ba:d1           01:40:96:ff:ff:c0   0x872d    Ethernet II
204   08:44:06.681842 172.22.25.76             172.22.24.14        ICMP      Echo (ping) request
205   08:44:06.681882 172.22.24.14             172.22.25.76        ICMP      Echo (ping) reply
206   08:44:07.457806 172.22.26.110            01:a0:f8:f0:f0:02   0x8781    Ethernet II
207   08:44:07.550249 AironetW_54:20:62        01:40:96:ff:ff:00   LLC       U, func=UI; SNAP, OUI 0x004096 (Unknown), PID 0x0000
208   08:44:07.616496 172.22.25.76             172.22.27.255       NBNS      Registration NB MDC-0654<00>
209   08:44:07.638535 172.22.25.76             224.0.0.22          IGMP      V3 Membership Report
210   08:44:07.669997 172.22.25.76             239.255.255.250     SSDP      M-SEARCH * HTTP/1.1
211   08:44:07.683047 172.22.25.76             172.22.24.14        ICMP      Echo (ping) request
212   08:44:07.683099 172.22.24.14             172.22.25.76        ICMP      Echo (ping) reply
213   08:44:08.241396 192.168.248.129          Broadcast           ARP       Who has 192.168.248.254? Tell 192.168.248.129
```

Figure 7.17: Preauthentication Monitor M1 Frames 188–213.

walking away from AP1 until the signal strength drops too low, and we begin the Discovery Delay. At this point, the STA associates with AP2. We see this Reassociation Delay in frames 203 to 215 in Figure 7.19.

Note the number of retransmitted frames (denoted by the letter "R" in the summary column) in Figures 7.19, 7.20, and 7.21, where sniffer M3 has captured the wireless traffic between the STA and the AP, which is the target of the roam. These retransmissions are likely due to a busy wireless environment resulting in increased interference and collisions. When a careful frame-by-frame analysis is performed, it is important to discount all but the final frame in a retransmission series, as the earlier transmissions were never received by the other end of the wireless hop. Applying this rule to frames 219 through 233 in Figures 7.19 and 7.20, we see that, unlike the tests described in Section 7.6.2, after associating with the new AP, there is no extensive EAPOL authentication here, which is because of the fact that the Authentication Delay was interleaved with the normal preroam association with AP1 and is thus separated from the actual roaming delays. The four-way EAPOL Key exchange is used to derive the PTK. It is possible to skip directly to this Key Management Delay phase as the PMK was already derived during the preroam Authentication Delay phase.

It is difficult to get a clear picture of what happens next in Figure 7.20 due to the fact that the wireless link is TKIP-encrypted following the four-way exchange. If we refer back to Figure 7.16, we see that the end of the authentication phase at frame 130 at time 08:44:11:151654 and the application pings *via the second AP* do not commence until frame 244 at time 08:44:43:643923 in Figure 7.18. This interval corresponds to the Key Management Delay followed by the Application Resumption Delay. At this point the roam is complete. It is important to note that the total number of frames exchanged between the STA and the AP2 in this preauthentication case is considerably less than that in normal 802.11i. The absolute time from reassociation to resumption of pings in the no-preauthentication case is the difference between frame 2529 (8:35:24:030368) in Figure 7.9 and frame 2584 (8:35:26:774851) in Figure 7.11, which is approximately 2.7 seconds.

The absolute time from reassociation for resumption of pings in the preauthentication case is the difference between frame 213 in Figure 7.19 (8:44:22.517366) and frame 525 in Figure 7.21 (8:44:33:604980), which is approximately 11.1 seconds. This observation seems counter-intuitive: If preauthentication is intended to reduce roaming latency, how is it possible that the observed results actually show larger values. The answer likely lies in the relative business of the wireless medium when the preauthentication tests were run and the large number of retransmissions that were observed. The fact remains that with preauthentication, fewer unique frames are required postroam before user traffic can resume.

The fact that with preauthentication, the Authentication Delay is incurred before the roam begins and clearly reduces the postroam frame count, as we have just seen. In the case of a

Chapter 7

No..	Time	Source	Destination	Protocol	Info
244	08:44:43.643923	172.22.25.76	172.22.24.14	ICMP	Echo (ping) request
245	08:44:43.644303	172.22.24.14	172.22.25.76	ICMP	Echo (ping) reply
246	08:44:44.425854	Cisco_8c:ba:d1	01:40:96:ff:ff:c0	0x872d	Ethernet II
247	08:44:44.641078	172.22.25.76	172.22.24.14	ICMP	Echo (ping) request
248	08:44:44.641311	172.22.24.14	172.22.25.76	ICMP	Echo (ping) reply
249	08:44:45.345563	AironetW_54:20:62	01:40:96:ff:ff:00	LLC	U, func=UI; SNAP, OUI 0x004096 (Unknown), PID 0x0000
250	08:44:45.568779	172.22.25.76	210.148.16.87	TCP	3652 > https [SYN] Seq=0 Ack=0 Win=16384 Len=0 MSS=1
251	08:44:45.569039	172.22.25.76	150.164.34.11	TCP	3653 > https [SYN] Seq=0 Ack=0 Win=16384 Len=0 MSS=1
252	08:44:45.569254	172.22.25.76	61.173.170.29	TCP	3654 > https [SYN] Seq=0 Ack=0 Win=16384 Len=0 MSS=1
253	08:44:45.569493	172.22.25.76	203.222.20.58	TCP	3655 > https [SYN] Seq=0 Ack=0 Win=16384 Len=0 MSS=1
254	08:44:45.569662	172.22.25.76	210.148.16.87	TCP	3656 > http [SYN] Seq=0 Ack=0 Win=16384 Len=0 MSS=14
255	08:44:45.569934	172.22.25.76	150.164.34.11	TCP	3657 > http [SYN] Seq=0 Ack=0 Win=16384 Len=0 MSS=14
256	08:44:45.570166	172.22.25.76	61.173.170.29	TCP	3658 > http [SYN] Seq=0 Ack=0 Win=16384 Len=0 MSS=14
257	08:44:45.572447	172.22.25.76	203.222.20.58	TCP	3659 > http [SYN] Seq=0 Ack=0 Win=16384 Len=0 MSS=14
258	08:44:45.641565	172.22.25.76	172.22.24.14	ICMP	Echo (ping) request
259	08:44:45.641793	172.22.24.14	172.22.25.76	ICMP	Echo (ping) reply
260	08:44:45.870534	172.22.25.76	62.219.55.96	TCP	3660 > 39645 [SYN] Seq=0 Ack=0 Win=16384 Len=0 MSS=1
261	08:44:45.871211	172.22.25.76	219.77.81.208	TCP	3661 > 37114 [SYN] Seq=0 Ack=0 Win=16384 Len=0 MSS=1
262	08:44:45.872024	172.22.25.76	212.179.13.42	TCP	3662 > 15307 [SYN] Seq=0 Ack=0 Win=16384 Len=0 MSS=1
263	08:44:45.872842	172.22.25.76	140.113.239.87	TCP	3663 > 12371 [SYN] Seq=0 Ack=0 Win=16384 Len=0 MSS=1
264	08:44:45.873616	172.22.25.76	222.164.150.112	TCP	3664 > 1273 [SYN] Seq=0 Ack=0 Win=16384 Len=0 MSS=14
265	08:44:45.874424	172.22.25.76	218.80.63.11	TCP	3665 > 51839 [SYN] Seq=0 Ack=0 Win=16384 Len=0 MSS=1
266	08:44:46.643320	172.22.25.76	172.22.24.14	ICMP	Echo (ping) request
267	08:44:46.643527	172.22.24.14	172.22.25.76	ICMP	Echo (ping) reply
268	08:44:47.380944	172.22.25.76	219.77.81.208	TCP	3666 > https [SYN] Seq=0 Ack=0 Win=16384 Len=0 MSS=1
269	08:44:47.381033	172.22.25.76	218.80.63.11	TCP	3667 > https [SYN] Seq=0 Ack=0 Win=16384 Len=0 MSS=1

Figure 7.18: Preauthentication Monitor M2 Frames 244–269.

Packet	Source	Destination	BSSID	Channel	Size	Absolute Time	Protocol	Summary
202	00:E0:B8:76:08:D8	00:03:7F:03:43:A7	00:E0:B8:76:08:D8	6	130	8:44:22.503954	802.11 Probe Rsp	FC=...R...., SN=3413, F
203	00:03:7F:03:43:A7	00:E0:B8:76:08:D8	00:E0:B8:76:08:D8	6	34	8:44:22.505018	802.11 Auth	FC=...R...., SN= 179, F
204	00:03:7F:03:43:A7	00:E0:B8:76:08:D8	00:E0:B8:76:08:D8	6	34	8:44:22.506815	802.11 Auth	FC=...R...., SN= 179, F
205	00:E0:B8:76:08:D8	00:03:7F:03:43:A7		6	14	8:44:22.507128	802.11 Ack	FC=.........
206	00:E0:B8:76:08:D8	00:03:7F:03:43:A7	00:E0:B8:76:08:D8	6	34	8:44:22.508403	802.11 Auth	FC=..., SN=3414, F
207	00:E0:B8:76:08:D8	00:03:7F:03:43:A7	00:E0:B8:76:08:D8	6	34	8:44:22.508987	802.11 Auth	FC=..., SN=3414, F
208	00:E0:B8:76:08:D8	00:03:7F:03:43:A7	00:E0:B8:76:08:D8	6	34	8:44:22.509558	802.11 Auth	FC=..., SN=3414, F
209	00:03:7F:03:43:A7	00:E0:B8:76:08:D8	00:E0:B8:76:08:D8	6	34	8:44:22.510447	802.11 Auth	FC=...R...., SN= 179, F
210	00:E0:B8:76:08:D8	00:03:7F:03:43:A7		6	14	8:44:22.510759	802.11 Ack	FC=.........
211	00:E0:B8:76:08:D8	00:03:7F:03:43:A7	00:E0:B8:76:08:D8	6	34	8:44:22.514585	802.11 Auth	FC=..., SN=3415, F
212	00:03:7F:03:43:A7	00:E0:B8:76:08:D8	00:E0:B8:76:08:D8	6	106	8:44:22.516256	802.11 Reassoc...	FC=...R...., SN= 180, F
213	00:03:7F:03:43:A7	00:E0:B8:76:08:D8	00:E0:B8:76:08:D8	6	106	8:44:22.517366	802.11 Reassoc...	FC=...R...., SN= 180, F
214	00:E0:B8:76:08:D8	00:03:7F:03:43:A7		6	14	8:44:22.517676	802.11 Ack	FC=.........
215	00:E0:B8:76:08:D8	00:03:7F:03:43:A7	00:E0:B8:76:08:D8	6	50	8:44:22.519151	802.11 Reassoc...	FC=..., SN=3416, F
216	00:03:7F:03:43:A7	FF:FF:FF:FF:FF:FF	00:E0:B8:76:08:D8	6	54	8:44:22.522696	802.11 WEP Data	FC=.F....W., SN=3417, F
217	00:03:7F:03:43:A7	FF:FF:FF:FF:FF:FF	00:E0:B8:76:08:D9	6	34	8:44:22.523310	LSAP-00	FC=.F...., SN=3418, F
218	00:03:7F:03:43:A7	FF:FF:FF:FF:FF:FF	00:01:F4:96:1B:4A	6	56	8:44:22.524889	802.11 WEP Data	FC=.F....W., SN=4085, F
219	00:E0:B8:76:08:D8	00:03:7F:03:43:A7	00:E0:B8:76:08:D8	6	157	8:44:22.527772	EAPOL-Key	FC=.F......, SN=3419, F
220	00:E0:B8:76:08:D8	00:03:7F:03:43:A7	00:E0:B8:76:08:D8	6	157	8:44:22.530084	EAPOL-Key	FC=.F.R...., SN=3419, F
221	00:E0:B8:76:08:D8	00:03:7F:03:43:A7	00:E0:B8:76:08:D8	6	157	8:44:22.532640	EAPOL-Key	FC=.F.R...., SN=3419, F
222	00:03:7F:03:43:A7	00:E0:B8:76:08:D8	00:E0:B8:76:08:D8	6	175	8:44:22.654763	EAPOL-Key	FC=T......, SN= 181, F
223	00:03:7F:03:43:A7	00:E0:B8:76:08:D8	00:E0:B8:76:08:D8	6	175	8:44:22.662421	EAPOL-Key	FC=T..R...., SN= 181, F
224	00:03:7F:03:43:A7	00:E0:B8:76:08:D8	00:E0:B8:76:08:D8	6	175	8:44:22.664279	EAPOL-Key	FC=T..R...., SN= 181, F
225	00:E0:B8:76:08:D8	00:03:7F:03:43:A7		6	14	8:44:22.664571	802.11 Ack	FC=.........
226	00:E0:B8:76:08:D8	00:03:7F:03:43:A7	00:E0:B8:76:08:D8	6	239	8:44:22.669344	EAPOL-Key	FC=.F.R...., SN=3424, F
227	00:E0:B8:76:08:D8	00:03:7F:03:43:A7	00:E0:B8:76:08:D8	6	239	8:44:22.671554	EAPOL-Key	FC=.F.R...., SN=3424, F

Figure 7.19: Preauthentication Monitor M3 Frames 202–227.

Packet	Source	Destination	BSSID	Channel	Size	Absolute Time	Protocol	Summary
228	00:E0:B8:76:08:D8	00:03:7F:03:43:A7	00:E0:B8:76:08:D8	6	239	8:44:22.676843	EAPOL-Key	FC=.F.R....,SN=3424,F
229	00:E0:B8:76:08:D8	00:03:7F:03:43:A7	00:E0:B8:76:08:D8	6	239	8:44:22.679062	EAPOL-Key	FC=.F.R....,SN=3424,F
230	00:E0:B8:76:08:D8	00:03:7F:03:43:A7	00:E0:B8:76:08:D8	6	239	8:44:22.681237	EAPOL-Key	FC=.F.R....,SN=3424,F
231	00:E0:B8:76:08:D8	00:03:7F:03:43:A7	00:E0:B8:76:08:D8	6	239	8:44:22.683439	EAPOL-Key	FC=.F.R....,SN=3424,F
232	00:E0:B8:76:08:D8	00:03:7F:03:43:A7	00:E0:B8:76:08:D8	6	239	8:44:22.685594	EAPOL-Key	FC=.F.R....,SN=3424,F
233	00:03:7F:03:43:A7	00:E0:B8:76:08:D8	00:E0:B8:76:08:D8	6	135	8:44:22.701884	EAPOL-Key	FC=T......,SN= 182,F
234	00:E0:B8:76:08:D8			6	14	8:44:22.702180	802.11 Ack	FC=........
235	00:03:7F:03:43:A7	FF:FF:FF:FF:FF:FF	00:E0:B8:76:08:D8	6	80	8:44:22.889627	802.11 WEP Data	FC=T..R..W.,SN= 183,F
236	00:E0:B8:76:08:D8			6	14	8:44:22.889788	802.11 Ack	FC=........
237	00:03:7F:03:43:A7	FF:FF:FF:FF:FF:FF	00:E0:B8:76:08:D8	6	84	8:44:22.890736	802.11 WEP Data	FC=.F....W.,SN=3433,F
238	00:03:7F:03:43:A7	FF:FF:FF:FF:FF:FF	00:01:F4:96:1B:4A	6	102	8:44:22.892145	802.11 WEP Data	FC=.F....W.,SN=4092,F
239	00:08:83:0C:88:00	00:03:7F:03:43:A7	00:E0:B8:76:08:D8	6	98	8:44:22.906507	802.11 WEP Data	FC=.F.R..W.,SN=3436,F
240	00:08:83:0C:88:00	00:03:7F:03:43:A7	00:E0:B8:76:08:D8	6	98	8:44:22.909254	802.11 WEP Data	FC=.F.R..W.,SN=3436,F
241	00:08:83:0C:88:00	00:03:7F:03:43:A7	00:E0:B8:76:08:D8	6	98	8:44:22.909777	802.11 WEP Data	FC=.F.R..W.,SN=3436,F
242	00:08:83:0C:88:00	00:03:7F:03:43:A7	00:E0:B8:76:08:D8	6	98	8:44:22.910509	802.11 WEP Data	FC=.F.R..W.,SN=3436,F
243	00:08:83:0C:88:00	00:03:7F:03:43:A7	00:E0:B8:76:08:D8	6	98	8:44:22.912509	802.11 WEP Data	FC=.F.R..W.,SN=3436,F
244	00:08:83:0C:88:00	00:03:7F:03:43:A7	00:E0:B8:76:08:D8	6	98	8:44:22.914286	802.11 WEP Data	FC=.F.R..W.,SN=3436,F
245	00:08:83:0C:88:00	00:03:7F:03:43:A7	00:E0:B8:76:08:D8	6	98	8:44:22.916294	802.11 WEP Data	FC=.F.R..W.,SN=3436,F
246	00:08:83:0C:88:00	00:03:7F:03:43:A7	00:E0:B8:76:08:D8	6	98	8:44:22.918356	802.11 WEP Data	FC=.F.R..W.,SN=3436,F
247	00:08:83:0C:88:00	00:03:7F:03:43:A7	00:E0:B8:76:08:D8	6	98	8:44:22.918750	802.11 WEP Data	FC=.F.R..W.,SN=3436,F
248	00:08:83:0C:88:00	00:03:7F:03:43:A7	00:E0:B8:76:08:D8	6	98	8:44:22.919111	802.11 WEP Data	FC=.F.R..W.,SN=3436,F
249	00:08:83:0C:88:00	00:03:7F:03:43:A7	00:E0:B8:76:08:D8	6	98	8:44:22.919636	802.11 WEP Data	FC=.F.R..W.,SN=3436,F
250	00:08:83:0C:88:00	00:03:7F:03:43:A7	00:E0:B8:76:08:D8	6	98	8:44:22.923211	802.11 WEP Data	FC=.F.R..W.,SN=3436,F
251	00:08:83:0C:88:00	00:03:7F:03:43:A7	00:E0:B8:76:08:D8	6	98	8:44:22.932660	802.11 WEP Data	FC=.F.R..W.,SN=3436,F
252	00:08:83:0C:88:00	00:03:7F:03:43:A7	00:E0:B8:76:08:D8	6	98	8:44:22.933291	802.11 WEP Data	FC=.F.R..W.,SN=3436,F
253	00:08:83:0C:88:00	00:03:7F:03:43:A7	00:E0:B8:76:08:D8	6	98	8:44:22.933653	802.11 WEP Data	FC=.F.R..W.,SN=3436,F

Figure 7.20: Preauthentication Monitor M3 Frames 238–253.

Packet	Source	Destination	BSSID	Channel	Size	Absolute Time	Protocol	Summary
525	00:03:7F:03:43:A7	00:60:E0:00:F8:B4	00:E0:B8:76:08:D8	6	112	8:44:33.604980	802.11 WEP Data	FC=T.....W.,SN= 261,F
526	00:E0:B8:76:08:D8	00:03:7F:03:43:A7		6	14	8:44:33.605609	802.11 Ack	FC=........
527	00:60:E0:00:F8:B4	00:03:7F:03:43:A7	00:E0:B8:76:08:D8	6	112	8:44:33.606939	802.11 WEP Data	FC=.F....W.,SN=3729,F
528	FF:FF:FF:FF:FF:FF	FF:FF:FF:FF:FF:FF	00:90:4B:34:B4:E0	6	102	8:44:33.630591	802.11 WEP Data	FC=.F....W.,SN=2614,F
529	00:03:7F:03:43:A7	00:E0:B8:76:08:D8	00:E0:B8:76:08:D8	6	28	8:44:33.762077	802.11 Data	FC=T...P..,SN= 262,F
530	00:E0:B8:76:08:D8	00:03:7F:03:43:A7		6	14	8:44:33.762103	802.11 Ack	FC=........
531	00:03:7F:03:43:A7	FF:FF:FF:FF:FF:FF	00:10:18:90:04:17	6	90	8:44:34.445061	802.11 WEP Data	FC=.F...DW.,SN=3779,F
532	00:03:7F:03:43:A7	00:60:E0:00:F8:B4	00:E0:B8:76:08:D8	6	112	8:44:34.602712	802.11 WEP Data	FC=T.....W.,SN= 263,F
533	00:E0:B8:76:08:D8	00:03:7F:03:43:A7		6	14	8:44:34.602750	802.11 Ack	FC=........
534	00:60:E0:00:F8:B4	00:03:7F:03:43:A7	00:E0:B8:76:08:D8	6	112	8:44:34.604709	802.11 WEP Data	
535	00:03:7F:03:43:A7	00:E0:B8:76:08:D8	00:E0:B8:76:08:D8	6	28	8:44:34.766359	802.11 Data	FC=T...P..,SN= 264,F
536	00:E0:B8:76:08:D8	00:03:7F:03:43:A7		6	14	8:44:34.766375	802.11 Ack	FC=........
537	00:03:7F:03:43:A7	00:08:83:0C:88:00		6	100	8:44:35.530463	802.11 WEP Data	FC=T.....W.,SN= 265,F
538	00:E0:B8:76:08:D8	00:03:7F:03:43:A7		6	14	8:44:35.530487	802.11 Ack	FC=........
539	00:03:7F:03:43:A7	00:08:83:0C:88:00	00:E0:B8:76:08:D8	6	100	8:44:35.530835	802.11 WEP Data	FC=T.....W.,SN= 266,F
540	00:E0:B8:76:08:D8	00:03:7F:03:43:A7		6	14	8:44:35.530856	802.11 Ack	FC=........
541	00:03:7F:03:43:A7	00:08:83:0C:88:00	00:E0:B8:76:08:D8	6	100	8:44:35.531042	802.11 WEP Data	FC=T.....W.,SN= 267,F
542	00:E0:B8:76:08:D8	00:03:7F:03:43:A7		6	14	8:44:35.531062	802.11 Ack	FC=........
543	00:03:7F:03:43:A7	00:08:83:0C:88:00	00:E0:B8:76:08:D8	6	100	8:44:35.531297	802.11 WEP Data	FC=T.....W.,SN= 268,F
544	00:E0:B8:76:08:D8	00:03:7F:03:43:A7		6	14	8:44:35.531316	802.11 Ack	FC=........
545	00:03:7F:03:43:A7	00:08:83:0C:88:00	00:E0:B8:76:08:D8	6	100	8:44:35.531494	802.11 WEP Data	FC=T.....W.,SN= 269,F
546	00:E0:B8:76:08:D8	00:03:7F:03:43:A7		6	14	8:44:35.531512	802.11 Ack	FC=........
547	00:03:7F:03:43:A7	00:08:83:0C:88:00	00:E0:B8:76:08:D8	6	100	8:44:35.531753	802.11 WEP Data	FC=T.....W.,SN= 270,F
548	00:E0:B8:76:08:D8	00:03:7F:03:43:A7		6	14	8:44:35.531773	802.11 Ack	FC=........
549	00:03:7F:03:43:A7	00:08:83:0C:88:00	00:E0:B8:76:08:D8	6	100	8:44:35.532006	802.11 WEP Data	FC=T.....W.,SN= 271,F
550	00:E0:B8:76:08:D8	00:03:7F:03:43:A7		6	14	8:44:35.532025	802.11 Ack	FC=........

Figure 7.21: Preauthentication Monitor M3 Frames 525–550.

pair of access points, this time saving occurs at a critical time of the roam at the expense of the slightly increased overhead incurred during the noncritical preroam phase. This saving is the crux of the argument in favor of preauthentication. In a real network having many access points, it is likely that the STA will find more than one roaming candidate during its scans. It is likely that the STA will attempt to preauthenticate with each of these candidate access points. The processing overhead on the RADIUS server per authentication is nontrivial; in the now-normal case that some TLS-related technology is used with a tunneled EAP method, this overhead is considerable. As the STA authenticates with many candidates and may not roam at all, there is a high chance of abuse of the RADIUS server's processing capabilities, possibly to a degree such that normal nonroaming users would have difficulty getting authenticated on the network.

7.8 Summary

In this chapter, we covered the 802.11 security staircase and the 802.11i standard's preauthentication. We then examined and analyzed a number of real-life secure roams. Our detailed analysis revealed where delays were occurring for the various security protocols. The results for WPA2-enterprise and preauthentication represent actual measurements taken with the test beds described. It would be overly pessimistic to imply that these results are indicative of the best that can be achieved with selected vendors' equipment, or some other tuning or enhancements designed to abbreviate some of the processes described here. One such shortcut is *Fast Session Resumption* which significantly reduces the number of EAP round trips to the AAA server. Further optimizations *beyond* the 802.11 standard itself will be discussed in the next chapter.

CHAPTER 8

Optimizing Beyond the 802.11 Standard

- 8.1 Introduction
- 8.2 Voice over Wireless IP Roaming
- 8.3 Opportunistic Key Caching
- 8.4 Centralized Wireless Switch Architectures
- 8.5 Summary
- References

8.1 Introduction

In this chapter, we will present the *Centralized Wireless Switch* architecture, which represents a paradigm shift in 802.11 networking. This new architecture, while not defined by the 802.11 standard, has created fertile ground for roaming enhancements that go beyond the evolution of 802.11 itself. We will discuss two such generic-roaming enhancements that are not part of the 802.11 standard but that which have been implemented by a number of vendors. They are *key caching* and *using tunnels to keep roams local*. However, before presenting these two topics, we will introduce the reader to the 802.11 application that has been most influential in forcing the industry to look for roaming performance superior to that currently offered by 802.11 implementations that are merely standards compliant: *Voice over Wireless IP*.

8.2 Voice over Wireless IP Roaming

8.2.1 Introduction

The most demanding application for 802.11 roaming that is currently practical and in widespread use is *Voice over IP (VoIP)*. Indeed, many of the enhancements discussed in this chapter were created either as stopgap measures to provide voice-capable roaming in the absence of standards, or, in some cases, to provide a capability for VoIP that is beyond what the 802.11 standards are expected to offer in the future. Our consideration of the vendor-specific solutions presented in this chapter will be against a backdrop of how they fare in different VoIP environments. In order to provide a context for these discussions, we first provide a brief primer on VoIP. The particular subset of VoIP running over 802.11 networks is important enough that new acronyms have begun to evolve to describe this subfield. These include *Voice over Wireless IP (VoWIP)* and *Voice over Wi-Fi* (denoted *VoWi-Fi* and also called *Voice over IP over Wi-Fi (VoFi)*).

8.2.2 Voice over Wireless IP Primer

In Figure 8.1, we show the basic components of an idealized VoWIP installation. A phone possesses not only the basic capability to transmit and receive voice signals but also control signals that may range from a simple collection of on-hook, off-hook, and ring signals, to a plethora of sophisticated calling features such as call waiting and caller ID. In an enterprise environment, these control signals occur as a dialog between the phone and the *Private Branch Exchange (PBX)*. The PBX provides a dial tone to the phone when it indicates a transition to an off-hook state. These traditional telephony concepts must map to equivalent functions in VoWIP. The control signals normally processed by a legacy PBX are provided by a *VoIP server* (sometimes called a *VoIP gateway*) or a *VoIP PBX*. In the interest of space, Figure 8.1 combines two different VoIP scenarios into a single diagram—the VoIP PBX scenario and the VoIP gateway with traditional PBX scenario.

If a VoIP call is to be made outside the enterprise LAN to another IP phone in another location in the enterprise, the VoIP server will know that the remote IP phone can be reached via the router shown connected to the Internet, as depicted in Figure 8.1. If the call is to a phone that is not an IP phone, then the call will be routed through the gateway shown in the figure. The conversion from IP to PSTN format will occur in the gateway. The normal digital or analog telephone line(s) that emanate from the gateway will either be connected to a traditional PBX, as shown in our diagram, or directly to the Public Switched Telephone Network (PSTN) if the enterprise has only VoIP phones. The VoIP server may also reside in a hybrid PBX, with traditional telephone lines connected to it as well. Such a hybrid PBX is

Optimizing Beyond the 802.11 Standard

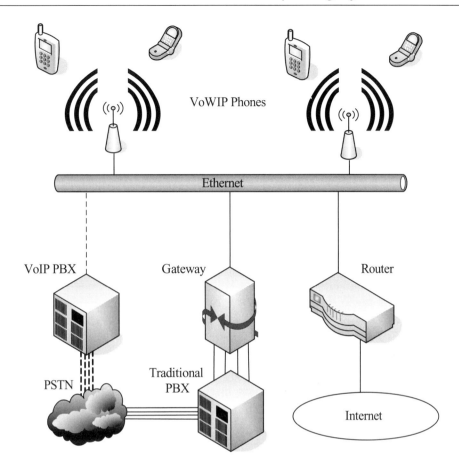

Figure 8.1: Generic VoIP Architecture.

shown as the VoIP PBX in Figure 8.1. In this case, the gateway functionality would also be built into the VoIP PBX. The dashed lines in the figure indicate that the VoIP PBX scenario is an alternative to the gateway scenario, as it would be unlikely that the two would exist in the same installation.

There are numerous commercial providers of VoIP equipment. This equipment includes many types of devices, including handsets, controllers, gateways, and VoIP PBXs. Major providers of VoIP equipment include Cisco, Nortel, Sonus, Tekelec, and Lucent. Alcatel and 3Com are also significant players. Avaya has a broad presence in many facets of the VoIP business due to its own strength in PBXs as well its relationships with companies such as Spectralink for Wi-Fi handsets, RIM for Blackberry devices, and Motorola for a dual-mode 802.11/GSM cellular handset that works with an IP PBX system. Companies that offer single-mode VoWIP product lines include Cisco, Broadvoice, Spectralink, and Vocera. Spectralink was a pioneer in VoWIP, and, as such, we will take a closer look at their technology in Section 8.2.5.

8.2.3 Voice over IP Protocols

There are a number of public and vendor-specific protocols in the VoIP family. One the most important family of protocols is the *ITU-T*'s (International Telecommunications Union Telecommunication Standardization Sector) H.323 recommendation. It was initially intended to provide a structure for generic multimedia traffic in LANs and has evolved to address many VoIP-specific requirements. H.323 prescribes that call establishment and tear down be accomplished via the H.225 and H.245 protocols. An alternative call setup/tear-down mechanism, intended by the IETF to replace H.323, is the *Session Initiation Protocol* (*SIP*) described in RFC 2543 [4] and RFC 3261 [7]. SIP is gradually overtaking H.323 as the dominant signaling protocol for VoIP. Detailed descriptions of the H.323 and SIP protocols are beyond the scope of this book. We refer the interested reader to [1] or [3] for more information.

The H.323 and SIP signaling protocols have the important task of identifying the IP addresses and port numbers that will be the endpoints of a VoIP call and then exchanging sufficient control information between those endpoints, so that VoIP data packets can begin to flow between them. Both H.323 and SIP specify the use of the IETF *Real-time Transport Protocol* (*RTP*) (RFC 3550 [8]) protocol as the layer-three transport protocol for the VoIP data packets themselves.

RTP is a relatively light-weight protocol, running on top of UDP. Unlike TCP, RTP supports no error-recovery mechanism, but it does support packet sequencing and time stamping. These characteristics lend themselves well to the support of VoIP, where some packet loss is tolerable but delays are not. The sequencing and time stamping of the RTP packets allows the receiver to reorder any out-of-order deliveries and to maintain de-jittering buffers that allow the smooth and timely delivery of the encoded-voice packets to the receiver. The exact packet rate for voice traffic will differ depending on the technology that is used, and it will range anywhere from 20 ms to 30 ms between voice packets.

Cisco's *Skinny Client Control Protocol* (*SCCP*) was developed as a vendor-specific but simpler alternative to the comparatively complex signaling of H.323. SCCP's "skinny" alternative to a generic "fat" H.323 client is accomplished by implementing the SCCP protocol in the VoIP phones themselves and by the use of an H.323 proxy server in the network infrastructure. Within the Cisco product nomenclature, this H.323 proxy is part of the *Cisco Call Manager*. The Cisco Call Manager processes the VoIP clients' call requests communicated via SCCP and translates them to H.323-compatible signaling for the generic VoIP infrastructure. While

SCCP is a Cisco technology, it has seen wide acceptance in the industry and is implemented by a number of VoIP vendors that interoperate with Cisco VoIP products.

8.2.4 Voice over Wi-Fi—VoIP's Newest Child

As of this writing, many VoWIP phones still offer limited roaming, QoS, and security features. The standards for these technologies lag behind the customer demands. While some of these phones do support more robust authentication, the most common identifier used for these VoWIP phones is still the MAC address, and, as such, this datum is the one on which authentication is most often based. As we mentioned earlier, MAC-based authentication is very weak, not the least because of the ease with which an MAC address can be spoofed. This weakness notwithstanding, these phones are being increasingly deployed, and, in most cases, the IT manager tolerates the weaker security in order to pass the benefits of VoWIP technology onto the users. As of this writing, VoWIP phones are usually 802.11b based, as the data rates for voice are so low that there is little justification for the higher speeds of 802.11g.

A new generation of VoWIP phones with improved battery life, security methods, and roaming performance is keenly awaited by the market and, as of this writing, a number of manufacturers have such phones available. *Softphones* consisting of a VoIP software application running on WPA-enterprise-enabled laptops or PDAs already have the most advanced security capabilities, unlike most dedicated VoWIP phones currently in use. Ironically, these very security capabilities adversely affect roaming performance.

One common approach to large 802.11 deployments of voice and data is to deploy the voice network on one channel and the data on another in order to reduce contention between voice and data, where nonprioritized voice would often suffer. In addition to mitigating contention problems for the VoWIP clients, this approach can also be used to impose different security policies for the voice and data services. A variant of this concept is where the physical AP offers multiple SSIDs on a single channel. One of the SSIDs can be dedicated to data applications, have the appropriate security measures in place, and map to a data-specific VLAN (with access through a firewall into the data LAN), whereas another SSID can be dedicated to voice and map to a different VLAN. In order to allow access for the security-poor VoWIP phones, the voice SSID can implement MAC-address security only, and force all traffic to go only to a voice controller, protecting the rest of the enterprise LAN from someone aspiring to attack the data network by pretending to be a voice device and spoofing the MAC address of a valid VoWIP phone. This approach does not address the contention problem for the voice service, but it does provide a hybrid approach to 802.11 security. For additional reading on VoWIP, we refer the reader to [6].

Chapter 8

8.2.5 Spectralink Voice Priority (SVP)

Spectralink was a pioneer in generating commercial enterprise-class voice over 802.11 solutions. They began in 1999 before any QoS-specific standardization work was being performed in the IEEE for 802.11. The company's goal was to provide untethered desk-set phone-equivalent service for enterprises, and, as such, they needed to provide very high-quality voice service, even in the presence of roaming. Indeed, the bar for roaming performance set by Spectralink in their certification program for infrastructure equipment, called the *Voice Interoperability Certification Program*, is higher than the service levels offered by the cellular phone providers in public networks. In order to provide this kind of call quality, Spectralink recognized that voice traffic would need very special treatment as compared to traditional data applications, which are more loss and delay tolerant. As no standards-based remedy was available at the time, Spectralink offered solutions to these voice-specific problems with their *Spectralink Voice Priority* (*SVP*) protocol.

In order to ensure that voice-quality goals are met, SVP specifies three areas for special treatment of voice packets in 802.11 networks. The first of these affects the manner in which access points transmit voice packets. The second is a call-admission mechanism that limits the number of simultaneous voice calls that an access point may support. The third is a *bunching* mechanism enforced by the SVP server (shown in Figure 8.2) that groups voice packets together before relaying them to the access point. It should be noted that like Figure 8.1, Figure 8.2 collapses both the traditional PBX and IP telephony server (VoIP PBX) scenarios into a single diagram, although it would be likely that only one or the other of these scenarios would exist in a single LAN.

A SVP-compliant access point must provide for priority queueing of voice packets over other data. The exact manner in which this queueing is done is left to the access-point manufacturer's discretion. SVP specifies the manner in which the voice-packet headers will be distinguished from nonvoice packets. The second SVP requirement on the access point is that when voice packets are waiting to be transmitted, the random-backoff algorithm, specified by 802.11's *Distributed Control Function* (*DCF*), is circumvented in a nonstandard manner such that the access point will force a backoff timer of zero in the case of a waiting voice packet. The intent is to ensure that in the case of a busy RF channel, in the battle for media contention, a retransmitted voice packet will always win against the other waiting transmitters on that channel. This approach would be dangerous if that BSS had other equipment contending for the channel, enforcing a similar nonrandom timer, but as Spectralink normally provides service in the more controlled IT environments of corporate enterprise, this situation can be avoided administratively.

The two rules just described provide absolute priority of voice packets over other data types. As SVP does provide absolute priority for voice packets at the expense of data packets, SVP

Optimizing Beyond the 802.11 Standard

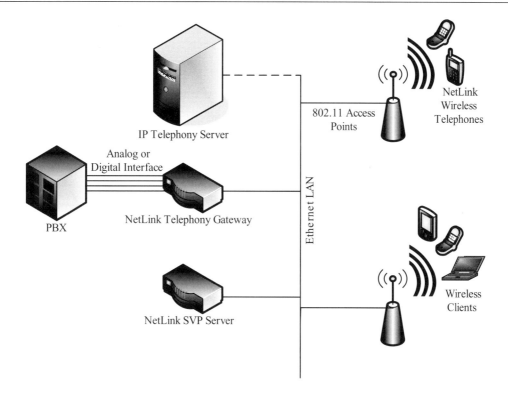

Figure 8.2: Spectralink Architecture.

stipulates that each access point provides a primitive *call-admission mechanism*, which enforces the maximum number of voice calls that may be active on an access point at any time. By limiting this number, the IT manager can effectively control the level of call quality that one wishes to guarantee to the VoWIP calls that are connected. Intuitively, it may appear that a second reason to limit the number of calls is to reserve bandwidth for nonvoice applications. However, this concern is not realistic as the bandwidth required for voice calls is negligible compared to that of most data applications.

One of the reasons that the maximum number of active voice calls should be kept small is that the DCF mutation prescribed by SVP can actually exacerbate contention for the channel as the overall number of voice calls across the multiple access points comprising the ESS increases. This fact is because the normal 802.11 DCF is designed to reduce post-collision contention through the random-backoff mechanism. While using a zero or otherwise artificially low backoff time to eliminate this entropy, one increases the likelihood of a post-contention collision between voice devices operating on different access points.

The SVP server depicted in Figure 8.2 coordinates the call with the IP telephony server (a VoIP server or VoIP PBX) and forces alignment of voice packets such that they are delivered to each access point in *bunches*. These bunches are organized to ensure that each individual voice stream's packets are delivered by the access point to the VoWIP phone within the tolerances required for Spectralink's service guarantees.

An additional aspect of SVP relates to how the access point handles 802.11 retries. Such retries are necessary when the 802.11 frame is not acknowledged. SVP requires that when there are multiple voice packets waiting to be transmitted, the access point does not "get stuck" retrying the packet at the head of the queue. Instead, other packets for other VoWIP phones will be transmitted. If multiple voice packets are waiting for retransmission, SVP stipulates that they receive round-robin service. As the wireless telephones may be highly mobile, this heuristic is important. Otherwise, during a roam, a packet waiting for transmission to a phone that has moved out of range could block the queue and delay transmission to phones that are still within the range of the access point. Such *head-of-the-queue blocking delay* is avoided by using SVP procedures.

While there has been widespread adoption of SVP by enterprise-class access-point manufacturers, Spectralink is committed to supporting public standards as they become available and broad enough to provide the level of support that their customers demand. The DCF modifications and priority queuing features of SVP have all now surfaced in similar forms in the IEEE 802.11e and *Wi-Fi Multimedia* (*WMM*) efforts.

Spectralink supports certification of WEP, WPA-PSK, and WPA2-PSK security modes in its certification *screen rooms*. A screen room is an RF-protected environment where certification tests can be performed in a controlled interference-free environment. A given vendor's equipment will only attain the Spectralink certification of a particular security mode if the certification criteria are met when the equipment is combined with Spectralink equipment in the controlled environment of the screen room. These test criteria are very stringent because the real-world environments of their enterprise customers will usually be more hostile to packet loss and delay than that encountered in the screen rooms. Of particular concern to us is the roaming aspect of the certification program, which requires that the roam be complete within 40 ms, corresponding to a voice-packet loss of no more than two packets. By adhering to this metric, the roam is accomplished without the user detecting the handoff in any way.

Spectralink acknowledges that the security modes currently supported are considered too weak for many enterprise environments, but considers the 802.1X-based WPA-enterprise security modes too protocol intense to meet their performance demands. As such, Spectralink is actively participating in the 802.11r work, which is expected to lessen the protocol overhead required for secure roaming. As of 2006, Spectralink's certification program was evaluating whether the intermediate step of Key Caching offered by some of its access-point partners represents an opportunity to provide WPA-enterprise level security and still meet

Spectralink's stringent performance goals without waiting for 802.11r. We examine Key Caching in the next section.

8.3 Opportunistic Key Caching

8.3.1 Introduction

In the earlier chapters, we presented and studied the IEEE 802.11i concept of preauthentication. The purpose of this complex mechanism is to place keying material in roaming candidate APs in advance of the roam itself. If we free ourselves from the bounds of the IEEE 802.11i standard and its laudable goal of wide interoperability, there are mechanisms that go beyond the standard to address this problem, which are simpler and sometimes more powerful than preauthentication. In particular, if a network implements a central management device controlling a *mobility domain* of a set of access points, it is possible for that central manager to make the 802.11i PMK available through each of the APs in that mobility domain. Such a centralized manager can be implemented via traditional *autonomous* APs using a vendor-specific inter-AP protocol, but it is the centralized wireless switch manufacturers that have been quickest to offer key-caching solutions. This solution is natural because with these architectures, the PMK is readily transferable to the *lightweight* APs under the control of the wireless switch. Indeed, in some centralized switch architectures, the PMK need never be sent to the AP at all as the encryption is performed entirely within the switch itself rather than in the AP. It should be noted that the term "fat" AP is sometimes used to denote an autonomous AP, and the term "thin" AP is often used to denote a lightweight AP. We will use the terms "autonomous" and "lightweight" in this chapter. We delve more deeply into the centralized wireless switch architecture in the next section.

It is worth noting that both the inter-AP protocol as well as the AP-to-wireless switch protocols could be standardized, but, to date, they have not been. The LWAPP and CAPWAP efforts that we will discuss in Section 8.4.3 are extant examples of such efforts. This fact is most likely because this area is where individual manufacturers believe that they can add value over and above the 802.11 standard and, hence, are reluctant to give away this advantage.

In whatever manner this key caching occurs, in order to benefit from it, the association exchange with an AP must denote that the AP possesses the PMK, which was obtained via one of these vendor-specific means, and thus the associating client understands that it can forego full authentication and that it may proceed directly to the four-way handshake which derives the PTK. This process is called *Opportunistic Key Caching* (*OKC*) as the process does not prescribe how the PMK reaches the target AP, just the manner in which the client will *opportunistically* take advantage of that situation.

OKC relies on a new 802.11i information element, PMKID, that was defined to support preauthentication. The 802.11i standard does mandate that the PMKID information element must be understood and correctly processed by compliant access points and STAs. The original intent in 802.11i was that the PMKID would identify a PMK that had arrived at the target AP, as a result of an earlier preauthentication. OKC is a matter of the client "guessing" that it is associated with a wireless switch, and therefore, instead of performing preauthentication with possible roaming candidate APs, it assumes on a roam that the target AP has knowledge of the current PMK. This information is encoded in the PMKID, which is sent in the Reassociation Request as an information element. If the PMK is known by the new AP, then authentication reduces to just the four-way handshake and the roam is fast; if the PMK is not known, then a full EAP reauthentication must take place. A wireless switch configured to do OKC will not advertise preauthentication capability; the absence of preauthentication stimulates the client to try OKC. The worst that can happen is that the OKC attempt will fail and full authentication will take place. In practice, this situation does not happen in an OKC-enabled mobility domain.

While OKC's performance equals that of preauthentication in terms of the reduced frame exchange at the time of the roam, it is far superior to preauthentication in terms of the authentication overhead on the AAA server. Recalling our discussion of preauthentication in Chapter 7, the preauthenticating client anticipates potential roams to a number of neighboring APs and completes full 802.1X-style authentications with each of them, including some to which the client may never roam. A common criticism of preauthentication is that it can actually increase the total authentication load on the AAA server as compared to performing a single full 802.11i authentication with each actual roam. OKC is superior to both of these in that there is a single 802.11i authentication when a client enters a mobility domain, and no further authentications as long as it roams within that domain. The only thing that would force a full reauthentication within that mobility domain would be the expiration of an AAA-session timeout, and these intervals are configurable by the IT manager and are typically quite long.

8.3.2 Cisco's Centralized Key Management (CCKM)

Cisco's Centralized Key Management (CCKM) is a key-caching technique that offers not only greater simplicity than preauthentication but a further reduction in roaming times as well as compared to generic OKC. CCKM technology has indeed proved to provide sufficiently low latency times, without sacrificing security, such that there are numerous implementations successfully using this technology for VoWIP roaming. A goal of CCKM was to decrease secure-roaming latency by minimizing the number of frames required for reauthentication following a roam. Cisco refers to this notion as "fast reauthentication" [2, 5].

We recall from Chapter 7 that the benefit of 802.11i preauthentication was to limit the post-roam frame exchange to the four-frame reassociation plus the four-frame key-exchange handshake, and this benefit was at the expense of considerable additional authentication traffic among the soon-to-roam client, multiple target APs, and the AAA server. While OKC greatly reduced this additional traffic, both preauthentication and OKC still require the four-way key exchange following reassociation in order to establish the new PTK and GTK. CCKM goes one step beyond this and reduces the post-roam association dialog to just the reassociation request and response frames, a reduction of four frames as compared to the 802.11i approach. The material which we will see from the 802.11r task group in Chapter 9 builds upon the CCKM model in achieving a more efficient standards-based post-roam frame exchange.

8.4 Centralized Wireless Switch Architectures

8.4.1 Introduction

The classical 802.11 deployment consists of a wired 802.3 switch with a number of intelligent APs connected to the switch providing the 802.11 radio access to the network infrastructure. In this traditional architecture, 802.11 intelligence is distributed among the intelligent APs. These APs are responsible for generating and processing 802.11 management frames, for maintaining and reporting frame statistics, and for handling security features, including authentication and encryption.

In the centralized architecture, some or all of these functions are moved back into a centralized wireless switch. Control of the association, authentication, and encryption processes may all be performed by the switch rather than by the AP. As mentioned in the previous section, the APs in this architecture are called *lightweight* APs, which, depending on the kind of *MAC Processing* used, may be little more than radios capable of 802.11 framing on one side and 802.3 framing on the other. The fact that the intelligence of scores of APs is concentrated in one switch is an enormous lever that can be used to the best advantage for *any* problem requiring coordination among APs. To be fair, such concentration can also present scaling challenges, but successful centralized switch manufacturers have proven that these are surmountable.

One of the best examples of such a problem requiring the coordination between APs is, of course, the hand-off of an STA from one lightweight AP to another AP during the course of a roam. This centralized architecture is attractive enough that a number of companies now offer 802.11 products based on this architecture. Indeed, the companies that make the boldest claims about fast-roaming support are those companies espousing this architecture; this greater speed follows directly from the simpler inter-AP coordination that the architecture allows. These products generally conform to the 802.11 standards on the radio side, although vendor-specific protocols are the norm for communication between the APs and the switch.

8.4.2 MAC Processing

Within the general centralized architecture family, there are very divergent architectures. One such area of divergence is in how much of the MAC layer is actually implemented on the AP.

This can range from all of the MAC layer, in the case of the traditional autonomous AP, to none of it, in the case of the pure-radio lightweight AP. When no MAC layer processing occurs on the AP, we call this a *local MAC* implementation. All packet processing, including QoS management and encryption-related functions, occurs only in the controller. As the APs are not involved in encryption at all, there simply is no need for the AP to possess the PMK ever, making OKC a natural fit for the local MAC architecture. Aruba is a prime example of a wireless switch manufacturer that uses local MAC processing.

In the *split MAC* architecture, a number of packet processing functions remain in the AP. These include frame encryption, handling beacon and probe response frames, frame acknowledgements, and the buffering and transmission of frames for STAs in power save mode. Functions that remain in the controller include 802.11 authentication and 802.11 association. In this kind of architecture, the keying information required for encryption needs to be pushed out to the APs from the controller, so split MAC architectures implement protocols to accomplish this in order for OKC to work. As of this writing, the market leader implementing Split MAC processing is Cisco, via its acquisition of Airespace in 2005.

Meru's *distributed MAC* architecture is best understood via a concept that they call a *virtual cell*. A virtual cell is not really a single cell in the normal 802.11 or cellular use of the term. As an example, if we have a virtual cell comprising ten APs, these ten APs will all operate on the same channel and have their RF characteristics constantly monitored and controlled by a centralized controller. The key characteristics of these APs are that they all have the same BSSID yet have ten independent, although coordinated, MAC layers. These MAC layers are coordinated in that once a STA is associated with one AP in the virtual cell, it may communicate with any of the APs in that cell. Any parameters related to security or QoS guarantees that were negotiated during the initial association are known to all the APs in that virtual cell. As all the APs share the same BSSID, the STA literally does not know to which of the member APs it is communicating because a frame that the STA transmits to that BSSID is actually directed to all the APs in that virtual cell. However, the member APs know which one of them is the one currently responsible for the communication with that client. This approach allows the system to "hand off" an STA from one AP to another without the STA perceiving that a roam has occured.

8.4.3 LWAPP, CAPWAP, and SLAPP

An early IETF effort to standardize a communication protocol between a wireless switch and a lightweight AP was led by the *Light Weight Access Point Protocol* (*LWAPP*) working group. LWAPP referred to the wireless switch as an *Access Router* (*AR*). While some vendors have built implementations around drafts of LWAPP, it was never ratified, and no real multi-vendor interoperability has resulted from this work. A more recent activity in the IETF to standardize such a protocol is, as of this writing, active under the auspices of the *Control and Provisioning of Wireless Access Points* (*CAPWAP*) *Working Group*. In this group, the wireless switch is referred to as the *Access Controller* (*AC*). CAPWAP is in its early stages and may face resistance from some of the manufacturers who have used the vendor-specific nature of that communication as a means of differentiating their products. CAPWAP has selected LWAPP as the basis for its standardization efforts.

Another effort, the *Simple Access Point Protocol* (*SLAPP*) was intended to be less ambitious and thus avoid some of the resistance faced by its predecessors [9], but this effort appears to have lost momentum, as of this writing. One of the reasons that these standardization efforts have been run in the IETF rather than the IEEE is the notion that the communication should be at layer three, so that the APs may be located more than a single network hop from the AC. The features covered by CAPWAP will offer real improvments in practical roaming performance across multiple vendors, so considerable importance is attached to this standardization effort.

8.4.4 Using Tunnels to Keep Roams Local

In the earlier chapters of this book, we emphasized that handoffs during a global roam will generally produce much greater latency than local roams. In fact, the expectation of the VoWIP industry is that calls *will* be dropped in the event of a global roam. Tunneling from the access points back to a centralized controller is one method that can greatly enhance the ability of an 802.11 network to shield the user from the adverse effects of roaming across subnets. No standard exists to allow this method in a multi-vendor environment, but a number of wireless switch vendors offer their own versions of this roaming enhancement.

In Figure 8.3, we show how tunneling can be used to turn global roams into local roams. In a straightforward enterprise 802.11 deployment, and in most of the examples provided thus far in this text, the IP subnet exposed to the 802.11 STA is the same subnet where the AP's wired interface is located. That is, from the point of view of the Ethernet switch to which the AP is connected, there are a number of MAC addresses that are reachable via the port on which the AP is present—one of which is the AP itself and the rest are the 802.11 clients that are currently associated with that AP. (This is not the case with some residential access point/routers, where the access point is actually built into a small router, but this situation is not the classical enterprise 802.11 access point where roaming is an issue.)

If we expand the traditional 802.11 model and create a tunnel from a centralized switch to the access point, it is possible to offer a totally different subnet (193.168.4.* in Figure 8.3) on the radio side of the access point independent from that of the wired subnet to which the access point is connected. Most importantly, it is possible to offer this same tunnel-based subnet

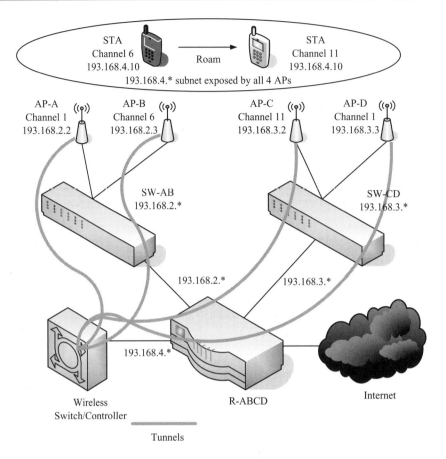

Figure 8.3: Local Roaming via Tunneling.

through different access points although they physically reside on distinct subnets. The tunnel may be layer-two or layer-three based. Most commonly, it is layer-three based as this allows the centralized controller to be more flexibly located with potentially one or more routers between it and the access points that it is managing.

In the example shown in Figure 8.3, the four access points reside on two distinct wired subnets: 193.168.2.* and 193.168.3.*. A roam from AP-B to AP-C would constitute a global roam were it not for the use of tunneling. By using the tunneling technique, an uninterrupted mobility domain of a third subnet, 193.168.4.*, is offered across the full set of four access points. The topology shown in Figure 8.3 is a realistic topology for a multibuilding campus that is offering VoWIP service with seamless roaming between adjacent buildings. In such a network, switch SW-AB would control the wired Ethernet access in one cluster of buildings, and the second switch SW-CD would control the subnet in an adjacent set of buildings. Multiple switches are linked via the router R-ABCD.

One downside of this approach is that the data path for the 802.11 clients' traffic is necessarily less direct than it would need to be in a straightforward deployment. In Figure 8.4, we show the convoluted data path for our client 193.168.4.10. The additional latency incurred by the extra hops through the wireless switch/controller is very small and not enough to perturb latency-sensitive traffic such as voice.

8.5 Summary

In this chapter, we have examined a number of methods that attempt to enhance the roaming experience for 802.11 users. Some of the technologies that we have covered in this chapter are accompanied by claims to have gone beyond the current standards while waiting for the ratification of IEEE 802.11r, or to have technology that addresses expected limitations of 802.11r.

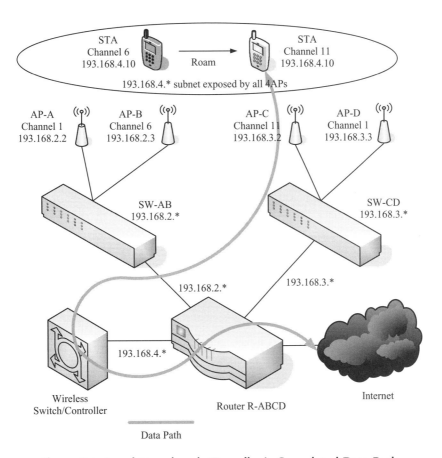

Figure 8.4: Local Roaming via Tunneling's Convoluted Data Path.

In all cases, the 802.11r standard represents a benchmark against which these vendor-specific solutions are measured. The 802.11r standard is the topic of our next chapter.

References

[1] Uyless Black, *Voice Over IP*, Prentice-Hall, 1999.

[2] Cisco Fast Secure Roaming Application Note, *Cisco Website [online]*, Internet, 2007.

[3] Daniel Collins, *Carrier Grade Voice Over IP*, McGraw-Hill, 2001.

[4] Mark Handley, Henning Schulzrinne, Eve Schooler, and Jonathan Rosenberg, SIP: Session Initiation Protocol, RFC 2543, Internet Engineering Task Force, 1999.

[5] Intermec, Cisco Compatible Extensions: Client Benefits of a Cisco WLAN, White Paper, Intermec Technologies Corporation, 2006.

[6] David Newman, Review: Voice over Wireless LAN, Network World, 2005.

[7] Jonathan Rosenberg, Henning Schulzrinne, Gonzalo Camarillo, Alan Johnston, Jon Peterson, Robert Sparks, Mark Handley, and Eve Schooler, SIP: Session Initiation Protocol, RFC 3261, Internet Engineering Task Force, 2002.

[8] Henning Schulzrinne, Stephen Casner, Ron Frederick, and Van Jacobson, RTP: A Transport Protocol for Real-Time Applications, RFC 3550, Internet Engineering Task Force, 2003.

[9] Techworld, *What's Behind the CAPWAP Flap? [online]*. Internet, 2007.

CHAPTER 9
The 802.11 Workgroups' Solutions for Fast Secure Roaming

- 9.1 Introduction
- 9.2 Overview of the 802.11r Standard
- 9.3 Detailed Concepts and Terminology of 802.11r
- 9.4 Protocol Exchanges in 802.11r
- 9.5 The 802.11k Standard Applied to Roaming
- 9.6 Concluding Remarks
- References

9.1 Introduction

There are two task groups under IEEE 802.11 whose efforts contribute strongly to improving the speed and security of the 802.11-roaming experience. The more obvious of these two is 802.11r, whose work is titled *Fast BSS Transition*; this task group is dedicated to fast secure roaming between access points within the same Extended Service Set. The work of the second group, 802.11k, is titled *Radio Resource Measurement*. While there are many applications of the measures and reports described by 802.11k that do not directly affect the roaming experience, there are significant potential roaming improvements that may be obtained through the use of 802.11k information. In particular, using 802.11k information, one can efficiently discover the best available target AP for a roam. This chapter will provide a tutorial for the complex mechanisms of 802.11r and a brief overview of the potential benefits that 802.11k may have for fast secure roaming in future.

9.2 Overview of the 802.11r Standard

The phrase "Fast BSS Transition" implies that the 802.11r standard attempts to define procedures and protocols that result in an 802.11 roam. In reality, 802.11r will be able to produce a *faster* security and QoS-sensitive BSS transition than would be possible without 802.11r. The standard does this by addressing the part of the delays in the hand-off procedures that lie in the areas of 802.11i security and 802.11e QoS. In this chapter, we will study how 802.11r proposes to reduce the roaming delays attributable to both of these factors. However, we should not lose sight of the fact that there are other components of roaming delay that are not addressed by 802.11r and will thus not be improved by it. In particular, the delays associated with detecting the need to roam and the selection of a target AP (Discovery Delay) as well as the resumption of downstream application data flow (Application Resumption Delay and Infrastructure Routing Delay) are all beyond the scope of 802.11r, yet contribute significantly to the roaming delays experienced by the user. As more time may be spent on these activities than is spent establishing the QoS and security contexts during the roam, the net roaming experience, even with full compliance with 802.11r, may still fall short of users' expectations.

The 802.11r standard defines many terms in an attempt to impose order on the chaos in the earlier discussion on 802.11 that centered around fast secure roaming. We will define a number of these new terms in the following sections, but for a complete reference we refer the reader to [1] of this chapter's references. Two of the most fundamental new terms are Mobility Domain (MD), which we first introduced in Chapter 4, and *first contact*. First contact for a given end-user station is defined to be the initial association of the STA with an AP in that MD. The procedures for this initial association are different from that for STA's subsequent association in the MD. These subsequent associations are the Fast BSS Transitions, which are the focus of 802.11r.

The bulk of 802.11r relates to reducing the security overhead during BSS transitions. The two most obvious security-related benefits are clarification of how the opportunistic key caching is accomplished and the elimination of the four-way handshake that traditionally followed reassociation. With regard to the postassociation four-way exchange, this elimination is accomplished by overloading the four-frame Authentication-Association exchange with new IEs that contain this information. In the case of OKC, the 802.11i standard did not stipulate the means by which the target AP obtained the keying information that the STA and AP developed during first-contact procedures. Instead, the term *opportunistic* was used because the first-contact keying information somehow "magically" reaches the target AP. (The term *magically* is commonly used in industry in relation to OKC.)

Opportunistic key caching assumed that the necessary keying information would be made available to the target AP either by some vendor-specific inter-AP protocol or, in the case of a wireless switch architecture, due to the fact that the four-way exchange was actually centrally controlled for all the access points in the ESS and that centralized controller was in possession of the PMK all along. The fact that the process was not dictated by the standard resulted in an inconsistent application of OKC. The 802.11r standard addresses this problem by defining a new key hierarchy and the concept of MDs as discussed subsequently.

A group of APs jointly form a single MD and in doing so gain access to a common key hierarchy. This can occur easily if the MD is designed to consist of a set of APs connected to a centralized controller or a set of controllers from the same manufacturer, but the specification of these mechanisms is beyond the scope of 802.11r. Such a centralized controller is sometimes referred to as the *Mobility Domain Controller* (*MDC*) in the 802.11r context. A centralized wireless switch fits naturally into this role.

During a roam, if the target AP advertises membership in the desired MD, the client may associate with it with the hope that the needed key hierarchy is present. It should be noted that 802.11r does not guarantee that the needed key hierarchy be present when the roam occurs, in which case unexpected latency may occur wherein the new AP needs to collect this information from its holder. When the STA performs its initial association and full authentication in that MD, it gains access to the 802.11r key hierarchy. This key hierarchy contains PMK-R0 material that is related to the 802.11i PMK. From this information a PMK-R1 key is derived, which is specific to that AP-STA pairing. The important fact is that a target AP in the same Mobility Domain can independently derive a new PMK-R1 equal to a new PMK-R1 derived by the client attempting to associate with the AP based on the client's BSSID and possession of the key hierarchy. This process can occur without resorting to the RADIUS server or any other external device, presuming that the AP has the key hierarchy.

The exact manner in which the target AP obtains the keying information is implementation dependent. As of this writing, no one anticipates the IEEE standardizing the inter-AP protocol. Such an effort would specify how this information could be exchanged between different manufacturers' APs. Indeed, the already formalized IEEE-802.11f work, targeted to standardize inter-AP protocols, has been *rescinded*. Because of this, as of this writing, it seems unlikely that IEEE standardization of such inter-AP protocols will be realized in the near future. It seems more likely that any related standardization will occur through the work of the IETF CAPWAP group, which focuses on communication between APs and the infrastructure switches that connect them. Interestingly, while we will see in the following sections that 802.11r defines roles and relationships in the key-exchange process, there remain enough unspecified functions (see [9]) that true vendor interoperability for Fast BSS Transitions is likely to remain an illusive goal.

A less significant portion of the 802.11r standard addresses the issue of reducing QoS overhead during a BSS transition. The standard does so by eliminating the two-frame 802.11e overhead related to call admission that would occur after the association and key exchange upon a roam. This two-frame removal is accomplished by overloading the effect of the 802.11e ADDTS into new IEs in the Authentication-Association exchange. While the elimination of these two call-admission frames has potential benefits as use of 802.11e expands in the future, its short-term benefits are likely to be more limited, which is attributed to the fact that the parts of 802.11e that are appearing in early commercial implementations (such as the Wi-Fi Alliance's WMM-compliant equipment) relate to traffic prioritization and novel non-DCF backoff mechanisms, rather than call admission.

The net effect of the combined security and QoS enhancements in 802.11r is that what could have been a $4+4+2$ frame exchange, even in the optimized case of OKC, is collapsed down to a four-frame exchange. While these remaining four frames grow in size as a result of the 802.11r extensions, the reduction in frame count is expected to reduce roaming latency. The benefits of the reduced frame count are multiplied by the elimination of the additional inter-frame spacing and frame acknowledgments that accompanied the now-eliminated frames.

9.3 Detailed Concepts and Terminology of 802.11r

9.3.1 Introduction

Figure 9.1 shows three different axes illustrating some of the facets of the 802.11r standard. All permutations along these axes are permitted by 802.11r. First, in the realm of security, three distinct paradigms need to be supported: no 802.11i security, 802.11i PSK mode, and

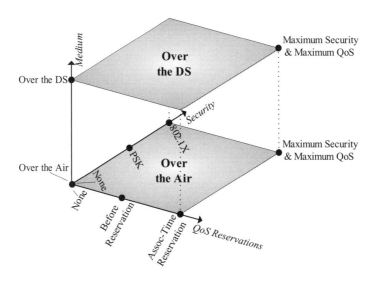

Figure 9.1: Prehandoff Possibilities Contemplated by the 802.11r Standard.

full 802.1X authentication. As for QoS, there may be no QoS reservation support, support for reservations prior to the handoff, and support for reservations during the handoff. Second, for any of the 802.11r exchanges that are permitted prior to association with the target AP, there are two possible media for these exchanges: Over the Air (*OTA*) or *Over the DS* (*ODS*). Both of these refer to dialogs between the STA and the target AP performed prior to association with the new AP.

The Over-the-Air dialog is a direct wireless communication between the STA and the target AP. As the only management frames that may be exchanged in this unassociated state are 802.11 authentication frames, the OTA dialog is entirely performed by overloading the desired information on 802.11 open-authentication frames without necessarily following them by the normal association request and response. The STA can remain associated with the current AP and still directly exchange information with the target via these open-authentication frames.

The other such preassociation exchanges between the STA and the target are those that were first described under 802.11i preauthentication. In this case, the dialog occurs in two hops. The first is the wireless hop to the current AP, and the next is done over the (normally wired) distribution service in the infrastructure that links the current and the target AP. The 802.11r ODS option is based on this precedent. In contrast to the OTA in the ODS case, the information is not placed in 802.11 authentication frames. Instead, a new *Frame Type* (*FT*) is created for 802.11r. This frame is called the *FT Action Request* frame. The FT Action Request frames are transmitted from the STA to the current AP and are relayed by the current AP to the target AP. These are two separate paradigms that different manufacturers may use and still remain compliant with 802.11r.

It is important to appreciate that OTA and ODS represent two options offering similar functionality. For that reason, if we refer back to Figure 9.1, we note that regardless of whether OTA or ODS is selected, all the permutations of security and QoS are still available within 802.11r. OTA and ODS represent two different conduits for the same set of information exchanges. The messages transferred over both the OTA and ODS paths are actually a harmonized end-to-end protocol exchange between peer SMEs. One reason that an implementation would choose ODS mode rather than OTA is to avoid delays resulting from switching channels to transmit the authentication frames in OTA mode.

As stated earlier, ODS mode retains some similarities to 802.11i preauthentication in that in both cases there is a dialog between the STA and the target AP via the current AP—however, the similarity ends there. The 802.11i standard preauthentication essentially duplicated the full 802.1X authentication protocol with the target AP, including the exchange with the AAA server. Preauthentication has long been criticized as unable to scale due to the potential overload of the AAA server from the many full 802.1X authentications that result from the roaming user in this mode. While 802.11i preauthentication would forward authentication

Chapter 9

requests to candidate target APs, it failed to specify precise mechanisms for this forwarding. The 802.11r standard specifies that the ODS relaying function be performed by the *Remote Request Broker* (*RRB*) in the current AP. The RRB is part of the SME function. Next, we will cover the most important of the 802.11r concepts and new terminology, and then, in Section 9.4, we will discuss eight distinct examples of how 802.11r achieves the Fast(er) BSS Transitions.

9.3.2 Architectural Elements of 802.11r

Figures 9.2, 9.3, and 9.4 show 802.11r-specific architectural components. Most of these components are not new. Part of the work of 802.11r has been to define these components and their relationships more precisely. One of the reasons for this effort is that earlier standards in the 802.11 family assumed that a particular functional component resided on a particular hardware component of the 802.11 architecture. As the hardware complexity has grown with 802.11r and the traditional fat-AP architecture has been blurred by a variety of wireless switch architectures, it became imperative to define functions precisely, allowing for the possibility that they may be implemented on different hardware components by different manufacturers.

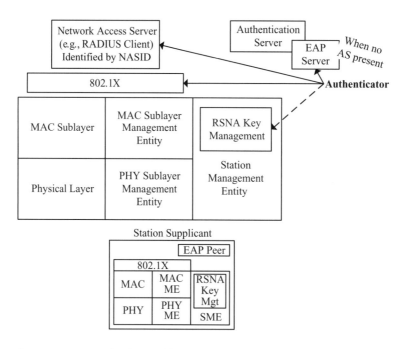

Figure 9.2: The 802.11r Standard's Architectural Entities Introduced. (Courtesy of the IEEE.)

240

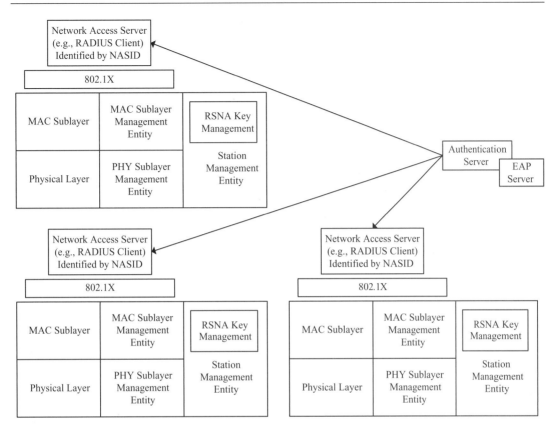

Figure 9.3: The 802.11r Standard's Entities in a Fat-AP Architecture. (Courtesy of the IEEE.)

The biggest dichotomy lies between the traditional fat-AP architecture which is shown in Figure 9.3 and the generic wireless switch architecture which is shown in Figure 9.4. Within the general wireless switch architecture shown in Figure 9.4, there can be many manufacturer-specific variants. For example, in some cases, the MAC entity exists entirely in the thin AP, whereas in others the MAC is actually split between the AP and the controller itself. We see the SME in all three figures. Inside the SME lies the *Robust Security Network Association* (*RSNA*) key management function. RSNA key management is the new terminology for what had been referred to as 802.1X key management in the early 802.11 security-related work. This key management, along with 802.1X port control on the AP, comprised the authenticator function described in those earlier works. The fact that these two functions may be on separate network devices (for example, the MDC and the AP) makes it important for 802.11r to define them clearly as distinct entities. More details on the components of RSNA Key Management are shown in Figure 9.5 and discussed in the following section.

Chapter 9

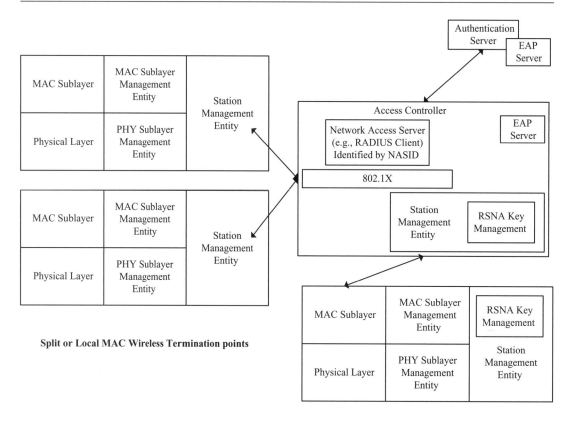

Figure 9.4: The 802.11r Standard's Entities in a Wireless Switch Architecture. (Courtesy of the IEEE.)

9.3.3 New Security Concepts

We recall from Chapter 4 that 802.11i defined a concept called the Pairwise Master Key (PMK) that was used to generate the dynamic session key called the Pairwise Transient Key (PTK). While the fundamental concepts surrounding these keys is not changed by 802.11r, there are refinements and new terminology relative to them. We direct the reader's attention to Figure 9.6.

This figure illustrates two fundamental shifts from the basic 802.11i key hierarchy. Similar to the 802.11i standard, the hierarchy begins with a key that is derived during the full 802.1X negotiation or, in the case of PSK mode, statically configured into the STA and AP. The key derived from the full 802.1X authentication is called the *Master Session Key* (*MSK*). In the case of Pre-Shared Keying (PSK) mode, the term PSK is used. In both cases, however, Figure 9.6 shows that this top-of-hierarchy key is not used directly to derive the PTK, which was essentially the case in 802.11i.

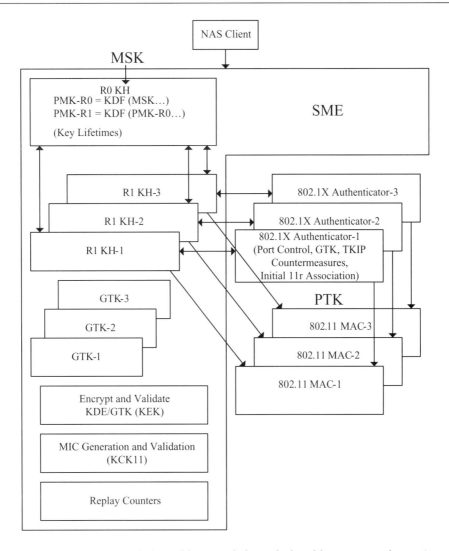

Figure 9.5: The 802.11r Keys, Their Holders, and the Relationships among Them. (Courtesy of the IEEE.)

Figure 9.6 shows that from the top-of-hierarchy, another level of the PMK is created. In the 802.11r standard, this key is called the *PMK-R0* and is distinguished from the PMK actually used to derive the PTKs. The PTKs are derived from a second level of the PMK called *PMK-R1*. As Figure 9.6 illustrates, multiple PMK-R1s will exist in a given MD—one for each BSSID. As there are now multiple PMKs in the domain, a new concept called *key holder* is created. A key holder of a PMK-R0 is called a *R0KH* and the key holder of a PMK-R1 is called a *R1KH*.

Chapter 9

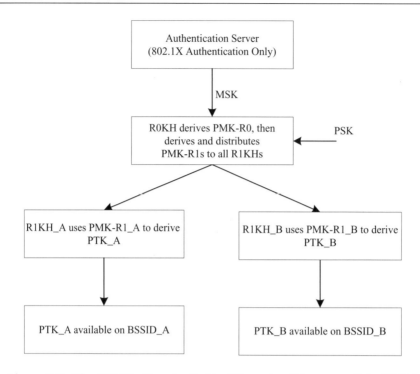

Figure 9.6: The 802.11r Standard's Key Hierarchy. (Courtesy of the IEEE.)

The 802.11r protocols themselves actually pass around identifiers of the PMKs and the key holders. These identifiers are called *PMKR0Name*, *PMKR1Name*, *R0KH-ID*, and *R1KH-ID*. The R0KH derives a PMK-R1 for each of the R1KHs in the MD and may eventually distribute them. For wireless switch architectures, it would be convenient if the PMK-R1 were the same for all the APs coming from the controller, but this creates security vulnerabilities. The reason that the PMK-R1 notion was created as distinct from the PMK-R0 was so that it could be AP-specific, and it must remain so to avoid a security weakness.

As an illustrative example, in a wireless switch architecture, the R0KH might reside in the wireless controller and the R1KHs might exist in the APs comprising the MD of that controller. The 802.11r standard uses concepts and terminology that allow its application to the full spectrum of wireless architectures available, including controller-free fat-AP architectures, thin APs that are in total control of the 802.11i processes, thin APs that defer the 802.11i processes to the controller, and MDs that span multiple wireless controllers, to name but a few.

As indicated earlier, Figure 9.5 graphically shows these 802.11r keys, their holders, and the relationships among them. Insights are provided into the data flow of how the keys move from one entity to another. The figure shows the more detailed components of RSNA key management that reside within the SME. It should be noted that the figure shows only the keys in the network infrastructure and does not show their counterparts in the STA. The term "authenticator" as used in 802.11i could encompass all four of the 802.11r functions consisting of the NAS client (NAS-ID), 802.1X port control (802.1X authenticator), the PMK-R0 Key Holder (R0KH), and the PMK-R1 Key Holder (R1KH).

This complexity can be a source of confusion, so the specific terms are used in 802.11r rather than the more general term "authenticator." In some implementations, these entities may exist on the same device and, in others, on separate devices. The R0KH will reside on the same physical device where the 802.1X-authenticator NAS-client functionality resides. The R0KH-ID is defined to be the 48 octet NAS ID of this NAS client, as defined by RFC 2865. The R1KH-ID is the six-octet MAC address of the physical entity that is the authorized holder of the PMK-R1 key (see [6]). The 802.11r-specific functions performed by each of these are as follows: NAS-ID receives the MSK from the Authentication Server, derives the PMK-R0 from the MSK, and transfers that PMK-R0 to the R0KH. The R0KH accepts that PMK-R0 and derives the multiple PMK-R1s from it and transfers the PMK-R1s to their respective R1KHs. (The mechanism by which the R0KH transfers the various PMK-R1s to their respective R1KHs is not specified by 802.11r.) The R1KH accepts that PMK-R1 and uses it during the four-way handshake with the STA, thus deriving the PTK on the AP side of the reassociation. All this happens within the SME(s), as shown in Figure 9.5.

The PTK is then transferred to the MAC entity. This transfer is frequently referred to as the *plumbing of the key* from the SME to the MAC entity. This term is used to imply the simple transfer of a complex object (the key) from its generator (SME) to the place where it is used (MAC). It should be noted that as the PTK is produced by the four-way exchange that occurs at reassociation time, the PTKs are *not* simultaneously alive at the different APs in the MD, although Figure 9.5 may suggest that is the situation.

The STA is able to derive the PMK-R1 specific to the target AP based on the information exchanged during the reassociation. This fact is because the STA remains in possession of the PMK-R0 information from its first-contact association. This complex infrastructure ensures that any STA-AP pairing in the MD will be able to derive unique PTKs using only the limited frame exchange allowed by 802.11r.

9.3.4 Resource Reservations

The resource-reservation procedures of 802.11e that we introduced in Chapter 4 use an ADDTS ("Add TSPEC") request following the association and security exchanges to request reservation of the resource specified in that TSPEC. Just as 802.11r collapses the 802.11i security exchanges into the authentication and reassociation messages, the 802.11r standard provides the ability to incorporate these TSPEC resource requests into this same exchange.

The protocol diagrams for QoS-sensitive roams shown in Section 9.4.4 illustrate that full resource reservation in Fast BSS Transitions requires a six-message exchange rather than the four-message exchange that suffices when no QoS is involved. This exchange is accomplished by adding two new open-authentication frame types, *authentication-ACK* and *authentication-Confirm*, for the third and fourth messages of the exchange. A given AP in a MD may require that the full six-message *FT reservation protocol* be used. Alternatively, it may provide the option to use the simpler four-message base exchange, placing the TSPEC request only in the reassociation frames. In the case of the four-message exchange, the STA does not know whether the reservation has succeeded or not until it has made the commitment of the reassociation itself.

The crux of the 802.11r reservation issue is that it takes four messages to know whether or not a reservation request will be granted by the target AP. If the STA needs assurance that it will have the needed resources upon association, then a four-message exchange must be completed prior to the reassociation itself, hence, a total of six messages for a true guarantee. We remind the reader that until the reassociation is begun, the STA may remain in active communication with the current AP and thus not have materially disturbed the ongoing user-data flow despite reserving resources at the target. It should be noted that in all cases, the AP's denial of the reservation request does not cause an otherwise successful reassociation attempt to fail. The STA always has the option of deciding whether or not to continue with the reassociation, despite the lack of short-term QoS guarantees.

Even with the full six-message FT reservation protocol, the AP may deny the reservation request in the fourth message. In this case, the STA has a number of options. The STA may temporarily delay the roam and try to reserve fewer resources at the same AP in the hope that a more modest request might be approved. The STA may abandon this AP and try to acquire the resource guarantees that it requires with another candidate. Finally, it may let the current reassociation complete without any resource reservations and issue an 802.11e ADDTS, following this reassociation with a modified resource request.

It is important to understand that with the six-message FT reservation protocol, the resources have been reserved at the end of the fourth message, but it is not necessary that the STA complete the roam immediately. As is shown in the figures of Section 9.4.4, this interval is limited by the deadline communicated by the AP in the response that it sends to the STA in the fourth message of the exchange. The AP imposes this deadline to prevent the situation where an aggressive STA wastes resources by reserving them long before the STA performs the reassociation and actually begins to use them.

There is an optional 802.11r procedure whereby the STA can ascertain the availability of resources without actually reserving them (see [11]). This resource *query* uses a *Resource*

Information Container (*RIC*) request similar to the one used for resource *reservation*, but no resources are reserved. This option eliminates the need for a deadline by the AP, as the request by the STA was only informational. This option can be very useful for a STA that will choose among alternate candidate target APs based on their responses to the query. The STA will only initiate the actual Fast BSS Transition procedures with an AP that has indicated that it currently has sufficient resources for the request that the STA is about to make.

The 802.11r standard permits the STA to execute the resource-reservation procedures up to but not including the reassociation exchange with multiple target APs in parallel. The STA may only associate with one, of course, as per the requirements of the base 802.11 specification. Any other reservations made with APs with which the STA did not associate with will expire after the reassociation-deadline time has elapsed. This process implies a potential waste due to oversubscription of resources, but only for a duration of reassociation-deadline time units. In the case that the reservation is denied, the AP may include in its response the resources that *could* be granted. This information can be used in a subsequent reservation request.

9.3.5 Information Elements

There are four principal information elements that are used by 802.11r to overload roaming, security, and QoS information onto existing 802.11 management frames. These are as follows:

- The *Fast Transition Information Element* (*FTIE*)

- The *Mobility Domain Information Element* (*MDIE*)

- The *Robust Secure Networks Information Element* (*RSNIE*)

- The *RIC Data Information Element* (*RDIE*) for resource reservations

Another, more minor IE that plays a supporting role in Fast BSS Transition is the *Timeout Information Element* (*TIE*). It should be noted that this acronym is different than "TIE" in "FTIE."

The Mobility Domain Identifier identifies a set of APs within the ESS that constitute a single Mobility Domain within which Fast BSS Transitions are supported. It should be noted that in general practice it is assumed that the MD will be the entire set of APs in an ESS. The MDIE is present in many 802.11r-related frames flowing between an AP and an STA. The role of the MDIE is to describe the capabilities of a given MD to which an AP belongs and an STA may choose to join. Regardless of the frame in which the MDIE appears, it should have the same value advertised by the AP in its beacon and/or probe-response frames, and they would normally be the same values for all the APs in that MD.

The MDIE contains the Mobility Domain Identifier, and the *Fast BSS Transition Capability* and *resource policy* value. The latter value includes only four bits; they describe the support of four basic features: the Fast BSS Transition Capability, Reservation Over the Air, Reservation Over the DS, and the Reserve Option. The Fast BSS Transition Capability bit will be set for an MD that supports Fast BSS Transitions, which would be the normal case for network equipment that has implemented 802.11r. The next two bits, *Reservation Over the Air* and *Reservation Over the DS*, indicate if reservations are permitted over each of those media, respectively. It should be noted that if both bits are set to zero then reservation requests should not be sent to the AP by an STA at all. When the final bit, Reserve Option, is set, the STA should first perform any QoS reservations prior to the reassociation. This setting dictates that the full FT reservation procedures must be used if any resources are to be reserved. This corresponds to a six-message reassociation sequence (as shown in the figures of Section 9.4.4). When the Reserve Option bit is not set, an STA has the option to use either the four-message base reassociation procedures or the full "reserve before associate" six-message procedures. Setting the Reserve Option bit would be normal for an AP whose support for QoS reservations could not be done "on the fly" because the AP requires prior processing of the requests.

The FTIE contains fields that permit the former 802.11i four-way EAPOL Key Exchange to be collapsed into the now multipurpose authentication and reassociation frames. As we will see in Section 9.4, the FTIE appears in a number of different messages defined by 802.11r procedures. It is a variable-length IE and may contain different pieces of information at various stages of the exchanges. The FTIE advertised in the beacon is its most simple form and only contains the BSS Transition Resource Value. In general, the closer to the end of a Fast BSS Transition in which the FTIE is used, the more information the FTIE contains. This information includes the security-related MIC, ANonce, and SNonce fields that we presented in Chapter 6. The FTIE may also contain the identity of the PMK-R0 Key Holder. Depending on the stage of the exchange, the FTIE might include the ID of the PMK-R1 Key Holder and the GTK. The PMK-R0 and PMK-R1 Key Holders were described in Section 9.3.3. Similar to the MIC, ANonce, and SNonce, the GTK is a familiar 802.11i concept presented earlier, which now finds a new carrier in the form of the FTIE.

The beacon and probe responses are not part of the Fast BSS Transition association procedures described by 802.11r, but they serve the important role of advertising the Fast BSS Transition capabilities of the MD of the AP that transmits them. Under the 802.11r standard, these frames will include both the MDIE and the FTIE. Early 802.11r work included additional security-related information in the FTIE used in the beacons and probe responses. This extra material created a *beacon bloat* problem, where the transmission medium would be increasingly occupied by what was originally intended to be simple beacon messages. As these messages are transmitted continuously, they represent a kind of necessary background tax that should be kept as small as possible.

The RSNIE draws it heritage from 802.11i and is re-purposed in new frames by 802.11r in order to support Fast BSS Transitions. Its original purpose was to advertise the security capabilities of an AP and an STA to each other during the association process. If the security levels were acceptable to each side, subsequent authentication would be performed in accordance with those negotiated bounds. The Opportunistic Key Caching discussed in Chapter 8 relied upon a PMKID field, which is also present in the RSNIE. Similar to OKC, the 802.11r standard exploits the PMKID field to communicate from an STA to an AP, the key that the STA wishes to use for the new association. The 802.11r standard improves upon OKC by overloading the four-way exchange, which is still required by OKC, into multipurpose open-authentication and association frames, thus reducing the roaming frame count.

The TIE is used to communicate various time intervals related to Fast BSS Transitions. There are currently two different time interval types defined for the TIE. These are the *Reassociation Deadline Interval* and the *Key Lifetime Interval*. The TIE is used only within the third message of the initial association procedures (see [7]). The reassociation-deadline timer represents the maximum allowable time that a STA may wait to reassociate after receipt of the authentication ACK message (the fourth message). After that timer expires, the STA must abandon the transition attempt. The Key Lifetime Interval specifies the maximum lifetime of the PTKs derived during transitions within that MD.

Resource-reservation requests are grouped in a RIC consisting of one or more data Information Elements, where each RDIE corresponds to a single resource request. Each resource request consists of an RDIE followed by one or more resource descriptors corresponding to the request. Although there may be multiple descriptors per resource, only one resource is reserved by a given RDIE. The RDIE allows the parsing of the container by including a length field for its resource descriptors. This parsing process is shown in Figure 9.7. The RDIE also includes a status bit, which is used in the response to the request. The value of this bit indicates whether the request was accepted or denied.

The resource descriptor itself conforms to the norms of 802.11e for such descriptions. Specifically, a resource descriptor consists of the TSPEC and optional TCLASS elements that are used in an 802.11e ADDTS request. The 802.11r standard has intentionally aligned the format of the resource descriptors with those of 802.11e for consistency. These Resource Information Containers are exchanged in request/response pairs which are called the *RIC request* and *RIC response*. The requests and responses share the same format. We recall that the 802.11e resource-reservation context is one where the STA is requesting that resources be reserved for it by the AP. When the RIC request is transmitted, the sender requests that the resources represented by all the requests be granted. The RIC response will be sent by the AP to the STA in the same format as the request, with the status bit in each RDIE indicating whether or not that resource request was granted.

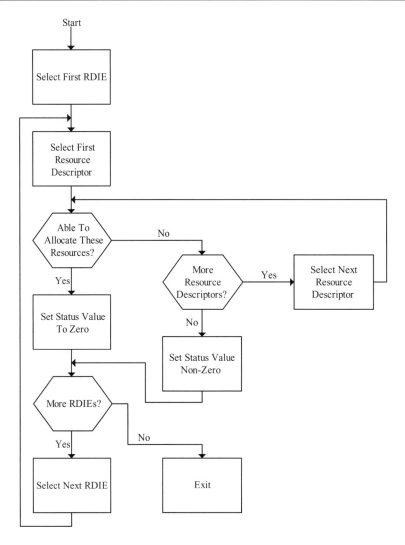

Figure 9.7: Resource Information Container Processing. (Courtesy of the IEEE.)

9.4 Protocol Exchanges in 802.11r

9.4.1 Introduction

In this section we will present eight different protocol diagrams showing the basic ways in which the 802.11r protocol exchanges may occur. If we refer back to Figure 9.1, it can be seen that there were a number of different prehandoff possibilities contemplated by the 802.11r standard. In the examples given in this section, we consider both the OTA and ODS cases. The Fast BSS Transition protocol exchanges are the same regardless of whether PSK or full

802.1X authentication is used. For those examples where resource reservations are made, we use the full FT reservation protocol to highlight the differences between the six-message paradigm and the simpler four-message case. In all the diagrams in this section, time flows from top to bottom. We assume in all instances that the initial association in the MD has already occurred, as shown in the upper portion of the diagram. The part of the dialog that occurs prior to the STA breaking the association is clearly delineated from the part that occurs after the STA disassociated from the current AP.

It is important to understand that the 802.11 open-authentication frames play a central role in how 802.11r reduces the frame count during a roam. The 802.11r standard exploits the fact that while a STA can only be associated with a single BSS at a given moment, the 802.11 standard does permit the exchange of 802.11 authentication frames with another BSS without breaking its association with the current AP. By enriching the kinds of information that the authentication frames can carry via new IEs, the 802.11r standard is able to set up security and QoS contexts before the roam, rather than paying this frame-count tax during the roam.

Once the 802.11r STA initiates the roam, the standard specifies that the STA should piggyback security and QoS information on the four-frame exchange that always occurs at this time, namely, the familiar authentication request/response and reassociation request/ response that occurs with every new association. A clear example of improved efficiency is the already mentioned fact that with 802.11i preauthentication, the four-frame authentication/ association exchange was followed by the four-way EAPOL key exchange. By overloading the authentication and association frames with the same information exchanged in the four-way EAPOL key exchange, we reduce an eight-frame exchange to merely four. The 802.11r standard offers a rich set of alternatives for supporting Fast BSS Transitions with 802.11-compliant security and QoS provisions.

9.4.2 Fast BSS Transition Over the Air, No QoS, and No Security

Figure 9.8 shows the most basic protocol exchange for an OTA Fast BSS Transition. The STA has already identified the target AP prior to the transmission of the initial open-authentication frame to the target. The initial authentication frame contains two fields particularly relevant to Fast BSS Transitions. These are denoted by the FT and MDIE attributes indicated in parentheses in the first frame shown in Figure 9.8. FT indicates that the authentication-algorithm field of the authentication frame is set to type *Fast BSS Transition Authentication*. MDIE indicates that this authentication frame includes a MDIE. This IE should be identical to the MDIE that was advertised by the AP in its beacon or probe-response messages. The remaining three frames in Figure 9.8 follow as they would in the normal non-802.11r case, with an authentication response followed by the reassociation-request/-response pair. The authentication response carries the same FT denotation and MDIE as the authentication request. The reassociation-request/-response pair carry the MDIE.

Chapter 9

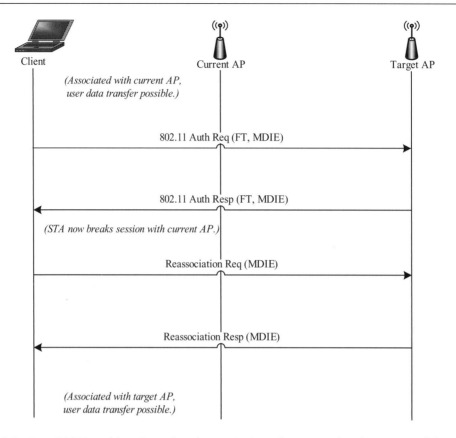

Figure 9.8: Fast BSS Transition Over the Air, No QoS, and No Security. (Courtesy of the IEEE.)

It should be apparent that in the case of an 802.11 roaming transition involving neither QoS nor security, which is the case for Figure 9.8, there is no substantive change with 802.11r. That is, moving to a new AP still involves a four-way frame exchange, just as it does without 802.11r. This fact is clear evidence that 802.11r is really a security and QoS enhancement standard. Both security and QoS protocols in 802.11 add significantly to the protocol exchange overhead during a roam, so if neither of those are involved in a transition, there is little benefit from using the 802.11r standard.

9.4.3 Fast BSS Transition Over the DS, No QoS, and No Security

Figure 9.9 shows the most basic protocol exchange for an ODS Fast BSS Transition. A fundamental concept for this section is that regardless of whether the Fast BSS Transition is accomplished by OTA or ODS, certain key pieces of information must be communicated in order for the 802.11r procedures to take place. Sometimes, this information is implicitly

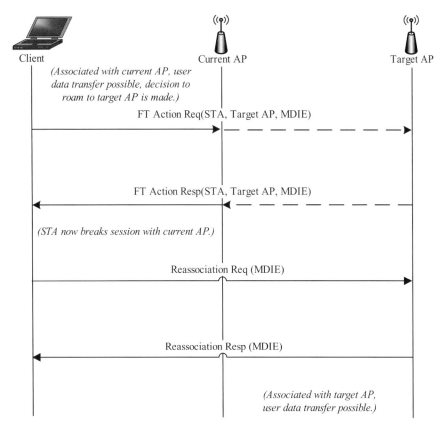

Figure 9.9: Fast BSS Transition Over the DS, No QoS, and No Security. (Courtesy of the IEEE.)

communicated as a result of the protocols in use, and at other times it needs to be explicitly communicated in the payload of the frames in the exchange. The first two frames in Figure 9.9 are excellent examples to illustrate this point.

The basic pattern that emerges regarding ODS with respect to OTA is that the Fast BSS Transition data which is communicated using the open-authentication frames in OTA mode are communicated via FT action-request frames in ODS mode. Clearly, *all* communication to or from an STA in 802.11 is done over-the-air, but in the case of FT Action Request Frames, the STA transmits these frames over-the-air to the *currently* associated AP which is *not* the target AP—the true destination of these frames. As the ultimate destination is the target AP, to be accessed via the distribution service, the identity of this AP must be communicated somehow.

In the case of OTA, the target AP was attained directly via the 802.11 transmission with an 802.11 DA set to the target's BSSID, whereas in the case of ODS, the target AP's address is communicated explicitly in the payload of the first FT action-request frame (see Figure 9.9).

This information will be used by the RRB in the current AP to relay the message to the intended recipient, the target AP. The dashed line in Figure 9.9, as well as in the remaining ODS diagrams in this section, denote the portion of the communications path done over the (usually wired) distribution service connecting the APs. There is no explicit notation in the FT action-request frame that this is a FT-type frame, as was the case in the authentication frames in OTA mode. This omission is okay because the FT action-request frame's very existence denotes FT mode, so there is no reason to allocate a special parameter to reinforce this fact. Conversely, it is necessary to include the STA and target AP addresses in these FT Action frames, whereas these data were communicated implicitly in OTA mode. The reassociation-request/-response pair that follows in Figure 9.9 is identical to the one in the OTA case shown in Figure 9.8.

9.4.4 Fast BSS Transition with QoS and Security

Figure 9.10 builds on Figure 9.8 and illustrates the additional protocol steps incurred when security is added. We note that the authentication exchange shown here now contains two new IEs—FTIE and RSNIE. The FTIE is used to carry the ANonce from the STA to the AP and the SNonce back from the AP to the STA in the response. This process is the overloading of the security elements of the 802.11i four-way key exchange onto the authentication and reassociation message exchanges in 802.11r. Referring back to the exposition of the four-way key exchange which we provided in Chapter 7, we see that the ANonce and SNonce were exchanged in the first two frames of the four-way key exchange, just as they were exchanged in the first two frames of the protocol exchange shown in Figure 9.10.

The RSNIE contains information regarding the security preferences and capabilities of the transmitting side. As we move on to the third and fourth frames of Figure 9.10, we see that the FTIE and RSNIE used to carry information formerly carried in the third and fourth messages of the four-way EAPOL key exchange. In addition to the ANonce and SNonce, the FTIE carries the message-integrity check. The RSNIE includes the PMKR1Name. The response from the AP includes the same information and the GTK, collapsing the group-key information into an existing message. A quick look back at Chapter 7 will confirm that this information is what was needed by 802.11i in the four-way EAPOL key exchange.

Figure 9.11 is the ODS counterpart of Figure 9.9, illustrating the additional protocol steps incurred when security is added to the base ODS case. The easiest way to understand Figure 9.11 is to contrast it with Figure 9.10. The security association resulting from the exchanges in these two figures is essentially identical; the only difference is that in the first case the first half of the exchange is done directly with the target AP via open-authentication frames Over the Air, and in the latter case the exchange is done indirectly

The 802.11 Workgroups' Solutions for Fast Secure Roaming

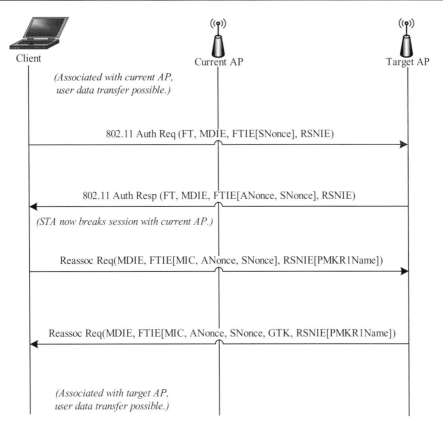

Figure 9.10: Fast BSS Transition Over the Air, No QoS, and Security. (Courtesy of the IEEE.)

Over the DS via the current AP. For this reason, the frames exchanged during the association phase are the same in both cases. The only difference we see in Figure 9.11 is that the security information corresponding to the first two frames of the 802.11i four-way key exchange is relayed in the FT action-request and FT action-response frames. Indeed, the information is carried in the same manner in the same FTIEs and RSNIEs, as we saw in Figure 9.10.

We now introduce the 802.11r exchanges that incorporate QoS reservations. The figures illustrating the QoS exchanges are built upon the corresponding nonQoS figures that preceded them. A quick comparison shows that Figure 9.12 extends Figure 9.8 by the addition of two new authentication frames following the initial request/response pair. These third and fourth frames are the new authentication-ACK and authentication-Confirm frames created for the 802.11r standard. They carry the RIC request and RIC response that are used to prereserve a set of resources, prior to the actual roam taking place.

Chapter 9

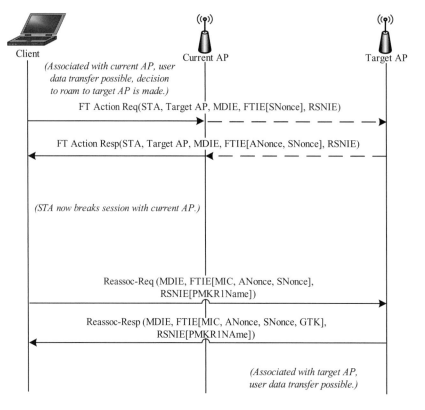

Figure 9.11: Fast BSS Transition Over the DS, No QoS, and Security. (Courtesy of the IEEE.)

The fact that the RIC request and RIC response messages are deferred to the third and fourth messages of the exchange results in the security association between the target AP and the STA being sufficiently established to afford the RIC request-and RIC response protection under the message integrity check. While this point is irrelevant in the case of a non-RSN (no security) implementation, the RIC exchange is still deferred to the third and fourth messages for consistency.

Figure 9.13 maps directly from Figure 9.12 in that each of the Information Elements carried in the authentication frames of Figure 9.12 is carried in a corresponding FT action frame in Figure 9.13. Whether performed in OTA or ODS mode, both these exchanges confirm that the required resources are reserved prior to reassociation. In both cases, the actual roam is reflected in the final two frames of the exchanges, where the reassociation request and response signal the completed reassociation with the target AP.

Figure 9.14 evolves from Figure 9.12, adding security to the OTA case with QoS. Figure 9.15 covers the same case as Figure 9.14 for ODS mode.

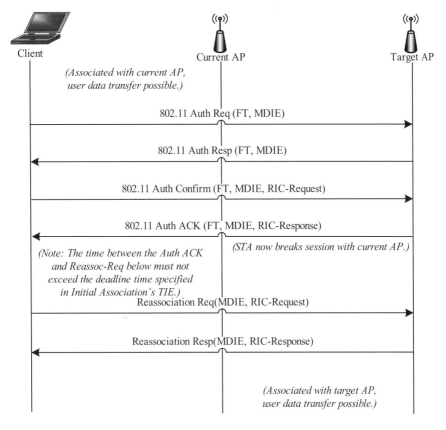

Figure 9.12: Fast BSS Transition Over the Air, QoS, and No Security. (Courtesy of the IEEE.)

9.5 The 802.11k Standard Applied to Roaming

9.5.1 Introduction

The 802.11k task group provides standards for measuring and reporting statistics about signal strength, channel usage, loading levels, and other important metrics about access points and end-user stations. This subgroup additionally defines protocols and MIBs for exchanging and reporting this information. We provided a general introduction to the 802.11k standard in Chapter 4. Radio Resource Measurement, as the work of the 802.11k task group is formally known, is one of the 15 architectural services of 802.11. The Radio Resource Measurement Service is intended to provide the following:

- Requesting and reporting of radio measurements in supported channels.
- Performing radio measurements in supported channels.

Chapter 9

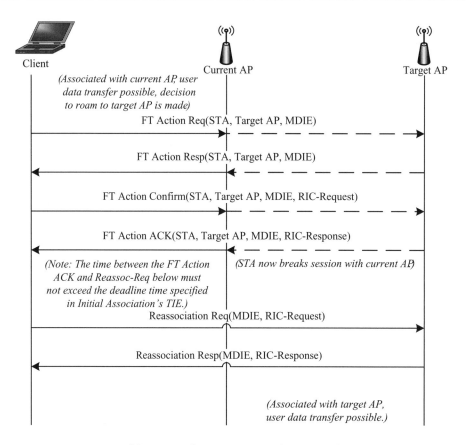

Figure 9.13: Fast BSS Transition Over the DS, QoS, and No Security. (Courtesy of the IEEE.)

- Providing an interface for upper-layer applications to access radio measurements using MLME primitives via Information Elements and/or MIB access.

- Providing information about neighboring APs.

These broad categories are broken down into the discrete reporting classes illustrated in Figure 9.16.

9.5.2 New Information Elements Defined by 802.11k

The 802.11k standard introduced a large number of new Information Elements in order to request measurements and to communicate the resulting measurement information. Beacon, probe, and association frames are candidates that carry these new IEs. Some of the measurement requests described in 802.11k result in repeated histogram reports from the recipient of the request back to the requestor. The 802.11k reporting is somewhat voluntary in that the STA or AP involved may judge itself too busy to spend the time "not working on user

The 802.11 Workgroups' Solutions for Fast Secure Roaming

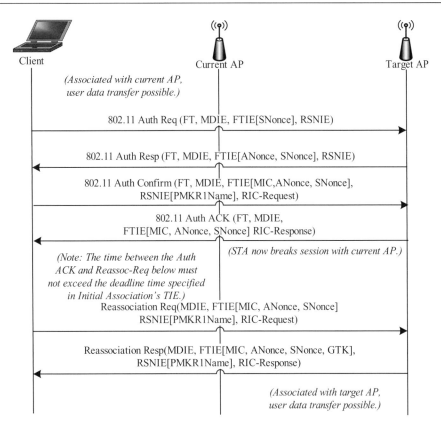

Figure 9.14: Fast BSS Transition Over the Air, QoS, and Security. (Courtesy of the IEEE.)

traffic" in order to gather the information requested. As this reporting may contend for computing resources that are needed for data forwarding or control, and as the reporting is considered of lower priority, the 802.11k information should be considered an *aid* for making more informed decisions about roaming, rather than the exclusive basis for those decisions. The details of what measures are defined by 802.11k, how these measures are initiated, and the results communicated and made available for subsequent SNMP-based interrogation are beyond the scope of this book. The interested reader is referred to [1].

If we focus exclusively on the individual statistics and reporting mechanisms defined by 802.11k, we run the risk of missing the potential ESS-wide import of an 802.11 system that widely implements 802.11k's features. Each individual AP or STA on its own has at best a partial view of the radio environment in the network. Relying solely on information that is directly received, it only has information about its immediate radio surroundings. In most instances where roaming is relevant, an 802.11 network consists of multiple APs and sets of STAs associated with them. This environment is where the true potential of 802.11k rests; by being able to interrogate neighboring STAs and APs about what they can measure about their

Chapter 9

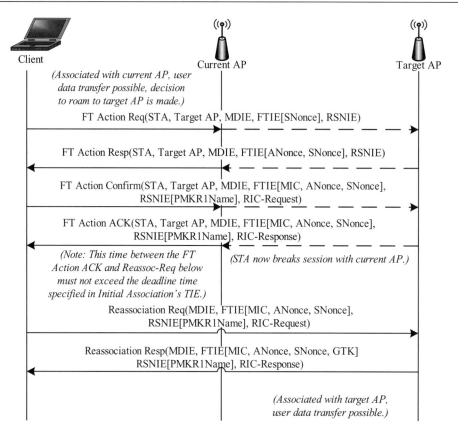

Figure 9.15: Fast BSS Transition Over the DS, QoS, and Security. (Courtesy of the IEEE.)

immediate RF environment and by relaying this information to their neighbors, each AP and STA can actually build an up-to-date image of the network. By constructing this network-wide picture about the RF environment, the location of STAs, and even AP loading, we can enable the ESS to behave in a globally optimized way, as opposed to its behavior being based on a disjointed series of locally optimized decisions, which are unlikely to result in as efficient an environment.

One obvious example that can help with a roaming decision is the 802.11k Neighbor Report. This report provides information about the APs that are neighbors of the current AP, and the report can provide information about the APs' current loading level (*BSS Load*). This load is the parameter that 802.11k uses to report the relative business of an AP. Intuitively, a less loaded AP should be more desirable as a roaming candidate than a heavily loaded alternative. This situation is particularly true when 802.11e resource reservations are being used, as it is not wise to waste resources requesting reservations from an AP, which is likely to refuse that request due to lack of resources. There is an effort to augment the 802.11k Neighbor

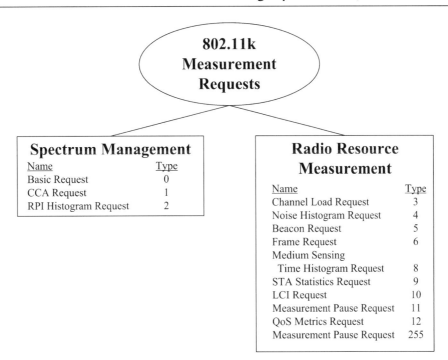

Figure 9.16: The 802.11k Standard's Type Definitions for Measurement Reports.

Report with 802.11r-specific information to make it more useful for roaming (see [4] for more details).

9.5.3 Utility of 802.11k

The 802.11k standard can help provide information about the surrounding APs so that the STA only scans on channels where APs are present. Scanning on an empty channel is a useless activity for the STA. While avoiding this waste is beneficial in 802.11b/g networks, it provides more advantage in 802.11a networks with its much greater number of nonoverlapping channels that need to be scanned. The ability to construct an informed picture of the wireless network, beyond what the STA or AP can directly measure, may have significant impact on how roaming occurs in future 802.11 networks. The 802.11k-location indicator measurements, given as a longitude and latitude, could be recorded as a function of time to capture the direction of movement of an STA relative to APs that this STA is likely to encounter later due to its own movement. We will consider two hypothetical examples of how such location-based information might affect roaming decisions.

Knowledge of Received Signal Strength (RSS) at other stations in the physical proximity of a predicted future location can help optimize a roaming decision. For example, in Figure 9.17

Chapter 9

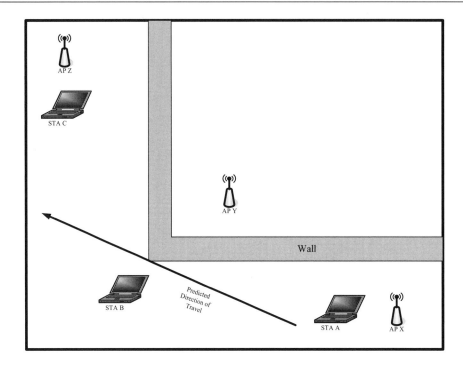

Figure 9.17: Roaming Enhanced by Knowledge Derived from 802.11k.

assume STA B is associated with an AP Y on channel 6 and is reporting marginal signal strength, and our roaming STA A currently associated with AP X on channel 11 has learned of the presence of a third AP Z on channel 1 through 802.11k. STA A by itself cannot know the location of AP Z because it cannot yet obtain a signal from AP Z, but it can know that location by using information relayed via 802.11k. To be clear, since we treat exclusively infrastructure mode in this book, STA B does not communicate the RSS information directly to STA A. Such information would be accumulated by 802.11k Measurement Reports by AP Y and then shared with neighboring APs.

Let us assume that STA A has also learned of a third STA C not too distant from STA B and still on A's predicted trajectory. The receipt of the signal of AP Z by STA C is much stronger than that of AP X by STA B. This information could permit STA A to approximate the network map that we see in the Figure 9.17. STA A might use such information to delay disassociation from its current AP in order to wait for the preferred AP Z which is soon to come within range, thereby avoiding the overhead of what is likely to be a marginally successful roam to the middle AP, AP Y.

Similarly, information about the QoS characteristics of AP Y in the previous example might also result in STA A deciding to defer its roam until it is within the range of AP Z.

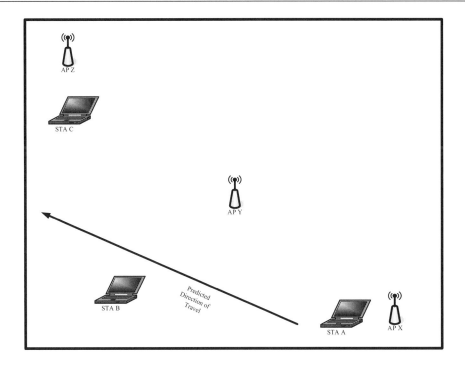

Figure 9.18: Roaming Enhanced by QoS Information Derived from 802.11k.

In Figure 9.18, a situation similar to that shown in Figure 9.17 is depicted but without a physical barrier (the wall) to provide RF interference. In this case, the occupancy measure (BSS Load) provided by 802.11k indicates that AP Y is already heavily loaded and not likely to be able to provide the performance desired by STA A, despite the fact that STA A receives a strong signal from AP Y. A similar measure of AP Z indicates that it is lightly loaded and despite its weaker signal, it represents a better roaming candidate than AP Y. Making the decision to roam to AP Z rather than AP Y is an example of *load balancing*. The fact that the 802.11k Neighbor Report can potentially allow the STA to make a location-sensitive and trajectory-sensitive decision, and thus arrive at a better outcome, suggests that future collaboration between the 802.11r and 802.11k task groups is desirable. It should be noted that this presumes much more intelligent roaming criteria than that exercised by 802.11 implementations, as of this writing.

As of now, a roaming decision is typically made because the error rate has risen unacceptably or the RSS from the current AP has fallen below a minimum acceptable threshold. To take advantage of the scenario just described, the decision to hold onto the current AP for a bit longer requires that this decision be made when the STA can still communicate acceptably with the current AP and have the ability to discern that the STA is moving toward the preferred AP. As long as the STA adheres to the basic 802.11 paradigm, of roaming only when about to lose the signal, there is limited ability to benefit from the potential location-sensitive sophistication allowed by incorporating 802.11k information into the roaming-decision process.

9.5.4 Limitations of 802.11k

There are a number of limitations to the amount of help that 802.11k will actually provide to roaming clients. Ideally, the information made available to a client via 802.11k about the surrounding network would provide a *global view* of the nearby 802.11 environment. The 802.11k information is derived by the neighboring APs gathering information and sharing it indiscriminately with their neighbors. Both the accuracy and the granularity of the information received from a neighbor are not well defined by the 802.11k standard. Part of the reason for this fact is that, with respect to accuracy and granularity, the standard is necessarily the least common denominator across the contributing companies. An example of this is BSS Load. In reality, the concept of BSS Load is quite vendor-specific and as such will be of little use in a multivendor environment. This situation is not surprising as AP vendors see the ability to add value in a single-vendor environment as one of the key means to differentiate themselves from their competitors. There is little incentive for a high-cost vendor to allow lower cost competitors to benefit from their more sophisticated measurement tools.

As another example, knowing which AP is receiving the client's signal more strongly is certainly part of the global view that would facilitate intelligent roaming decisions. While technically, the 802.11k standard could have done so, as of this writing, the standard does not provide information about a client's RSS. Limitations such as this restrict 802.11k from providing as global a view as it otherwise could.

9.6 Concluding Remarks

The security tightening of 802.11i and the QoS enhancing of 802.11e complicated basic 802.11 roaming and increased the frame counts involved in roaming. Addressing the resulting increases in delay has been the focus of 802.11r. In the post-802.11r environment, other more fundamental causes of roaming delay will receive more attention. With the exception of round-trip delay to the RADIUS server, higher frame counts between the STA and the AP do not add to roaming latency as significantly as the following items:

1. The client taking a long time to decide to leave the current AP.
2. The client taking a long time to decide on a target AP.
3. The client making a poor selection of the target AP.
4. Data packet loss during the roam.

The first three of these are components of the discovery delay introduced in Section 3.3.3 and the fourth occurs during the Application Resumption Delay and Infrastructure Routing Delay phases. The first of the four items involves not being aggressive enough about switching from the current AP. The reason for not being aggressive enough is a fear of being too aggressive in this decision, which leads to a common problem known as *ping-ponging*, where the STA enters a state of hysteresis—vacillating between two APs when it would do better not to roam at all and stick with one of them. While (1) relates to how long it takes for an STA to decide to roam, point (2) relates to how the STA decides on the next AP once that decision is taken. It is clearly evident that by gathering additional information with increased scanning, better informed decisions can be made, but consuming much time exacerbates the handoff delay.

It is clearly not productive for an STA to spend time scanning (a) nonchannels where the available APs on those channels are under high-load conditions or (b) on other channels where there are lightly loaded APs, but they are out of effective range for error-free, high-speed communication (the further the range, the more the speed adapts to a lower speed and the more the air medium is full of traffic, thereby producing more collisions).

We explained earlier that of these four issues, combining 802.11r with 802.11k may address both points (2) and (3). The final issue (4), however, may be more important than all the others. The mere fact that the STA is associated, authenticated, and all the resources are reserved is of little importance to the user at the STA, if the user's data packets are still being forwarded to the preroam AP because the infrastructure has still not reacted to the fact that the STA has moved to a new AP. This important area for standardization may be addressed by the CAPWAP-protocol work but is beyond the scope of 802.11r and 802.11k.

Part of the material and many of the figures in this chapter come from the following IEEE references: [2], [3], [5], [8], [10], [12], [13], [14], and [15].

Figures 9.2–9.15 from Draft IEEE Standard, IEEE P802.21/D01.00, March 2006, Copyright IEEE 2006, Draft Admendment to Standard for Information Technology IEEE P802.11r/02.1, May 2006, Copyright IEEE 2006, Interim Contribution 802.11-06/0825r1 2006-05-31, Copyright IEEE 2006, and Interim Contribution 802.11-06/0566r2, Copyright IEEE 2006, all rights reserved.

References

[1] Draft Amendment to Standard for Information Technology—Telecommunications and Information Exchange Between Systems—LAN/MAN Specific Requirements; Part 11: Wireless Medium Access Control (MAC) and Physical Layer (PHY) Specifications: Amendment 7: Radio Resource Measurement, IEEE P802.11k/D2.2, New York, July 2005.

[2] Draft Amendment to Standard for Information Technology—Telecommunications and Information Exchange Between Systems—LAN/MAN Specific Requirements; Part 11: Wireless Medium Access Control (MAC) and Physical Layer (PHY) Specifications: Amendment 8: Fast BSS Transition, IEEE P802.11r/D2.1, New York, May 2006.

[3] Interim Contribution, IEEE 802.11-06/171r0, New York, 2006.

[4] Interim Contribution, IEEE 802.11-06/282r0, New York, 2006.

[5] Interim Contribution, IEEE 802.11-06/323r0, New York, 2006.

[6] Interim Contribution, IEEE 802.11-06/566r2, New York, 2006.

[7] Interim Contribution, IEEE 802.11-06/605r0, New York, 2006.

[8] Interim Contribution, IEEE 802.11-06/624r0, New York, 2006.

[9] Interim Contribution, IEEE 802.11-06/637r0, New York, 2006.

[10] Interim Contribution, IEEE 802.11-06/650r0, New York, 2006.

[11] Interim Contribution, IEEE 802.11-06/685r0, New York, 2006.

[12] Interim Contribution, IEEE 802.11-06/695r0, New York, 2006.

[13] Interim Contribution, IEEE 802.11-06/792r0, New York, 2006.

[14] Interim Contribution, IEEE 802.11-06/825r1, New York, 2006.

[15] Interim Contribution, IEEE 802.11-06/948r1, New York, 2006.

CHAPTER 10

Roaming between 802.11 and Other Wireless Technologies

10.1 Introduction

10.2 Ideal Roaming Experience

10.3 IEEE 802.16: WiMAX

10.4 IEEE 802.15.1: Bluetooth

10.5 Relevant Standards Bodies and Industry Organizations

10.6 Third Generation Partnership Program

10.7 UMA: A Transitional Step for 3GPP

10.8 Third Generation Partnership Program 2

10.9 The 802.21 Standard

10.10 Summary

References

10.1 Introduction

As 802.11 is not the only wireless communications technology in widespread use, it is natural for users to demand devices with multicommunication radios that are capable of bridging these different technologies. Such devices are already on the market, and it is expected that there will be a rapid growth in the number of such devices in consumers' hands in the near future. In Chapter 8, we discussed the dedicated VoWIP phones that have had some success in the market. These devices have enjoyed success related to improved productivity and collaboration, and also because they have sometimes provided the mobility of the cellular phone with superior quality and feature sets without the need to pay service dollars to a network provider on an ongoing basis [3].

While the advantages just mentioned have been sufficiently attractive to sustain this industry, future growth of these markets is hampered by the expectation by enterprise IT managers that the dual-mode Wi-Fi cellular handsets will shortly replace the dedicated VoWIP phones. As network operators continue to push their third-generation and *second-generation-plus* (*2G-plus*) services, as the single source for both data and voice connectivity, the limited data rates and higher costs of these solutions will lead purchasing managers to look for hybrid solutions. We see evidence of this direction in Qualcomm, a leading manufacturer of 3G CDMA-based data and voice chipsets, who joined the Wi-Fi Alliance in order to ensure that its chipsets are compatible with Wi-Fi technologies as well [9]. Another piece of evidence is Qualcomm's purchase of Airgo in 2007. The advent of the roaming-capable, dual-mode, Wi-Fi cellular phone is at hand.

10.1.1 Vertical Versus Horizontal Roaming

There are many cases where the same device may be at times connected via a physical wire and at other times wirelessly. It is, therefore, important to note that the concept of roaming begins to encompass wired access as well. The blurring of these media boundaries is an example of the much-hyped *Fixed Mobile Convergence* (*FMC*) that is rapidly undergoing transition from the conceptual stage to reality. The interested reader can pursue the latest developments online. In this section, we describe *vertical* and *horizontal* roaming.

Earlier in this book, we distinguished local roaming from global roaming. The complexities of a global roam form a superset of those of local roaming. Most often, global roaming occurs because the user switches from one infrastructure provider to another. As a result of this context switch, many conditions may change. At the least, we would expect the IP subnet to change, potentially disrupting IP-based user applications. Other impacts include the need to authenticate the user with the new provider. In the case of an enterprise network, this authentication may be for security reasons alone. In the case of a public network provider, the authentication is usually required both for security reasons and for the purpose of charging the approved user for the service. When a user roams between 802.11 and other wireless technologies, these cross-technology roams represent yet another degree of complexity, as compared with global roams. These kinds of cross-technology roams may include all the complexities of global roaming, in addition to those related to a switch in physical-layer technology.

This discussion leads us to define two new terms—*horizontal* and *vertical* roaming. As shown in Figure 10.1, horizontal roaming occurs when the type of underlying access technology does not change during the roam. The figure illustrates that horizontal roaming is exemplified by roaming between two GSM base stations or between two 802.11 access points. However, when the user roams between an 802.11 access point and a GSM base station, this event is called a *vertical roam*, in which case we consider the figure along the vertical axis, as opposed to horizontal axis.

Just as our global-roam concept was a superset of the local roam, it can be easily understood that our existing definitions of global and local roams are subsumed by the vertical-roam class. Continuing with this same line of reasoning, a vertical roam would encompass a superset of the problems of the global roam. However, the use of Mobile-IP technology can cloud these otherwise clear boundaries. With Mobile IP, it is possible for the mobile station to change underlying service providers, and even underlying access technologies have not yet changed its IP address *as seen by the application layer*. However, in fact, the MS has indeed roamed to a new subnet and acquired a new IP address, thereby fulfilling that part of our definition of a global roam. The MS uses Mobile-IP procedures to establish an IP tunnel that allows the application layer to be shielded from the change in the IP subnet.

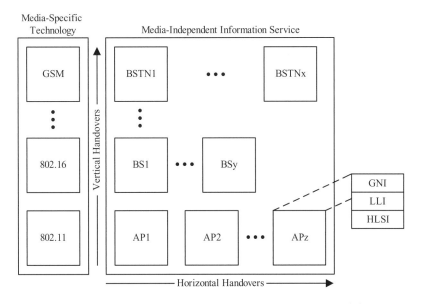

Figure 10.1: Horizontal and Vertical Handovers. (Courtesy of the IEEE.)

10.2 Ideal Roaming Experience

10.2.1 Introduction

As we consider roaming between 802.11 and other access technologies, what will be our future users' real expectations? Think of a typical user, call her Kai, of a roaming cellular phone. Kai does not want to know, and usually does not know, when she roams locally within her provider nor when she roams globally from her provider to an alternate provider—Kai merely wants to be able to make and maintain her phone conversation without being aware of these details. This level of confidence is the benchmark that we should use as we consider roaming between 802.11 and different wireless technologies. It is a benchmark that will be very hard to reach for some time, but it provides a useful measure against which the current state-of-the-art can be compared.

The wireless industry still segregates its users into two categories: The users of data applications such as Web and email, and the voice users. The industry is much closer to meeting the ideal expectations of the cross-technology roaming users for data applications than it is for voice. In the next two sections, we state criteria for the ideal roaming experiences in each of these situations.

10.2.2 Enterprise Data User

We propose that the ideal roaming experience for the enterprise data user would meet the following criteria:

1. Within the enterprise, the user would automatically use an 802.3 wired connection, if it is available and provides the desired connectivity. As authentication, encryption, and other security-related features become more commonplace with 802.3 connections, we expect that the selection and configuration of the 802.3 connection will incorporate much of the complexities that we have seen for 802.11 connections. We assume that the wired connection is preferable in terms of bandwidth, security, and QoS, as compared with a wireless connection. As of this writing, this remains a relatively safe assumption, but this assumption could change in the future.

2. If there is no 802.3 connection, the user will connect to the correct 802.11 network and access point in accordance with the user's established profile.

3. If the user is roaming outside the enterprise, the user's profile will determine whether any acceptable public 802.11 providers are within range, and if not, will make the decision to establish a cellular connection soon enough that no disruption in service is experienced. This step is a nontrivial technical challenge, as even a pedestrian can move out of the range of an 802.11 network in a matter of seconds.

4. If the user is roaming outside the enterprise, the system itself will be intelligent enough to judge if a VPN connection needs to be established and will do so automatically without intervention from the user.

5. In all cases, the roaming user will expect the system to be intelligent enough to make roaming decisions itself, so that it does not switch from one service to another at such a high frequency as to affect the service provided, yet still switch often enough to provide an acceptable level of service. The notion of what is acceptable needs to be both application specific and user specific and will likely be encoded in one or more profiles that help guide the roaming decision.

10.2.3 Voice User

The ideal roaming experience for the *voice* user must achieve all the goals described in Section 10.2.2 and also achieve handoff times short enough such that voice service is not disrupted. Achieving this goal in the cross-technology roaming experience is likely to remain challenging for some time to come. In this chapter, we examine pairings of different wireless technologies where intertechnology roaming is often proposed. Of all of these intersecting technologies, the pairing that currently attracts the most interest is Wi-Fi to cellular network. A major portion of this chapter will focus on the current state-of-the-art of this topic. Since the topics of roaming between 802.11 and 802.16, and between 802.11 and Bluetooth often arise, we will briefly try to clarify how these wireless technologies relate to 802.11 and identify whether roaming solutions between 802.11 and each of these is likely to materialize.

10.3 IEEE 802.16: WiMAX

IEEE 802.16 is a *Wireless Metropolitan Area Network* (*Wireless MAN* or *WiMAN*) standard that was first ratified in 2002. The standard is often referred to as WiMAX [24]. It was updated with 802.16d in 2004 and with a mobile version, 802.16e, in 2006. Both 802.16d and 802.16e use OFDM. The 802.16 standard has received the strongest commercial promotion from Intel who is betting heavily on the success of this technology in its communications-related semiconductor strategy. The 802.16 standard's supporters tout its very high data rates and long transmission ranges. Indeed, an 802.16 cell may have a radius of 30 miles, as compared to the few hundred feet of an 802.11 cell. The transmission rates may reach 70 Mbps under ideal circumstances. These characteristics support the idea of using 802.16e as an alternative to cellular technology.

Not surprisingly, the reality of 802.16 is more muted than what some of the claims imply. Intel's data shows that the 70-Mbps transmission rate is achievable for distances less than four miles. The Intel data also shows that as the distance to the user terminal increases, the maximum achievable speed drops to 50 Mbps between four and six miles from the base station, and further drops to 25 Mbps, as the range extends out to the maximum of 30 miles. It is also important to note that 802.16, similar to 802.11, is a shared access medium. Just as in 802.11b, the theoretical maximum rate of 11 Mbps is never achieved by any single user; the individual users of the 802.16 link must share that aggregate bandwidth and also pay an overhead penalty for the MAC mechanism.

The Media Access Control mechanism in 802.16 is different from the CSMA/CA mechanism in 802.11. Unlike the randomized collision avoidance mechanism of CSMA/CA, the 802.16 standard uses a deterministic scheduling algorithm that assigns precise times for the 802.16 terminals to access the medium. Such an approach has the dual benefits of easier support of QoS guarantees than a stochastic scheduling mechanism such as 802.11 as well as the ability to deliver those QoS guarantees with a higher percentage of channel loading than 802.11 could ever achieve. However, the approach has a drawback: *Randomized Arbitrary Bit Rate (ABR)* traffic will usually use the medium less efficiently under such deterministic scheduling than it would with a randomized MAC such as 802.11.

The supporters of 802.16 generally fall into two separate camps: first, those who view it primarily as a fixed-wireless backhaul technology and, second, those who view it as a high-bandwidth alternative to cellular network. Broadly speaking, the first camp is based on using 802.16d, whereas the second requires the mobility features of 802.16e. The technical differences between 802.16d and 802.16e are significant, since limiting support to a fixed terminal, as is the case for 802.16d, eliminates the numerous problems related to roaming that we have seen in this book. We will now look at how 802.11 might interact with 802.16 in the view of each of these two camps.

10.3.1 Interactions between 802.11 and 802.16

The most obvious and commonly cited use of 802.16 with 802.11 is the fixed wireless application of 802.16. In this application, 802.11 access points may be placed in locations that may be difficult to reach with a normal wired distribution service. Such a wired distribution service is most often seen as coming from an Ethernet, a cable, or a DSL-based infrastructure, but using 802.16 as a wireless backhaul DS eliminates the challenges of reaching locations, where no such wired DS alternative exists. However, in this application, we have no 802.11 to 802.16 roaming issues to discuss, as the access technology in these examples remains exclusively restricted to 802.11. From a roaming perspective, the more exciting application of 802.16 is its possible application as an alternative to or an augmentation of the services provided by cellular network operators. In this event, the idea of roaming from 802.11 to 802.16 really becomes just an extension of the roaming from 802.11 to cellular network. The work of IEEE 802.21 that we will discuss in Section 10.9 includes 802.16, along with 802.11,

3GPP, and 3GPP2, as one of the primary *Radio Access Technologies* (*RATs*) that is being considered.

As we mentioned earlier, IEEE 802.16 is often referred to as WiMAX. In fact, this name is an abbreviation for *Worldwide Interoperability for Microwave Access* Forum. WiMAX is a consortium of companies formed in 2001 to encourage industry adoption of IEEE 802.16 through conformance testing and a resulting certification mark. Conforming products could bear the WiMAX mark. The relationship between 802.16 and the WiMAX Forum is analogous to the relationship between 802.11 and the Wi-Fi Alliance.

10.4 IEEE 802.15.1: Bluetooth

10.4.1 Introduction

IEEE 802.15.1 [20] defines a *Personal Area Network* (*PAN*) standard. As it is a *personal area* network standard, the transmission distances involved are very small. Depending on the 802.15.1 power class involved, the transmission ranges are between a single meter and one hundred meters. During 802.15.1's early years, there was confusion in the market about the overlap between this technology and 802.11. Clearly, at its high power limits, the overlap is significant. Perhaps due to the proliferation of 802.11, the commercial success of 802.15.1 has been at the low end of its transmission range. The very low power required for such small ranges eliminates many of the interference problems that would otherwise result from simultaneous use of 802.15.1 and 802.11 devices in close proximity. This fact is particularly important as both 802.15.1 and 802.11 may operate in the same 2.4-GHz frequency range. Limiting the transmission distances to a few meters has created a market niche for 802.15.1 in the realm of exchanges among devices that are likely to be located nearby a user.

Unlike the Wi-Fi Alliance or WiMAX, the name *Bluetooth* [20] commonly associated with this technology is not an acronym built from a collection of technical terms. As 802.15.1 defines an MAC which requires negotiation for gaining access to the medium, the inventors of 802.15.1 technology chose the name "Bluetooth" after the Danish king known as Harald Bluetooth—famed for getting warring parties to negotiate with each other. Common examples of when Bluetooth is used include linking desktop peripherals such as a printer or a mouse to the desktop, or connecting a cell phone ear piece to a belt-carried mobile phone.

10.4.2 Bluetooth's Relationship with 802.11

Discussions related to roaming between 802.11 and Bluetooth seem unlikely to leave the conceptual stage. It is certainly technically achievable, but as long as Bluetooth remains faithful to its roots as a PAN, roaming applications are unlikely. This statement is true because the kinds of devices that people use with Bluetooth remain in close physical proximity with the targets of their communications. Imagine the common case of a wireless mouse using Bluetooth to communicate with the PC that it is controlling: one can easily move that mouse and use it with another PC, but no one would expect (or want) that mouse to continue to control the original PC from its new home.

Bluetooth does not incorporate the concept of a cell surrounding a centralized base station. Thus, the very concept of roaming as we have defined, is foreign to Bluetooth. This observation is not to say that devices will not have both Bluetooth and 802.11 technology present at the same time. In fact, this case is already common. For example, a Bluetooth mouse could easily be in control of an 802.11-enabled laptop; it is unlikely that the mouse would dynamically switch from Bluetooth to 802.11 to reach the laptop. It is difficult to conceive of realistic 802.11 to Bluetooth roaming applications becoming a commercial reality, but stranger things have happened, and we cannot predict the types of devices which might be available in the future.

10.5 Relevant Standards Bodies and Industry Organizations

The networking field is cluttered with acronyms and organizations. In this section, we clarify the relationships among a number of standards bodies. Figure 10.2 shows four major international bodies whose work includes standardization of roaming across different wireless technologies. As the majority of future communications applications, whether data, voice, or multimedia, will most likely be IP based, the IETF plays an important coordinating role.

The IETF's contributions are made at layer three and above, and any such contribution should be independent of the physical- and access-layer protocols of the layers below. The most relevant contribution of the IETF to multitechnology roaming is in the area of Mobile IP, which we introduced briefly in Chapter 3. All the other three organizations shown in Figure 10.2 have a primary focus on one or more basic transmission technologies.

The IEEE 802.21 is focused on interworking between IEEE 802-family technologies as well as interworking among those technologies and non-802-family technologies. Such non-802-family technologies are most commonly understood to mean 3G cellular telephone technologies. We discussed these cellular technologies in Chapter 2. The 3G world is split between the Universal Mobile Telecommunications System (UMTS) technology dominant in Europe and in many parts of the world, and the Code Division Multiple Access (CDMA2000) technology in widespread use in North America and parts of Asia. UMTS is based on *Wideband Code Division Multiple Access* (*W-CDMA*). The W-CDMA standard is distinct from the well-known North American CDMA standard and its successor CDMA2000, resulting in the separate and incompatible 3GPP [22] and 3GPP2 [23] efforts.

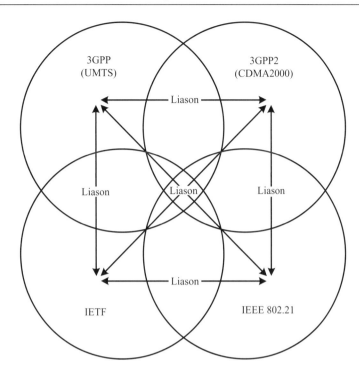

Figure 10.2: The Major Organizations Involved in Intertechnology Roaming.

CDMA2000, formally IS-2000, is an updated version of the original IS-95 version of CDMA with enhancements to support 3G cellular requirements. The umbrella organization for UMTS-based 3G technology is the 3GPP established in 1998; for CDMA2000, it is 3GPP2, which was also established in 1998. These two organizations have similar scope and goals, each specific to its target technology. However, the published set of 3GPP specifications are more extensive than those of 3GPP2.

Just as 3GPP is the standardization group for UMTS, and UMTS is the 3G heir to the GPRS and GSM technologies that preceded it, 3GPP2 is the standardization group for CDMA2000—the 3G heir to the 2G CDMA technologies that evolved before it. Third-generation multimedia services have been designed from the outset as IP based, which has created a strong dependency among the 3GPP, the 3GPP2, and the IETF organizations. Each of IEEE 802.21, 3GPP, and 3GPP2 has recognized the need to develop inter-operability standards with others, and as IP applications are central to all three, all three of them coordinate with the IETF, although to varying degrees.

These interorganizational liaisons are indicated by the arrows in Figure 10.2. These liaisons can be quite formal, where periodic liaison reports are issued and made available through each organization's Website, or informal, through the cross-referencing of each other's specifications. The 3GPP working group TSG CT W63 is charged with "Interworking with External Networks." Early specifications from this group list IETF dependencies but not IEEE dependencies. However, this situation may change as IEEE 802.21 gains momentum. Another organization, *Unlicensed Mobile Access* (*UMA*), which was started in 2004, gained an early prominent role in the area of accessing GSM/GPRS services over unlicensed spectrum technologies such as 802.11. While the term UMA is still seen frequently in the literature, the work of this group was fully transferred to 3GPP in 2005. We will discuss how UMA is used as a transitional technology for 3GPP-WLAN interworking in Section 10.7.

In a fast-moving field such as wireless, companies with leading-edge technology often find the standards process far too slow to meet their business needs. Recall that the Wi-Fi Alliance, which we discussed in Chapter 4, was created due to the need to commercialize a subset of the still-evolving 802.11i standard. As intertechnology wireless communication is likely to have enormous commercial impact, it is not surprising that a number of different groups of vendors have emerged to try to gain an early foothold in this fecund area. Some of these, while retaining a commercial focus, may resemble a public standards body. Others are more accurately called "clubs" of companies who have complementary technology and band together in order to bring solutions to market before the emergence of real standards. Examples of the latter are the *Seamless Converged Communication Across Network* (*SSCAN*) Forum and the *Mobile Integrated Go-to-Market Network IP Telephony Experience* (*MobileIGNITE*) [19].

Organizations such as SSCAN and MobileIGNITE are usually formed by collections of companies with an immediate short-term view to the benefits of collaboration; they sometimes do not bear the test of time.

A larger organization, the *IP Multimedia Subsystem* (*IMS*) Forum, is focused on seamless mobility for IP-based applications across broadband wireless and wireline connections. The IMS Forum was known earlier under the name *Wireless Wireline Convergence Group*. This group has a greater application-based focus than 3GPP, 3GPP2, and IEEE 802.21, and fosters the commercialization of the IMS functions, protocols, and network elements defined by 3GPP.

10.6 Third Generation Partnership Program

10.6.1 Introduction

Network equipment vendors generally fall into the category of those specializing in selling to network operators or to the enterprise market. The IEEE 802 task groups are heavily populated by companies that tend to benefit from the proliferation of the related technology, regardless of whether or not there are service revenues associated with the sale of the technology. Conversely, the 3GPP [22] specifications are driven by network operators and the equipment manufacturers that live off of them. Technological advances and increased network efficiency may be to their advantage in some ways, but the ultimate goal of most of the participants is to attract more customers to public for-charge networks [2].

This dichotomy results in the distinct approaches of 3GPP and the IEEE 802 workgroup to the problem of 3GPP-WLAN interoperability. The 3GPP discussion of WLAN roaming is invariably presented as 802.11 being "just another" RAT available to access 3GPP core services; using the 802.11 access technology for any reason *other* than accessing the 3G core network is not addressed by the 3GPP specifications. Indeed, the actual deployment of IMS-capable User Equipment (*UE*) in 3G networks is likely to be constrained by the network operators who do not want the flexibility allowed by IMS (for example, VoIP calls that circumvent the 3G network) to erode their control over service revenues [2].

10.6.2 Vertical Roaming Issues

The problem of vertical roaming has so many dimensions that it is daunting to try to decompose it into any single set of architectural diagrams or specifications. This complexity is exacerbated by the fact that the individual technologies are evolving such that even purely horizontal roaming remains a moving target. Because of this situation, simplifying assumptions must be made to establish an architectural framework to discuss the problem of vertical roaming. Early 3GPP work made such assumptions by portraying a future where voice calls are just a component of a universal IMS-based infrastructure that was entirely IP based. This approach simplifies the vertical-roaming problem for voice to switching VoIP calls over different RATs.

Initially, these RATs included 3GPP's native UMTS *Universal Terrestrial Access Network* (*UTRAN*) and eventually would be extended to other RATs, notably IEEE 802.11. These simplifying assumptions permitted the early definition of an IMS-based WLAN-3GPP interoperability architecture. IMS is defined by 3GPP but based on IETF-defined protocols. The central signaling protocol in IMS is SIP, and it has been extended to support session establishment specifically for 3GPP networks.

As the signaling is SIP based and both 3GPP and 802.11 networks natively transmit IP packets, it is straightforward to base the 3GPP vertical-roaming architecture on these principles. We graphically illustrate basic IMS-based roaming between a UTRAN RAT and an 802.11 RAT in Figure 10.3. The signaling shown in Figure 10.3 is all SIP based. The voice traffic itself follows the *bearer* path shown in the figure. In this pure IMS example, the voice traffic is carried in IP packets regardless of whether the call originates from the 802.11 RAT or the UTRAN RAT. It should be noted that the terms *signaling path* and *bearer path* are

Chapter 10

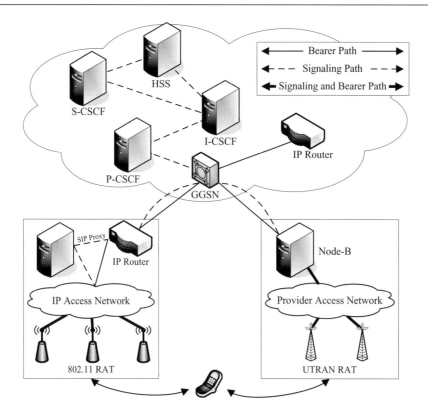

Figure 10.3: IMS-Based Roaming between Cellular and 802.11.

analogous to the terms *control path* and *data path* of the computer-networking vernacular. Both signaling and bearer IP packets first traverse a *GPRS Gateway Support Node* (*GGSN*), which is a specialized IP router. In a pure IMS network, the bearer traffic remains in IP-packet form and is switched by IP routers.

The signaling functions are handled by three kinds of *Call Session Control Function* (*CSCF*) nodes. All three types provide SIP signaling services. They are as follows:

- The *Proxy CSCF* (*P-CSCF*)
- The *Interrogating CSCF* (*I-CSCF*)
- The *serving CSCF* (*S-CSCF*)

The *Home Subscriber Server* (*HSS*) holds user authorization and profile information. The Home AuC derives the IMS-AKA quintet of authentication vectors and then distributes them to one or more HSSs. In practice, the HSS and AuC functions are often combined within a single network device. Figure 10.3 does not show all the network elements of the IMS architecture, nor does it show the full details of the connections of these elements. As such details are beyond the scope of this book, we refer the interested reader to references [11] and [12] provided at the end of the chapter.

10.6.3 Interworking between 3GPP and 802.11

There are a number of 3GPP specifications and technical reports that specifically deal with interworking between 3GPP and 802.11. Specifically, much of the material appearing in this section is drawn from the 3GPP documents, [22.234], [22.934], and [23.234], which are references [18], [10], and [13], respectively, provided at the end of this chapter. In Release 6 of 3GPP, the specifications introduced interoperation with WLAN networks, although only to a limited degree. The Release 7 versions of these specifications begin to address some of the more detailed problems related to roaming between heterogeneous technologies. First, it is worth noting that, in general, 3GPP provides more of a system-level set of architecture documents than the IEEE specifications that have been the earlier focus of this book. IEEE specifications tend not to "reach" to the same degree, as they are less about network architectures and more about a precise description of one network element that needs to function independently of the layers above and below.

There is a core 3GPP architecture, which ultimately encompasses specifications for all layers in the communications hierarchy, from PHY radios to the protocols to be used for encoding voice. At some of these layers, 3G contemplates multiple competing technologies for that layer, all of which fit within the architecture. An example of this point is the different RATs included in the current 3GPP specifications. These different RATs include the following:

- UTRAN

- The *GSM EDGE Radio Access Network* (*GERAN*)

- The *Long Term Evolution* (*LTE*) RAT

The UTRAN is the UMTS Terrestrial Radio Access Network composed of *Node-B's* (UMTS's term for base station) and *Radio Network Controllers* (*RNCs*). GERAN is an enhancement to 2G GSM-access technology. We recall from Chapter 2 that GSM radio access technology is based on TDMA. GERAN is a technology bridging RAT, which is intended to be replaced by W-CDMA. LTE is largely conceptual at this stage and is intended to describe the evolution of the 3GPP RAT from W-CDMA toward a high data rate, low latency, and packet-optimized, radio access technology.

The 3GPP IMS-related interworking specifications such as [22.234] emphasize that 3G's rationale for interworking with WLAN technology is to permit the WLAN-enabled device to access 3G core services. A specific interworking instance that would be *outside* the scope of 3GPP would be the following scenario: an enterprise user with a dual-mode phone is on a VoWi-Fi call using the corporate WLAN to access the corporate IP-telephony services to call a colleague in the same corporation in another city. This user leaves the office and walks to a car in the parking lot. From a technical viewpoint, the goal is clear—the user wishes for the dual-mode phone to establish a cellular connection before the WLAN signal is lost and then immediately to connect this call to the corporate IP telephony services via the 3G network.

In the case just described, the 3G network plays the subordinate role of an access network to the core IP-telephony services being provided by the enterprise. Addressing such a scenario will *not* be the focus of 3GPP. One may wonder, if the applications are IP based, what does it matter whether the application is presented by the 3G core or some service provided by an enterprise outside of the network operators? From a pure technology perspective, this question is really of no significance, but the reality is that 3GPP is a network-operator driven standard, and the actual implementations of 3GPP systems will push users toward chargeable provider-based services. Such a goal may be accomplished by technical or administrative means, or both.

The 3GPP work supports both legacy *Circuit Switched* (*CS*) voice services and 3GPP-native *Packet Switched* (*PS*) voice services. The PS voice services are part of the broader IP Multimedia Services 3GPP component. In Release 6, interworking between 3GPP and WLAN is limited to the PS services. Release 7 contemplates roaming from a CS-based 3GPP voice call to a PS-based voice call [8], which we will discuss in more detail shortly. The [22.234] specification stipulates that dual-mode devices be able to make or receive calls from the CS domain via the 3G RAT without interfering with the device's connection to the WLAN. More generally, the dual-mode device should be able to be simultaneously connected to both the WLAN and PS domain for different services, if required. A potential scenario for such access splitting would be access to a high-bandwidth IMS data application over the WLAN and using the PS domain for access to a low-bandwidth but QoS-sensitive IMS application (for example, VoIP).

10.6.4 Five Levels of Interoperability

In this section, we deal with the five basic levels of interoperability that 3GPP defines for roaming between a native 3GPP RAT and 802.11 [7]. The first and most basic of these five levels is *common billing and customer care*. This one is purely an administrative level of interworking. As this level merely implies that the customer receives one bill for both 3GPP and WLAN services, this level does not imply any protocol interoperability between 3GPP and WLAN. This point distinguishes it from the second level, *3GPP System-based Access Control and Charging*, which presumes that the user authentication and charging systems used by the WLAN provider employ the same authentication and charging servers as the 3GPP system. This assumption implies that the WLAN will use a SIM-based authentication system and some back-end interface into the 3G authentication and charging systems. It additionally implies that the WLAN uses AAA mechanisms compatible with those used in the 3GPP system, specifically the HSS that we saw in Figure 10.3. It should be noted that this will require the use of the DIAMETER AAA protocol that we introduced in Chapter 6, rather than the RADIUS protocol which is currently in common use in 802.11 networks.

The third level of interoperability is *Access to 3GPP System IMS-Based Services*. This level corresponds to an 802.11 user with a Wi-Fi enabled device being able to access an IP-based data or voice application supplied by a 3G provider. Assuming the availability of IMS-capable Wi-Fi devices, this level of interoperability should be achievable with only minor technical hurdles as the SIP signaling on which IMS services are based is easily and natively supportable by the 802.11 network. Interoperability challenges here should be limited to integration problems such as use of compatible versions of SIP. 3GPP strongly suggests the use of IPv6 for IMS services rather than IPv4, but early IMS implementations will support IPv4 user equipment. Access to data applications provided by 3G will likely occur sooner than that for voice as the QoS capabilities of the WLAN network may not be sufficient for voice calls. As 802.11e-capable WLANs proliferate, the ability to support WLAN access to 3G voice services should improve.

It should be noted that level three only implies *static* access to the 3G services from the WLAN—levels four and five contemplate *roaming* between the WLAN and 3G. These two layers are called *Service Continuity* and *Session Continuity*, respectively. The interoperability of level four can be achieved to some degree in Release 6 of 3GPP, but much of the implementation details are vendor-specific and thus there is not widespread use of these technologies so far. The IMS-capable infrastructure deployments are already advanced to the point where there are not enough IMS-capable handsets to exploit the infrastructure fully. The expectation is that twenty-percent penetration of IMS-capable mobile phones will not occur until 2009 [1], and that only ten percent of IMS-capable phones will be dual-mode cellular 802.11 by 2011 [1].

It is important to distinguish IMS-capable phones, a central component of the 3GPP architecture, from SIP-capable non-IMS phones, sometimes called "Naked SIP" phones. Naked-SIP phones implement the correct IP protocols to permit access to the IP infrastructure offered by the 3G providers without requiring that they use the provider-offered services. This means that a SIP-signaled VoIP call may be placed via the 3G service provider and hence over the public Internet, and so escape service-provider fees for the voice call. The IMS-capable phones are based on the same technology, but will restrict the user to the IP-based services offered, and charged for, by the provider.

The real difference between level four and level five is one of degree: The roaming provided at level four acknowledges that the user may be aware that the roam has occurred. There may be obvious delays in the handover between 3GPP and WLAN, and vice versa, but some data applications may be able to tolerate such latency. For example, an *Instant Message* (*IM*) application may be able to tolerate a roam that takes five seconds without causing the user much consternation. In general, voice is not considered a good candidate for level-four interoperability between 3GPP and WLAN. At least as of this writing, the technical deficiencies of WLANs, in particular their QoS limitations, keep our roaming horizon at level four.

The holy grail of WLAN-3G roaming is, of course, the seamless service continuity of level-five interoperability, which future improvements in WLAN technology and in dual-mode devices should allow. Feasibility studies of such level-five interoperability are discussed in [8]. Meridith also discussed the related problem of *Voice Call Continuity* (*VCC*), or *seamless roaming*, between CS-based voice calls over UMTS and IMS-based voice calls over WLAN. This work expands the roaming challenge past the simplifying assumption stated at the beginning of this section, where we limited the 3GPP-WLAN roaming problem to maintaining IMS-based service across the roam. In fact, the majority of voice calls are likely to be CS-based for a long time to come, so a practical consideration of voice roaming between these technologies needs to contemplate CS to PS voice roaming. In addition to [8], early work in this area appears in [14], which we discuss further in the following text. An additional level of interoperability being contemplated includes providing IMS service access to CS supplementary services, such as PSTN/ISDN emulation.

10.6.5 Seamless Roaming between CS and PS Services

A significant amount of work regarding CS-IMS voice-call continuity has been compiled in [14], a 3GPP technical report that considers different approaches to providing CS-IMS voice-continuity service. Rather than a specification, it is more of an evaluation or feasibility document—capturing the current state of discussions for VCC service. The document is important to our focus on interworking with 802.11 as the IMS-based RAT most often used in the examples in [14] is 802.11. There are two different models promulgated in [14] for the voice-call continuity service. These are the *IMS Controlled with Static Anchoring* (*ICSA*) model and the *Original Domain Controlled* (*ODC*) model. Despite pros and cons for both, 3GPP has subsequently converged on the ICSA model, which is the model that we discuss in this section.

The UE shown in Figure 10.4 is assumed to be a dual-mode IMS-CS device. In terms of extant technologies, we will consider a dual-mode 802.11-GSM handset to be such a device. There are many advanced features such as call waiting and multiparty calling that need to be handled by any model supporting VCC. In this discussion, we will limit ourselves to the simple factors of the current *point of attachment* of the UE, the destination service of the call, and whether or not the call is placed *from* the UE or *to* the UE to consider.

Roaming between 802.11 and Other Wireless Technologies

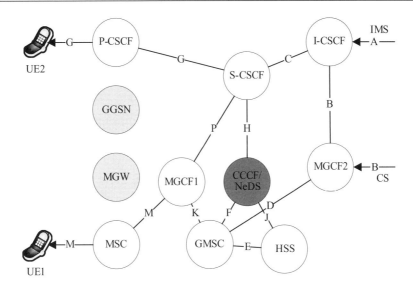

Figure 10.4: CCCF-Anchored Voice Call Continuity Signaling in 3GPP.

The first factor is whether the device is IMS-registered via a PS-access technology such as 802.11, or CS-attached via GSM or UMTS, or both. This will be manifested both in the radio access technology itself as well as the use of different signaling paradigms. Another factor is whether the call ends up accessing IMS services or CS services. This access will depend on the ultimate destination of the call. The final factor is the direction of the call. Depending on the particular subset of these factors that is selected, we end up with a specific instantiation of the *call chain*. For any call, the call chain is the set of network elements that the signaling path traverses. A call chain exists for any active call.

One of the core elements of ICSA is that of *static anchoring* of the call through a newly proposed function—the *Call Control Continuity Function* (*CCCF*). Within the CS-IMS VCC architecture, the CCCF is the function that ultimately controls the bearer path for both CS calls and IMS sessions. The CCCF remains an *anchor* of the call in that even if the UE roams from a CS-access network to PS-access network, the signaling path from the UE into the core network will still traverse the original CCCF. It is called a *static anchor* because the CCCF remains on the call path regardless of how many times the UE switches from one access network to another. The roaming scenarios contemplated in [14] include not only switching to a new access network but potentially between the CS signaling and bearer paradigm to the PS signaling and bearer paradigm. When a call is placed from the UE, the decision about the access network has already been made by the UE, well before the call reaches the CCCF.

An outbound call placed toward the UE is less straightforward. Another new function, *Network Domain Selection* (*NeDS*), is defined to select the domain that is to be used in routing a call to the UE. The NeDS is aware of where the UE is attached and of specific preferences or prohibitions that the operator or UE may have regarding which domain to use. This information is available for use by the NeDS when consulted by the CCCF at the time that the initial call chain is being established. This information is also used by the NeDS when the UE is roaming between access networks. The CCCF and NeDS are defined as separate functions so that a given implementation may instantiate them in separate devices, but it is reasonable to imagine cases where the CCCF and NeDS function are implemented in a single network device. This is how we depict them in Figure 10.4.

In the ICSA model, the CCCF/NeDS function needs to be aware of the current point of attachment of the UE. When the UE accesses the network via a 3G RAT, this information becomes available at the time that the UE registers for IMS services. This registration occurs with a CSCF function. When the access is via a GSM RAT, it registers with the MSC connected to the GSM RAT. The ICSA model stipulates that when this registration occurs, the CSCF or MSC must communicate this information to the CCCF/NeDS function. Once this communication occurs, the CCCF/NeDS function is aware of the current point of attachment(s) of the UE. The UE is permitted to be attached simultaneously through both the CS- and IMS-access technologies.

We will now walk the reader through different permutations of the factors described previously by referring to the specific call chain segments shown in Figure 10.4.

If a call is directed to the UE from the IMS domain, regardless of where the CCCF/NeDS decides to route it, the call will arrive directly at an I-CSCF (segment A in Figure 10.4).

If a call arrives from the CS domain, it arrives first at a *Media Gateway Control Function* (*MGCF*), and then the I-CSCF is consulted (segments B-B). In both cases, the call ultimately needs to reach the CCCF/NeDS.

If a call is routed via IMS, the CCCF/NeDS is reached by routing the call via a S-CSCF (segments C–H).

If a call is routed via CS, the call is routed through a *Gateway Mobile Switching Center* (*GMSC*) (segment D), next the HSS is consulted (segment E), and then the GMSC contacts the CCCF/NeDS to determine where this call should be routed (segment F). Whether the call originated from the CS or IMS domain, at this point the incoming-call signaling has reached the CCCF/NeDS anchor.

In the event that the NeDS determines that the call should be routed via the IMS-network domain, the signaling will eventually follow segment G, keeping the CCCF/NeDS on the call chain in anticipation of possible future roaming. In the case that the IMS-destined call came in from CS, the path involves more signaling back from the GMSC through the correct MGCF (segments K–P) before it intersects with segment G.

If the CCCF/NeDS determines that the call should be routed via the CS network domain, it consults the HSS (segment J). The CCCF/NeDS then signals the S-CSCF (segment H) (or, if CS-originated, the MSC (segment K)) to route the call through the identified MGCF (segment P), which follows normal CS call-setup procedures through the MSC to which the UE is currently attached (segments M–M). The MGCF, through the *Media Gateway* (*MGW*) it controls, converts from IMS-session setup to CS signaling. An MGW is a translation function that can convert between different bearer services such as TDM for PSTN, ATM for GSM, and IP for IMS.

If a call is IMS originated from the UE, the signaling path for this all-IP path to the CCCF/NeDS function is the straightforward path through the P-CSCF and S-CSCF (segments G–G–H) shown in Figure 10.4.

For the case where the call is CS originated from the UE, there are three alternatives described in [14]. At a high level, all three alternatives first consist of a normal call from the UE into the MSC (segment M). We will not attempt to trace the call chain segments for this permutation as Figure 10.4 is overly simplistic to permit this sequence. The MSC then sends a message to the CCCF/NeDS function to inform it of the call. Then, depending on the particular alternative, either the MSC or CCCF prompts the MGCF to forward the call to the CCCF via an I-CSCF. The CCCF thus terminates the incoming leg and then initiates a session to the originally called party via the S-CSCF via normal call-setup mechanisms.

Our discussion of ICSA signaling is intended to provide the reader with an intuition for the signaling flow of the ICSA solution for VCC services. Figure 10.4 does not provide complete and precise details for a reader who is interested in the precise details of the proposed signaling. For a detailed description of these exchanges, we refer the reader to [14].

We have used the terms *roaming*, *handoffs*, and *handovers* interchangeably throughout this text based on our definitions of these terms. The definition of *handover* in 3GPP vernacular is a radiolayer transition controlled by the RAT, and *roaming* 3GPP is related to changing provider networks. The VCC transitions that we have just described are not RAT controlled, so [14] states that CS-IMS handovers are *not* supported. These VCC transitions certainly do fall into the broader definition of roaming used in our text.

The work that provides support for VCC service will not address all the problems related to providing mobile multimedia services roaming between a 3G access network and 802.11. For example, a data application such as online gaming will require a secure IPSec tunnel between the UE and the 3G network. The mobility provided by the CCCF function and the ICSA model does not address the maintenance of this tunnel as the mobile UE *drags* the tunnel across different access networks. We will see that 3GPP2 plans to address this problem using a new IETF protocol called *MOBIKE* [6], which we will discuss briefly in Section 10.8.

10.7 UMA: A Transitional Step for 3GPP

The advent of UMA technology came about through a consortium of companies that were independent of 3GPP but anxious to see near-term deployments of dual-mode 802.11-GSM handheld devices. Considerable specification work was accomplished by this group, and commercial products are now available. Nevertheless, compared with the sweeping scope of 3GPP-WLAN interworking as undertaken by 3GPP itself, UMA developed a niche transitional solution for 802.11-WLAN interoperability. As we stated earlier in this chapter, the UMA consortium no longer exists as a separate group; its work has all been folded into 3GPP.

While UMA is a transitional technology, this transitional phase may be quite durable, as of this writing, UMA is a commercial reality. A number of deployments of UMA services for public Wi-Fi access are already planned by some established network operators, most notably the following: TeliaSonera, Telecom Italia, France Telecom, and British Telecom [2]. Nokia, Samsung, and Motorola have UMA-capable Wi-Fi/GSM handsets. Kineto offers commercial UMA Network Controllers.

UMA manufacturers claim that seamless roaming between GSM RATs and 802.11 RATs for access to the GSM core network is feasible. The *UMA Network Controller* (*UNC*) is the key interface between the WLAN and the GSM/GPRS core network, as shown in Figure 10.5. In the figure, we see that the UNC connects to the MSC similar to the way in which the GSM base station connects to the MSC. The UNC appears as a network controller to the MSC. The network controller is responsible for radio-layer handovers within a GSM RAT. The HLR function stores a database of user information. The AuC processes SIM-card authentication requests. These two 2G functions were subsumed by the 3G HSS function mentioned earlier. In the same manner that the base station needs to interface with the HLR and AuC functions as part of call setup, the UNC will also provide this service for the UMA mobile station.

On the user side of the UNC, both bearer traffic and signaling are transferred via an IPSec tunnel between the UMA-enabled handset and the UNC. It should be noted that this signaling is *not* compatible with the SIP-based signaling envisioned for use over WLANs in the IMS-based 3GPP specifications discussed earlier, which is shown in Figure 10.3. This drawback is one of the reasons that UMA will ultimately remain a commercialized yet transitional technology.

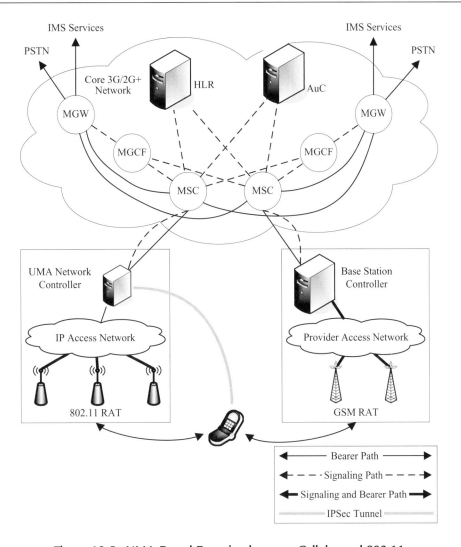

Figure 10.5: UMA-Based Roaming between Cellular and 802.11.

10.8 Third Generation Partnership Program 2

In addition to and separately from 3GPP, active work is proceeding on independent 3GPP2 specifications for interworking with WLAN [23]. Of particular interest are the following three documents: [15], [16], and [17]. In [17], a description of the general requirements and goals for interworking between 802.11 and 3GPP2 is given. Not surprisingly, these broadly align with the goals set out by 3GPP for WLAN interworking. Five levels of interoperability are defined, similar to those described previously for 3GPP.

Chapter 10

The 3GPP2 specification makes very clear the fact that the interworking goal includes session continuity for circuit-switched CDMA2000 voice calls to roam back and forth between WLAN-based VoIP calls. It also mandates session continuity between the various packet-switched CDMA2000 services, including voice, and those same services over the WLAN. A number of requirements related to *network selection* are stipulated in [17]. The CDMA2000 network operator may automatically provision a *preferred* or *forbidden list* of WLAN operators. The network selection may be based on combinations of criteria such as cost or QoS characteristics, and the selection may ultimately be performed by either manual or automatic means. Finally, the user should be notified when the network selection process results in a switch between access networks. As [17] is a requirements document, specific solutions or implementation are not proposed.

In [15], detailed procedures are provided for achieving the second level of interworking between 802.11 and 3GPP2, offering common billing and customer care between the two access technologies. Specific relationships and protocol exchanges between the *WLAN AAA Server* (*W-AAA*), *Broker AAA Server* (*B-AAA*), and the *Home AAA Server* (*H-AAA*) as well as the relationship between the H-AAA and the *Home Location Register/Authentication Center* (*HLR/AuC*) are described. These exchanges are based on Mobile-IP technology. In [15], specific details are also provided on the authentication methods and EAP exchanges that must occur between the MS and the H-AAA in order to authenticate the roaming CDMA2000 user over the WLAN. The two EAP methods that are recommended are EAP-TLS with preshared key and EAP-AKA, both of which we introduced in Chapter 6. Details on the actual access to the CDMA2000 PS services appear in [16]. This level of interworking corresponds to the third level of the five-level hierarchy.

In [16], we see a detailed description of tunnel management and mobility procedures related to accessing the data-packet services in a CDMA2000 network. In order to access the data-packet services of the CDMA2000 network, an IPSec tunnel is established to carry the MS traffic through the WLAN network and Internet to the *Packet Data Interworking Function* (*PDIF*). The PDIF is an IP gateway into a CDMA2000 network, similar to the GGSN function in 3GPP. The IPSec tunnel provides the Mobile-IP functionality of exposing a consistent IP address to the MS across the roam and the authentication and privacy benefits of IPSec protection of that tunnel.

Much of [16] is dedicated to the detailing of the IPSec and AAA-protocol exchanges between the MS, PDIF, V-AAA, B-AAA, and H-AAA entities. The *serving CDMA2000 network* is distinguished from the *home CDMA2000 network*. If the active CDMA2000 user was already in a global-roam situation when the roam to the 802.11 WLAN was initiated, then an IP tunnel will be established between the MS and the serving CDMA2000 network to which the roaming user was most recently connected. The tunnel will *not* extend to the user's home

CDMA2000 network. Access to the home CDMA2000 network is limited to authentication purposes, where the H-AAA and HLR are consulted during the tunnel establishment procedures. If at the time of the roam to the WLAN the user was connected to the user's own home CDMA2000 network, the serving CDMA2000 network and the home CDMA2000 network are the same.

In [16], considerable detail is provided regarding the authentication issues surrounding the establishment of the IPSec tunnel as well as information about how MOBIKE or Mobile IP will be used to maintain a consistent IP address during roaming. This specification does restrict itself to accessing packet-data services and, in this respect, it appears that 3GPP has a more mature set of specifications with regard to roaming between PS-based voice calls in 802.11 networks and CS-based voice calls in 3G networks.

Active liaison work is underway between 3GPP2 and IEEE 802.21. Anecdotal observation indicates that there is less activity happening between 802.21 and 3GPP than between 802.21 and 3GPP2. Although at a high level, the 3GPP2 and 3GPP bodies look at interworking with WLAN similarly, the obvious difference being that the cellular-side technology is CDMA2000 versus one of the native 3GPP radio access technologies.

Based on the history of 2G and 3G cellular deployments and the maturity of the 3G in general, it seems safe to assume that the 3GPP and 3GPP2 specifications will eventually be sufficiently robust to permit roaming at all five levels of interoperability between the RATs that are a native part of the 3G family. As we extend the RATs contemplated by 3GPP and 3GPP2 to include IEEE 802 families, we encounter problems at the level of details. Differences in roaming paradigms render it impossible to simply map the 3GPP RAT interface specifications onto the IEEE 802 interfaces. This situation is not only because of differences in the paradigms but also due to the fact that the 802-interface definitions are not really sufficiently rich to permit intertechnology roaming to the same degree permitted by the 3G specifications.

The IEEE 802.21 standard is the IEEE's effort to create a generic intertechnology roaming interface into which 3GPP RATs, IEEE 802-family RATs as well as yet-to-be-defined access technologies can all present a uniform upper-layer interface to the applications affected by roaming as well as those controlling roaming decisions.

10.9 The 802.21 Standard

The IEEE 802.21 standard, which as of this writing is still in draft form [4], began with the premise that many of the roaming problems have been solved independently by existing access methods. For example, the *Preferred Roaming List* (*PRL*) is a generic concept pioneered in cellular networks, yet foreign to 802.11. The 802.21 standard stipulates a generic set of Information Elements to communicate information similar to the PRL for any underlying network type. As of this writing, existing 802-family networks do not have that information, but the definition in 802.21 identifies the need for that enhancement and provides a standard homogeneous mechanism for communicating that information. It is challenging to homogenize an interface to access technologies that support four fundamentally different paradigms of handover:

- *MN initiated*
- *MN controlled*
- *Network initiated*
- *Network controlled*

We have seen that 802.11 handovers are MN initiated and MN controlled. Cellular handovers are traditionally network initiated and network controlled. The 802.21 standard is an attempt to create a set of interface definitions that accommodate this variety of paradigms.

The 802.21 standard is a classic IEEE standard that clearly defines its scope and attempts to define a complete solution within that scope. Functions that lie outside that scope but may be necessary for a complete solution are often not addressed. This approach may be viewed as either a weakness or a strength of the IEEE. It is certainly different from 3GPP. The scope of 802.21 is captured in the following excerpt from the IEEE 802.21 PAR [5]:

The scope of this project is to develop a standard that shall define media access independent mechanisms that enable the optimization of handover of handover-capable upper layer entities (for example, Mobile IP sessions) between homogeneous or heterogeneous media types both wired and wireless. The standard shall also specify a means to the detection and selection of network attachment points. The scope of the standard must address the full range of upper layer handover strategies in common use. Consideration will be made to ensure compatibility with the 802 architectural model including at least 802, 802.2, 802.1D, 802.1Q, and 802.1X. Consideration will be made to ensure that compatibility is maintained with 802 security mechanisms including 802.1X and that existing security is not compromised. Neither security algorithms nor security protocols shall be defined in the specification. The purpose of the project is to enable mobile devices to handover between 802 networks whether or not they are of different media types, including both wired and wireless, where handover is not otherwise defined and to make it possible for mobile devices to perform seamless handover where the network environment supports it. A further purpose is to provide mechanisms that will also be usable by non 802 access networks, enabling handover between 802 and non 802 networks. This will improve the user experience of mobile devices by improving the available network coverage through the support of multiple media types and preventing the interruption of upper layer sessions during handover.

The 802.21 standard is intended to address part of the problem of *vertical roaming*. The sample underlying access technologies offered in 802.21 are 802.11, 802.16, 802.3, 3GPP2, and the 3GPP family. Future support for other access technologies is assumed. The work is based on the presumption that the decision of when and where to roam is based on many factors, especially when vertical roaming is contemplated. These include QoS, cost, and security. Indeed, the relative importance of these factors may vary from application to application. Solving this problem completely for vertical roaming will be very hard, and a solution is likely to evolve over many years. A solution will also involve vendor-specific steps at different stages of this evolution. This issue is part of what drives the scope of 802.21 to be as narrow as it is.

The 802.21 standard defines generic sets of the following items:

- *Command Services* (*CS*)
- *Asynchronous Event Notification Services* (*ES*)
- *Information Services* (*IS*)

that are media-independent services. These services are accessed through the *Media Independent Handover* (*MIH*) function shown in Figure 10.6. There are multiple possible

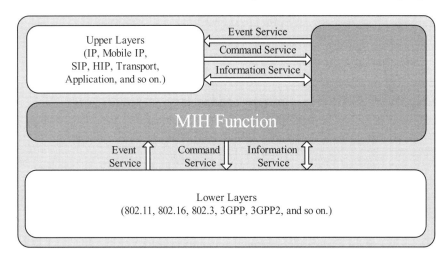

Figure 10.6: MIH Function Location and Key Services. (Courtesy of the IEEE.)

MIH users interfacing to the MIH function. These include application, transport, network, Mobile IP, and network-selector functions.

Although the network-selector function is not listed in Figure 10.6, it is an important upper-layer user of the MIH. The idea is that by presenting such a media-independent interface to a higher layer *network selector* entity, this entity can be application specific when necessary and even improve over time in its ability to select the best network at a given moment for the user. The vendor-specific technology mentioned earlier is most likely to occur inside the network-selector entity. While the interfaces presented to the higher layer network selector are generally media independent, the lower layer interfaces to the different access methods are definitely media dependent.

Figure 10.7 focuses in greater detail on the upper- and lower-layer interfaces into the MIH function. In this figure, we see the example of two 802-family members: 802.16 and 802.11. The 3GPP2 access technology CDMA2000 and the three alternative access technologies of 3GPP: GERAN, UMTS, and LTE are also shown. A *Service Access Point* (*SAP*) is a grouping of service primitives and is often simply called an *interface*. The *MIH-SAP* is the interface to the MIH function from the upper layer entities such as the network-selector entity. The lower layer interface, which in the aggregate is called the *MIH-LINK-SAP*, is shown in Figure 10.7 in its component media-specific SAPs:

- MLME-SAP and LLC-SAP for 802.11

- CS-SAP, C-SAP, and M-SAP for 802.16

- MIH-3GLINK-SAP for both 3GPP and 3GPP2

While Figure 10.7 does not show the 802.3 media type, the 802.21 standard does address 802.3. The media-dependent interface to this 802.3 is *Logical Service Access Point* (*LSAP*). LSAP is also used as an interface to the 802.11 media type. While each of these access

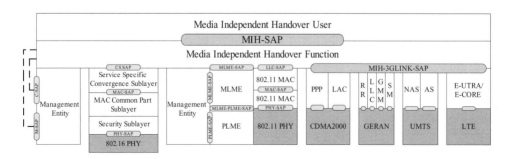

Figure 10.7: MIH Reference Model for Mobile Stations with Multiple Protocol Stacks. (Courtesy of the IEEE.)

Roaming between 802.11 and Other Wireless Technologies

technologies presents some uniqueness in its interface to the MIH, Figure 10.8 shows that they fall into two broad categories, which align with the 802 family and the 3GPP(2) family.

In the 802 family, the IS, CS, and ES services that are media dependent, flow consistently through a clearly defined management interface and then through the appropriate 802 MAC and PHY and communicate with their peer functions one 802 "hop" away. As 802.21 proposes reuse of existing 3GPP solutions when possible, information about layer-two mapping for 3GPP is not as consistent or complete as that provided for the 802 family. On the mediaindependent side of the MIH function, the two families align, both using a higher layer transport such as IP to communicate into the network infrastructure the existing MIH-related network entities such as an MIH Server Controller and MIH Information Database. The functions of an infrastructure-based MIH entity would likely overlap with the CCCF/NeDS functions that we presented in Section 10.6.5.

To the extent that functions such as the CCCF/NeDS and the MIH Server Controller overlap, it is desirable that harmonization efforts be undertaken so that they collapse into a single function. Indeed, the 802.21 standard emphasizes reuse of existing 3GPP and 3GPP2 technology

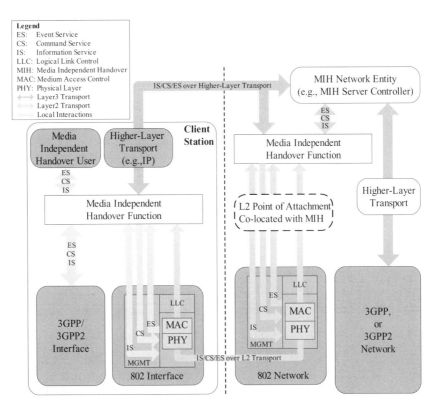

Figure 10.8: MIH Reference Model across Different Networks. (Courtesy of the IEEE.)

293

when it already provides a vertical-roaming function. This reuse minimizes development on 3GPP and 3GPP2 platforms required for the sake of 802.21 compliance. Similarly, the reuse of extant 802-family solutions is desirable. For example, Neighbor Reports that 802.21 defines as part of its information service are based on the Neighbor Reports of 802.11k that we described in Chapter 9. To the extent that the required radio-resource management functions already exist within one of those media interfaces, the 802.21 standard will provide a mapping to that existing function. In those cases where the underlying technology does not yet offer such an interface, the definition provided by 802.21 provides a direction for future enhancements of that interface type. As existing technologies gradually implement the support primitives required for both *horizontal* and *vertical* roaming, the 802.21 standard will provide a common set of interfaces to which they should conform.

10.10 Summary

In this chapter, we have examined the current realities and specification work related to the interworking between 802.11, 3G, and other existing wireless technologies. This kind of interworking falls into a somewhat vague, yet exciting new area broadly called 4G [21]. A variety of names are associated with 4G, including *Fourth Generation Mobile Systems* and IEEE's *3G and Beyond*. Fourth generation includes not only the interworking between existing technologies but also radio transmissions providing digital data rates much higher than those offered now as well as new channel-access schemes and network topologies such as mesh networking. Our book concludes in the next chapter with a look at a few samples of the kinds of research that may transport us beyond our current horizons.

Figures 10.1 and 10.6–10.8 from Draft IEEE Standard, IEEE P802.21/D01.00, March 2006, Copyright IEEE 2006, Draft Amendment to Standard for Information Technology IEEE P802.11r/02.1, May 2006, Copyright IEEE 2006, Interim Contribution 802.11-06/0825r1 2006-05-31, Copyright IEEE 2006, and Interim Contribution 802.11-06/0566r2, Copyright IEEE 2006, all rights reserved.

References

[1] Dean Bubley, The Evolution of SIP- and IMS-Capable Mobile Handsets *End-User Battleground for IMS Services and Disruptive New Entrants*, Disruptive Analysis, Ltd., 2006.

[2] Glen Campbell, J. Feng, and F. Chen, *Telecommunications: 2006 Global Phone Book—Convergence or Collision?* Merrill Lynch Industry Overview, 2006.

[3] Ellen Daley, *Companies Want Wi-Fi/Cellular Calling*, Forrester Research, 2005.

[4] Draft IEEE Standard for Local and Metropolitan Area Networks: Media Independent Handover Services. IEEE P802.21/D01.00, New York, March 2006.

[5] *802.21 Project Authorization Request (PAR) [online]*, Internet, 2007.

[6] Pasi Eronen, IKEv2 Mobility and Multihoming Protocol (MOBIKE), Internet Engineering Task Force, 2005.

[7] Fermin Galan Marquez, Miguel Gomez Rodriguez, Tomas Robles Valladares, Tomas de Miguel, and Luis Angel Galindo, *Interworking of IP Multimedia Core Networks Between 3GPP and WLAN*, IEEE Wireless Communications, 2005.

[8] John Meredith, Specification Corner: Voice Call Continuity in 3GPP, CompactPCI and AdvancedTCA Systems, 2006.

[9] *Qualcomm Joins the Wi-Fi Alliance [online]*, Internet, 2007.

[10] The 3rd Generation Partnership Program, Technical Specification Group Services and System Aspects, *Feasibility Study on 3GPP System to Wireless Local Area Network (WLAN) Interworking (Release 6)*, Technical Report 3GPP TR 22.934 V6.2.0 (2003-9), 3GPP Organizational Partners, Valbonne, France, 2003.

[11] The 3rd Generation Partnership Program, Technical Specification Group Services and System Aspects, *Network Architecture (Release 6)*, Technical Specification 3GPP TS 23.002 V6.4.0 (2004-3), 3GPP Organizational Partners, Valbonne, France, 2004.

[12] The 3rd Generation Partnership Program, Technical Specification Group Services and System Aspects, *IP Multimedia System (IMS) (Release 6)*, Technical Specification 3GPP TS 23.228 V6.5.0 (2004-3), 3GPP Organizational Partners, Valbonne, France, 2004.

[13] The 3rd Generation Partnership Program, Technical Specification Group Services and System Aspects, *3GPP System to Wireless Local Area Network (WLAN) Interworking, System Description (Release 6)*, Technical Specification 3GPP TS 23.234 V6.3.0 (2004-12), 3GPP Organizational Partners, Valbonne, France, 2004.

[14] The 3rd Generation Partnership Program, Technical Specification Group Services and System Aspects, *Voice Call Continuity between CS and IMS Study (Release 7)*, Technical Report 3GPP TR 23.806 V7.0.0 (2005-12), 3GPP Organizational Partners, Valbonne, France, 2005.

[15] The 3rd Generation Partnership Program 2, *CDMA2000 Packet Data Services: Wireless Local Area Networking (WLAN) Interworking: Access to the Internet*, 3GPP2 X.S0028-100-0 V1.0, Organizational Partners of 3GPP2, 2006.

[16] The 3rd Generation Partnership Program 2, *CDMA2000 Packet Data Services: Wireless Local Area Network (WLAN) Interworking: Access to Operator Service and Mobility*, 3GPP2 X.P0028-200 V0.3 Draft, Organizational Partners of 3GPP2, 2006.

[17] The 3rd Generation Partnership Program 2, *CDMA2000 Packet Data Services: Wireless Local Area Network (WLAN) Interworking: Stage 1 Requirements*, 3GPP2 S.R0087-A V1.0, Organizational Partners of 3GPP2, 2006.

[18] The 3rd Generation Partnership Program, Technical Specification Group Services and System Aspects, *Requirements on 3GPP System to Wireless Local Area Network (WLAN) Interworking (Release 7)*, Technical Specification 3GPP TS 22.234 V7.4.0 (2006-6), 3GPP Organizational Partners, Valbonne, France, 2006.

[19] *Wi-Fi/Cellular Roaming Overview [online]*, Internet, 2007.

[20] Wikipedia, *Bluetooth [online]*, Internet, 2007.

[21] Wikipedia, *4G [online]*, Internet, 2007.

[22] Wikipedia, *3GPP [online]*, Internet, 2007.

[23] Wikipedia, *3GPP2 [online]*, Internet, 2007.

[24] Wikipedia, *WiMAX [online]*, Internet, 2007.

CHAPTER 11

Future Directions

11.1 Introduction

11.2 Survey of Ongoing Work Related to 802.11

11.3 A Mobility Model for Studying Wireless Communication

11.4 Conclusions

References

11.1 Introduction

In the earlier chapters of this book, we dealt with 802.11 secure-roaming technology as well as many possible additions and extensions to 802.11. We covered many important historical points and technical details relating to the 802.11 standard. We indicated how rapidly this field of technology has been emerging and addressing new issues. Throughout the book, we have also referred to ongoing work in this field. A great deal of research is taking place related to WLAN technology that attempts to provide a radical leap forward in roaming capabilities. In this chapter, we provide a survey of some recent work.

We provide a list of about a dozen references that describe research in the following three areas: mobile 802.11, security, and Quality of Service. Our goal is to provide the reader with a flavor of the ongoing work. The interested reader will want to pursue these references and also explore the bibliographies contained therein. In order to remain current in this field, one must keep up-to-date with the latest academic research, and perhaps the best way to do this is to conduct your own research on the Web and to attend conferences in this area.

Following the survey of ongoing work, we present a mobility model for studying wireless communications. This model is much more theoretical in nature than the material described previously in the book. The model provides a theoretical framework for examining important

Chapter 11

questions related to the mobile-computing environment. We list examples of the types of questions that are worth exploring in the model. The objective of including such questions is to expose yet another aspect of the ongoing research related to wireless networking. On the one hand, such models must abstract away all but the most important details in order to remain usable. On the other hand, over simplification by such a modeling effort means that the results ascertained from the model have little to no practical implications. This model seems to achieve a good balance.

We end the chapter with a brief section devoted to conclusions.

11.2 Survey of Ongoing Work Related to 802.11

11.2.1 Introduction

In this section, we survey a number of interesting research papers related to 802.11; they cover a wide range of topics. These papers are good to begin with for the reader who is interested in delving more deeply into the latest work in this field. It should be remembered that there is a vast amount of ongoing work, and with this short survey we are only to able to touch on a small fraction of this research.

For each paper we provide a reference, brief remarks, and an abstract. We encourage the interested reader to pursue the paper and the references included in their bibliographies.

11.2.2 Mobile 802.11

In this section, we include a number of papers related to mobile 802.11.

1. Sangeetha Bangolae, Carol Bell, and Emily Qi. *Performance Study of Fast BSS Transition Using IEEE 802.11r*, pages 737–742, Proceedings of the International Conference on Communications and Mobile Computing. Vancouver, British Columbia, Canada, 2006. © [2006] ACM, Inc. Included by permission. doi.acm.org/10.1145/1143549.1143696.

    This paper [1] relates to material that we covered in Chapter 9. The paper observes that roaming quickly is crucial to supporting real-time applications such as VOIP. The authors developed a prototype of the 802.11r standard, where they implemented the security features of the standard. They did not support QoS issues from 802.11e in their prototype. Under their assumptions, the experiments that they conducted demonstrated that the Fast BSS Transition time for a roam in their partial implementation of 802.11r was significantly faster than that of a roam in 802.11i. The conclusion is that 802.11r supports real-time applications such as VoIP better than earlier standards due to faster roams. The authors noted that their work "is the first ever implementation and evaluation of the 802.11r pre-standard."

> *Abstract:* The mass deployment of IEEE 802.11 based wireless local area networks (WLANs), and increased sales in portable devices supporting WLAN, have resulted in an urgent need to support real time applications while wireless users are on the move. This critical support necessitates research into current WLAN roaming capabilities. This paper discusses how WLAN roaming capabilities are affected by new standards such as security (IEEE 802.11i) and quality of service (IEEE 802.11e), and describes the new IEEE 802.11r standard, which was developed to address issues faced by real time applications that implement the security and quality of service enhancements. The performance evaluation of 802.11r prototype and the 802.11i baseline mechanisms show that a voice application using 802.11r can achieve significantly shorter transition time and reduced packet loss during AP-AP transition, and can therefore realize a noticeable improvement in voice quality.

2. Richard Good and Neco Ventura. *A Multilayered Hybrid Architecture to Support Vertical Handover Between IEEE 802.11 and UMTS*, pages 1295–1300, Proceeding of the International Conference on Communications and Mobile Computing. Vancouver, British Columbia, Canada, 2006. © [2006] ACM, Inc. Included by permission. doi.acm.org/10.1145/1143549.1143601.

This paper [4] relates to material that we presented in Chapter 10. The paper explores the issue of vertical handovers in a heterogeneous network that incorporates both SIP and Mobile IP. An architecture is proposed so that the advantages offered by both SIP for real-time applications and by Mobile IP for nonreal time applications can be utilized. A framework was constructed to test the architecture. Although many simplifying assumptions were required, the system which was successfully developed exposed many of the important issues relating to this type of hybrid network, as pertaining to vertical handovers.

> *Abstract:* Telecommunications advances have created the need for a high speed, ubiquitous network capable of catering for diverse application domains. This Next Generation or 4G network can be achieved through the interworking of several existing architectures to form a seamless global network. An important issue involved in interworking is vertical handover. This paper reviews mobility protocols, Mobile IP and the Session Initiation Protocol (SIP), and compares their ability to implement vertical handovers.

Chapter 11

> *A multilayered, hybrid architecture is proposed and described, which incorporates both SIP and Mobile IP. A Mobile IP framework is introduced and evaluated, based on its ability to implement vertical handover.*

3. Michael Hempel, Hamid Sharif, Ting Zhou, and Puttipong Mahasukhon. *A Wireless Test Bed for Mobile 802.11 and Beyond*, pages 1003–1008, Proceedings of the International Conference on Communications and Mobile Computing. Vancouver, British Columbia, Canada, 2006. © [2006] ACM, Inc. Included by permission. doi.acm.org/10.1145/1143549.1143750.

This paper [6] describes a simulation model to study WLANs where users are traveling on trains. The trains travel at several different velocities. The paper describes an actual implementation where APs were established along a stretch of train tracks in Nebraska. The testbed was used to study maximum obtainable network throughput, as well as questions relating to roaming and handoffs. The initial results suggest that 802.11b performs well for mobility considerations. The authors also plan to test other standards.

> **Abstract:** *In this paper, we present our approach of establishing a wireless test bed as a part of our collaborations with the Federal Railroad Administration (FRA) for studying the performance of current and upcoming wireless technologies in a mobile railroad environment. The focus is on studying the impact of mobility on the wireless system throughput for moving trains with different velocities. We describe details of our test bed design including selection of the location, equipment as well as system topology and performance evaluations approach. We also present and discuss test results obtained from our test bed.*

4. Yong Liao and Lixin Gao. *Practical Schemes for Smooth MAC Layer Handoff in 802.11 Wireless Networks*, pages 181–190, Proceedings of the International Symposium on World of Wireless, Mobile and Multimedia Networks. Niagara Falls, Buffalo, NY, 2006.

This paper [8] is related to our discussion in Chapter 8. The authors describe a couple of new algorithms for reducing packet delay and loss during handoffs. These algorithms divide the scanning of channels into subgroups, for example, rather than scanning n channels at once, n/g channels are scanned; there is a period of transmission; and then the next group of channels is scanned; and so on. Parameters relating to RSS are adjusted so that scan operations begin a little earlier than they usually would. The authors have implemented and tested their algorithms in software. The experiments indicate that packet loss and delay are significantly reduced. For applications such as VoIP, their algorithms seem very promising. It is also worth noting that no changes or updates are needed to APs but only to clients. This fact makes deployment of this technology much more feasible.

> *Abstract:* The limited service range of the access points demands mobile wireless stations to handoff frequently between different cells in the IEEE 802.11 infrastructure networks. However, the handoff scheme used in the current 802.11 infrastructure networks is far from graceful. In this paper, we propose a smooth MAC-layer handoff scheme and a greedy smooth MAC-layer handoff scheme. Our schemes scan channels in a smooth manner so that the handoff can have less impact on upper-layer applications. In order to limit the frequency of channel scanning, an adaptive mechanism is used to dynamically adjust the threshold triggering the channel scanning operations. We have implemented our handoff schemes using commodity 802.11 devices and extensive experiments have been conducted to evaluate the performance. The experimental results demonstrate that our schemes reduce packet delay and loss during handoff. Our handoff schemes are implemented in the client side only and do not require changes to access points. Therefore, our schemes can be deployed in the existing 802.11 networks.

5. Vivek Mhatre and Konstantina Papagiannaki. *Using Smart Triggers for Improved User Performance in 802.11 Wireless Networks*, pages 246–259, Proceedings of the 4th International Conference on Mobile Systems, Applications and Services. Uppsala, Sweden, 2006. © [2006] ACM, Inc. Included by permission. doi.acm.org/10.1145/1134680.1134706.

This paper [10] expands on notious that we discussed in Chapters 5 and 7. The goal is to develop handoff algorithms that continuously monitor "The quality of the links of all the APs operating on the client's current and overlapping channels." This technique is in contrast to most systems where handoffs are triggered just by loss of connectivity or poor performance. The idea of this work is to use more information to make more intelligent decisions about which AP to roam to next. The authors implemented their system and ran a series of experiments on a campus network. The results reported are promising, and the authors noticed a reduction by more than a factor of two in the average handoff delay. Plans are underway to extend the experiments to test overlapping channels rather than just the client's current channel.

> *Abstract:* The handoff algorithms in the current generation of 802.11 networks are primarily reactive in nature, because they wait until the link quality degrades substantially to trigger a handoff. They further rely on instantaneous signal strength measurements when choosing the best AP. This

approach leads to handoff delays on the order of 1–2 seconds that are unacceptable for delay sensitive applications such as VoIP. We propose a fundamentally new approach to handoffs that is based on continuous monitoring of wireless links. In our approach, a client measures the beacon strengths of all the APs operating on the current, and the overlapping channels, and makes its handoff decisions based on the long-term, and short-term trends in these signals. We show through experiments in a campus wireless network that our proposed algorithms result in more than 50% reduction in average handoff delays, while having the potential to improve overall user performance. Our algorithms have been implemented in today's hardware, and unlike other proposed roaming algorithms in the literature, need no infrastructure support.

6. Anthony J. Nicholson, Yatin Chawathe, Mike Y. Chen, Brian D. Noble, and David Wetherall. *Improved Access Point Selection*, pages 233–245, Proceedings of the 4th International Conference on Mobile Systems, Applications and Services. Uppsala, Sweden, 2006. © [2006] ACM, Inc. Included by permission. doi.acm.org/10.1145/1080730.1080732.

This paper [11] expands on concepts that we discussed in Chapters 5 and 7. The idea is to develop an algorithm for access-point selection that incorporates more information than just RSS. In fact, the authors ran one set of experiments which showed that selection of APs by RSSs performs about the same as random selection. The authors implemented their selection algorithm in a system called Virgil, which takes into account *Round Trip Times (RTTs)* to the reference server, estimated bandwidth of the reference server, and port availability. By caching information that was discovered earlier with regard to APs, the overhead of the system drops to an acceptable level.

> ***Abstract:*** *This paper presents Virgil, an automatic access point discovery and selection system. Unlike existing systems that select access points based entirely on received signal strength, Virgil scans for all available APs at a location, quickly associates to each, and runs a battery of tests to estimate the quality of each AP's connection to the Internet. Virgil also probes for blocked or redirected ports, to guide AP selection in favor of preserving application services that are currently in use. Results of our evaluation across five neighborhoods in three cities show Virgil finds a usable connection from 22% to 100% more often than selecting based on signal strength alone. By caching AP test results, Virgil both improves performance and success rate. Our overhead is acceptable and is shown to be faster than manually selecting an AP with Windows XP.*

7. Ahmed H. Zahran, Ben Liang, and Aladdin Saleh. *Signal Threshold Adaptation for Vertical Handoff in Heterogeneous Wireless Networks*. Mobile Networks and Appliances **11**(4):625–640, August 2006.

This paper [12] relates to material that we presented in Chapter 10. Vertical handoffs between 3G cellular and WLANs are examined. The authors deal with *Moving Out* (*MO*) of and *Moving Into* (*MI*) the preferred network. This paper includes a good survey of related work.

> *Abstract: The convergence of heterogeneous wireless access technologies has been envisioned to characterize the next generation wireless networks. In such converged systems, the seamless and efficient handoff between different access technologies (vertical handoff) is essential and remains a challenging problem. The heterogeneous co-existence of access technologies with largely different characteristics results in handoff asymmetry that differs from the traditional intranetwork handoff (horizontal handoff) problem. In the case where one network is preferred, the vertical handoff decision should be carefully executed, based on the wireless channel state, network layer characteristics, as well as application requirements. In this paper, we study the performance of vertical handoff using the integration of 3G cellular and wireless local area networks as an example. In particular, we investigate the effect of an application-based signal strength threshold on an adaptive preferred-network lifetime-based handoff strategy, in terms of the signaling load, available bandwidth, and packet delay for an inter-network roaming mobile. We present an analytical framework to evaluate the converged system performance, which is validated by computer simulation. We show how the proposed analytical model can be used to provide design guidelines for the optimization of vertical handoff in the next generation integrated wireless networks.*

11.2.3 Security

In this section, we include two of the papers related to security in wireless networks as we have focused a great deal of attention on the security aspect of the roaming problem.

1. Ezedin S. Barka, Emad Eldin Mohamed, and Kadhim Hayawi. *End-to-end Security Solutions for WLAN: A Performance Analysis for the Underlying Encryption Algorithms in the Lightweight Devices*, pages 1295–1300, Proceeding of the International Conference on Communications and Mobile Computing. Vancouver, British

Columbia, Canada, 2006. © [2006] ACM, Inc. Included by permission. doi.acm.org/10.1145/1143549.1143809.

Although this paper [2] does not specifically address the roaming problem, it does address security issues in wireless networks. The authors examine the performance characteristics of RC4 and AES in terms of processing power, memory consumption, and time on portable devices. The conclusion drawn from the experiments that the authors conducted is that RC4 requires less processing power and memory than AES; therefore, RC4 is the preferred choice on systems where there are very limited resources.

> ***Abstract:*** *The advances in the wireless technology, represented by the improved computational capabilities of third generation (3G) and fourth generation (4G) wireless devices and the wider bandwidth wireless networks, make it possible to support a variety of security-sensitive applications such as m-banking and m-commerce. This, however, requires increasingly robust End-to-End security solutions. The contribution of this paper is twofold. First, it presents a design and an implementation of a light weight application-level security solution for handheld devices in wireless LAN. Second, it analyzes the impact of the processing power, time, and memory on the performance of two of the widely known encryption algorithms—RC4 and AES—used by the lightweight handheld devices in the wireless network environment. The work in this paper uses pure Java components to provide end-to-end client authentication and data confidentiality and integrity between wireless J2ME-based clients and J2ME-based servers.*

2. Ulrike Meyer, Jared Cordasco, and Susanne Wetzel. *An Approach to Enhance Interprovider Roaming through Secret Sharing and Its Application to WLANs*, pages 1–13, Proceedings of the 3rd ACM International Workshop on Wireless Mobile Applications and Services on WLAN Hotspots. Cologne, Germany, 2005. © [2005] ACM, Inc. Included by permission. doi.acm.org/10.1145/1080730.1080732.

This paper [9] introduces a new EAP type based on key splitting. The goal is to provide more secure interprovider roaming through secret sharing. This task is accomplished by the Home Network retaining a part of the secret key for each Foreign Network. Then "An authentication based on the roaming key can thus not only prove FN's authorization by HN to MD but also prove FN's identity to MD." The authors illustrate how their techniques can be applied to billing in micropayment schemes. The paper includes a survey of research on "InterProvider Roaming in Public WLANs" and an extensive list of references.

Abstract: In this paper, we show how secret sharing can be used to address a number of shortcomings in state-of-the-art public-key-based inter-provider roaming. In particular, the new concept does not require costly operations for certificate validation by the mobile device. It furthermore eliminates the need for a secure channel between providers upon roaming. We demonstrate the new approach by introducing a new protocol, EAP-TLS-KS, for roaming between 802.11i-protected WLANs. In addition, we show that the properties of EAP-TLS-KS allow for an efficient integration of a micropayment scheme.

11.2.4 Quality of Service

In this section, we include two papers related to Quality of Service considerations in wireless networks.

1. Ozgur Ekici and Abbas Yongacoglu. *A Novel Association Algorithm for Congestion Relief in IEEE 802.11 WLANs*, pages 725–730, Proceeding of the International Conference on Communications and Mobile Computing. Vancouver, British Columbia, Canada, 2006. © [2006] ACM, Inc. Included by permission. `doi.acm.org/10.1145/1143549.1143694`.

   Although this paper [3] is not directly applicable to roaming, it addresses technology at the MAC layer, which influences roaming that will occur in a given ESS. This work introduces the *predictive association algorithm* as a way of associating STAs to APs. The algorithm attempts to associate STAs to APs in a more intelligent manner than relying solely on RSS. The idea is to maximize available data rates and do some load balancing on APs. The authors conducted a series of experiments, in which their algorithm was compared with existing algorithms. The results indicate that the new algorithm is very promising.

   Abstract: Many wireless local area network (WLAN) performance estimations are done with the assumption of uniformly distributed stations (STAs). In practice, on the contrary STAs are distributed unevenly among access points (APs), causing hot-spots and under utilized APs in a wireless network. Considering a WLAN is made up of multiple APs, having some APs carrying excessive loads (i.e. hot-spots) degrades both the considered APs as well as the overall network performance. The system performance can be improved by associating incoming STAs effectively throughout the network, in a sense to balance the network load evenly between APs and relieve the hot-spot congestion. Currently employed user association method in IEEE 802.11 WLANs considers only the received signal strength of APs at STAs, and associates STAs to the closest (in signal strength sense) AP, ignoring its load and interference value. Novel user association algorithms are required for congestion relief and network performance improvement. In this work, a new distributed association algorithm taking into consideration not only the received signal strength of the APs at STAs but also AP loadings and interference is proposed. A new AP load calculation method acknowledging the interference between STAs and APs is presented. Our simulations demonstrate that the proposed algorithm can improve the overall system throughput performance more than 50% and offers a better load distribution across the network compared to conventional association algorithm.

2. Ming Li, Hua Zhu, Imrich Chlamtac, and Balakrishnan Prabhakaran. *End-to-end QoS framework for Heterogeneous Wired-Cum-Wireless Networks*, Wireless Networks **12**(4):439–450, 2006.

This paper [7] proposes a new protocol for flow reservation and admission control in wireless networks. A large number of simulations using the ns-2 network simulator were conducted to evaluate the protocol, including an interesting analysis of the number of handoffs and percentages of handoffs blocked during roaming. As expected, when MDs are blocked from roaming to a heavily loaded AP, the QoS of those already at the AP continues at the same level and does not degrade. Of course, depending on the density of APs in the ESS, if an MD is blocked and traveling, then that MD could completely lose connectivity. It would be interesting to see how the authors' protocol behaves in the real world.

> ***Abstract:*** *With information access becoming more and more ubiquitous, there is a need for providing QoS support for communication that spans wired and wireless networks. For the wired side, RSVP/SBM has been widely accepted as a flow reservation scheme in IEEE 802 style LANs. Thus, it would be desirable to investigate the integration of RSVP and a flow reservation scheme in wireless LANs, as an end-to-end solution for QoS guarantee in wired-cum-wireless networks. For this purpose, we propose WRESV, a lightweight RSVP-like flow reservation and admission control scheme for IEEE 802.11 wireless LANs. Using WRESV, wired/wireless integration can be easily implemented by cross-layer interaction at the Access Point. Main components of the integration are RSVP-WRESV parameter mapping and the initiation of new reservation messages, depending on where senders/receivers are located. In addition, to support smooth roaming of mobile users among different basic service sets (BSS), we devise an efficient handoff scheme that considers both the flow rate demand and network resource availability for continuous*

QoS support. Furthermore, various optimizations for supporting multicast session and QoS re-negotiation are proposed for better performance improvement. Extensive simulation results show that the proposed scheme is promising in enriching the QoS support of multimedia applications in heterogeneous wired-cum-wireless networks.

11.3 A Mobility Model for Studying Wireless Communication

This section was coauthored with Dr. Sanapawat Kantabutra, and much of this material appears in [5].

11.3.1 Introduction

In this section, we describe a communications model for a *mobile network*, that is, a wireless network in which *the access points themselves may be moving* and the distribution service interconnecting the access points is wireless based. Such networks are of great importance in supporting relief efforts for natural disasters such as earthquakes, tsunamis, tornadoes, and hurricanes. In addition to these purely humanitarian uses, such networks are of great interest to the military for conducting field operations. We believe that additional uses of mobile networks will emerge over time, as technology continues to improve. We first present the model's definition. An instance of the model is then illustrated by an example, and a few interesting research problems related to this model are defined. This presentation should give the reader a feel for the type of modeling research that takes place in academia.

11.3.2 The Mobility Model

We will refer to the model for mobile wireless communications as the *mobility model*. On the one hand, we want to be able to model actual mobile networks. Thus, the model needs to be sophisticated enough to model complex real-life situations. On the other hand, we need to abstract out the key features of a mobile network so that the model is feasible to study and apply. The model strikes a suitable balance among the model's complexity, its ability to describe real-world situations, and its simplicity and ease of use. Once the model has been defined, we present a communication protocol (see page 309) to interpret the model. Let $\mathbb{N} = \{1, 2, 3, ...\}$. The set S^k for $k \in \mathbb{N}$ is the k-fold Cartesian product of S. Following that we present an instance of the model.

Definition of the Model

The model is defined to operate on a 2-dimensional grid. A *mobility model* is an 8-tuple $(\mathcal{S}, \mathcal{D}, \mathcal{U}, \mathcal{L}, \mathcal{R}, \mathcal{V}, \mathcal{C}, O)$, where

Chapter 11

1. The set $S = \{s_1, s_2, \ldots, s_m\}$ is a finite collection of *sources*, where $m \in \mathbb{N}$. The value m is the *number of sources*. Corresponding to each source s_i, for $1 \leq i \leq m$, an *initial location* (x_i, y_i) is specified, where $x_i, y_i \in \mathbb{N}$.

2. The set $\mathcal{D} = \{000, 001, 010, 101, 110\}$ is called the *directions*, and these values correspond to no movement, east, west, south, and north, respectively.

3. The set $\mathcal{U} = \{u_1, \ldots, u_p\}$ is a finite collection of *mobile devices*, where $p \in \mathbb{N}$. The set $\mathcal{U}$ is called the set of *users*. The value p is called the *number of users*. Corresponding to each user u_i, for $1 \leq i \leq p$, an initial location (x_i, y_i) is specified, where $x_i, y_i \in \mathbb{N}$.

4. The set $\mathcal{L} = \{l_1, \ldots, l_t\}$ is a finite collection of "bit strings," where $t \in \mathbb{N}$ and $l_i \in \mathcal{D}^t$ for $1 \leq i \leq t$. Each group of three bits in l_i, beginning with the first three, defines a step in a given direction for the user u_i's movement or no movement at all if the string is 000. The value t is called the *duration of the model*.

5. Let $t(i) \in \mathbb{N}$ for $1 \leq i \leq m$. The set $\mathcal{R} = \{r_1, r_2, \ldots, r_m\}$ is a finite collection of "bit strings," where $r_i \in \mathcal{D}^{t(i)}$ for $1 \leq i \leq m$. Each group of three bits in r_i beginning with the first three defines a step in a given direction for the source s_i's movement or no movement at all if the string is 000. The set $\mathcal{R}$ is called the *random walks* of the mobility model.

6. The set $\mathcal{V} = \{v_1, v_2, \ldots, v_m\}$ is a finite collection of numbers, where $v_i \in \mathbb{N}$. The value v_i is the corresponding number of steps from r_i per unit time that s_i will take. This set is called the *velocities*.

7. The set $C = \{c_1, c_2, \ldots, c_m\}$ is a finite collection of lengths, where $c_i \in \mathbb{N}$. The value c_i is the corresponding diameter of the circular coverage of source s_i. This set is called the *coverages*.

8. The set $O = \{(x_1, y_1, x_2, y_2) \mid x_1, y_1, x_2, y_2 \in \mathbb{N}, x_2 > x_1, \text{ and } y_2 > y_1\}$ is a finite collection of rectangles in the plane. This set is called the *obstacles*.

Several remarks are in order about the definition. The model is defined on a 2-dimensional grid for simplicity, but it would certainly be interesting to extend the model to the 3-dimensional case. The sources in S correspond to wireless access points. They are broadcasting and receiving signals. Although real mobile sources do not move in discrete steps, by using a sufficiently fine grid, we lose little information by assuming that the sources are always at grid point locations.

The set $\mathcal{D}$ represents the usual four possible directions for movement in the grid, plus no movement at all. The set $\mathcal{U}$ represents users with mobile devices. We have modeled the movement of the users by random walks contained in the set L. Although we have assumed that all the walks have the same length, this convention is not really a restriction as we can simply pad out shorter walks using the no-movement bit string 000 from $\mathcal{D}$. For the sake of simplicity, we have assumed that all users travel at the same velocity. The users move to their new locations in unit steps instantaneously.

The movement of the sources is modeled by random walks contained in the set $\mathcal{R}$. To accommodate for different velocities, the walks in $\mathcal{R}$ have different lengths. In real-life situations, mobile-access points move around at different speeds, for example, a Hummer may be traveling at speeds in excess of 100 kilometers per hour, whereas an elephant working his way through dense brush may be moving at 1 kilometer per hour. The relative speeds of the sources are represented by natural numbers contained in the set $\mathcal{V}$. Of course, a given source may not always travel at a constant velocity. It would be worth examining an extension of the model where any source's speed can change over time.

Different sources will broadcast at different signal strengths depending on a variety of factors, the main one being the amount of power available. The various signal strengths are represented by specifying the diameter of a circle c_i for each source indicating where its signal can be received. This region is called the *coverage area*. As buildings and other obstacles may interfere with signal transmission, the model incorporates a set of obstacles O. For simplification, we only permit rectangular obstacles.

We now turn the discussion to the communication protocol, which will allow us to illustrate how the model is used. The sources are always on; they are always broadcasting and listening. Users with mobile devices are moving in and out of the range of each other and various sources. Mobile devices would like to communicate (send and receive messages) with one another. We specify the manner in which they may communicate in the following. Let $k > 2$ and $k \in \mathbb{N}$.

- At a given instance in time, any two sources with overlapping coverage areas may communicate with each other in full duplex fashion as long as the intersection of their overlapping coverage area is not completely contained inside obstacles. We say that these two sources are *currently in range*. A series $s_1, s_2, \ldots, s_k$ of sources are said to be *currently in range* if s_i and s_{i+1} are currently in range for $1 \leq i \leq k-1$.

- Two mobile devices cannot communicate directly with one another.

- A mobile device D_1 always communicates with another mobile device D_2 through a source or series of sources as defined subsequently. The mobile devices D_1 at location (x_1, y_1) and D_2 at location (x_2, y_2) *communicate through a single source s* located at (x_3, y_3), if at a given instance in time the lines between points (x_1, y_1) and (x_3, y_3), and points (x_2, y_2) and (x_3, y_3) are within the area of coverage of s and do not intersect with any obstacle from O. The mobile devices D_1 at location (x_1, y_1) and D_2 at location (x_2, y_2) *communicate through a series of sources* s_1 at location (a_1, b_1), s_2 at location (a_2, b_2), ..., and s_k at location (a_k, b_k) that are currently in range if the line between points (x_1, y_1) and (a_1, b_1) is inside s_1's coverage area and does not intersect any obstacle from O, and the line between points (x_2, y_2) and (a_k, b_k) is inside s_k's coverage area and does not intersect any obstacle from O.

The model captures both the mobility of the sources and the users. This reflects the situation in a real mobile network where access points and users are moving around. For simplicity, the model implicitly assumes that all users are moving at the same rate of speed, whereas we explicitly modeled sources moving at different velocities. It should be noted that the model can handle users moving at different rates of speed by having some users remain stationary, whereas others are moving at each step. In the next section, we present an example of the model.

A Sample Instance of the Model

To illustrate the mobility model, we subsequently provide a specific instance. Let $M = (S, \mathcal{D}, \mathcal{U}, \mathcal{L}, \mathcal{R}, \mathcal{V}, C, O)$ be defined as follows:

1. Let $S = \{s_1, s_2, s_3, s_4\}$ with initial locations $(2,5)$, $(5,5)$, $(6,4)$, and $(5,2)$, respectively.

2. Let $\mathcal{D} = \{000, 001, 010, 101, 110\}$.

3. Let $\mathcal{U} = \{u_1, u_2, u_3\}$ with initial locations $(3,4)$, $(2,1)$, and $(6,2)$, respectively.

4. Let $t = 3$ and $\mathcal{L} = \{l_1, l_2, l_3\}$, where $l_i = (000, 000, 000)$ for $1 \leq i \leq 3$.

5. Let $\mathcal{R} = \{r_1, r_2, \ldots, r_4\}$. For clarity, the figure only illustrates $r_1 = (101, 001, 101)$ and omits the other r_i's, which we assume are all $(000, 000, 000)$ except for r_2, which is twice as long.

6. Let $\mathcal{V} = \{1, 2, 1, 1\}$.

7. Let $C = \{2, 2, 2, 4\}$.

8. Let $O = \{(2, 1, 4, 2)\}$.

Figure 11.1 illustrates this model instance M graphically. In this case, there are three stationary users. There are four sources s_1, s_2, s_3, and s_4 who are located at $(2,5)$, $(5,5)$, $(6,4)$, and $(5,2)$, respectively. An obstacle in this figure is the rectangle defined by the lower-left coordinate

Future Directions

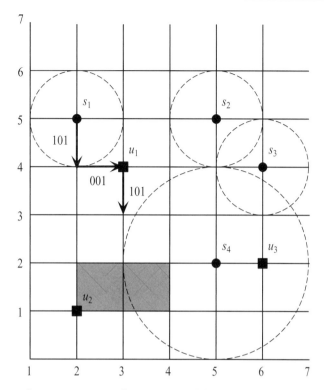

Figure 11.1: Sample Instance of the Mobility Model.

$(3,2)$ and the upper-right coordinate $(5,3)$. Sources s_1, s_2, and s_3 each have a coverage with a diameter 2, and s_4 has a coverage with a diameter 4. The steps of s_1 at initial location $(2,5)$ are defined by r_1. In this case, s_1 moves south in the first step, east in the second step, and south in the third step. The moves are made with a velocity of $v_1 = 1$ or one step per unit of time.

It should be noted that initially, for example, sources s_2 and s_3 are currently in range, sources s_2, s_3, and s_4 are a series of sources currently in range, and sources s_1 and s_2 are not currently in range. Initially, users u_1 and u_3 cannot communicate either by a source or a series of sources. After three steps, u_1 can communicate with u_3 through the series of sources s_1 and s_4.

11.3.3 Problem Definitions

In this section, a few interesting research problems related to the mobility model are defined.

User Communication Problem

INSTANCE: A mobility model $(\mathcal{S}, \mathcal{D}, \mathcal{U}, \mathcal{L}, \mathcal{R}, \mathcal{V}, \mathcal{C}, \mathcal{O})$, two designated users u_a and u_b from $\mathcal{U}$, and a time k.

QUESTION: Can users u_a and u_b communicate at time k?

Sources Reachability Problem

INSTANCE: A mobility model (S, D, U, L, R, V, C, O), two designated sources s_a and s_b from S, and a time k.

QUESTION: Are sources s_a and s_b in range at time k?

Access-Point Location Problem

INSTANCE: A mobility model (S, D, U, L, R, V, C, O), two designated users u_a and u_b from U, an access point diameter d, and a natural number k.

QUESTION: Can users u_a and u_b communicate if k or fewer access points of diameter d are appropriately placed in the grid?

Access-Point Placement Problem

INSTANCE: Two mobility models $M = (S, D, U = \{u_1, u_2\}, L, R, V, C, O)$ and $M' = (S', D, U = \{u_1, u_2\}, L, R', V', C', O')$.

QUESTION: Can u_1 and u_2 communicate for more steps in model M than they can in model M'?

Obstacle Removal Problem

INSTANCE: A mobility model (S, D, U, L, R, V, C, O), two designated users u_a and u_b from U, and a natural number k.

QUESTION: Can u_a and u_b communicate throughout the duration of the model if at most k obstacles are removed?

Including a discussion of the computational complexity of these decision problems is beyond the scope of this book. It is sufficient to say that one can make many revealing statements about the time required to solve numerous other interesting issues related to the mobility model by solving problems such as these.

11.4 Conclusions

This book has covered a lot of ground relating to the 802.11 standard and safe roaming. The research material in this chapter certainly indicates that the field is healthy and will continue to evolve rapidly. With the ever increasing number of deployments of 802.11 networks, we see an even greater need for more efficient roaming. At no time in the past have we ever seen a decrease in the need for computer security, and we fully expect to see an even greater emphasis and effort placed on the ability to roam securely in 802.11 networks over time. The future of this field looks bright with many interesting problems that are yet to be solved and many interesting modeling challenges that remain to be developed.

A query entered into Google of the title of this book

Roaming Securely in 802.11 Networks

in March of 2007 turned up 72,500 hits. Interestingly, the same query with the word "in" deleted turned up a whopping 479,000 hits, and a query of

Roaming 802.11 Networks

turned up 894,000. These results indicate that there is a lot of past and current work taking place relating to 802.11. We hope that you will continue to explore the roaming issue in 802.11 and make your own contributions to the field, just as we have tried to make ours in this work ...

References

[1] Sangeetha Bangolae, Carol Bell, and Emily Qi, *Performance Study of Fast BSS Transition Using IEEE 802.11r*, pages 737–742, Proceedings of the International Conference on Communications and Mobile Computing. Vancouver, British Columbia, Canada, 2006. `doi.acm.org/10.1145/1143549.1143696`.

[2] Ezedin S. Barka, Emad Eldin Mohamed, and Kadhim Hayawi, *End-to-end Security Solutions for WLAN: A Performance Analysis for the Underlying Encryption Algorithms in the Lightweight Devices*, pages 1295–1300, Proceeding of the International Conference on Communications and Mobile Computing. Vancouver, British Columbia, Canada, 2006. `doi.acm.org/10.1145/1143549.1143809`.

[3] Ozgur Ekici and Abbas Yongacoglu, *A Novel Association Algorithm for Congestion Relief in IEEE 802.11 WLANs*, pages 725–730, Proceeding of the International Conference on Communications and Mobile Computing. Vancouver, British Columbia, Canada, 2006. `doi.acm.org/10.1145/1143549.1143694`.

[4] Richard Good and Neco Ventura, *A Multilayered Hybrid Architecture to Support Vertical Handover Between IEEE 802.11 and UMTS*, pages 1295–1300, Proceeding of the

International Conference on Communications and Mobile Computing. Vancouver, British Columbia, Canada, 2006. doi.acm.org/10.1145/1143549.1143601.

[5] Raymond Greenlaw and Sanpawat Kantabutra, *A Mobility Model for Studying Wireless Communications*, The 15th International Conference of Forum for Interdisciplinary Mathematics on Interdisciplinary Mathematical & Statistical Techniques, Shanghai, China, 2007.

[6] Michael Hempel, Hamid Sharif, Ting Zhou, and Puttipong Mahasukhon, *A Wireless Test Bed for Mobile 802.11 and Beyond*, pages 1003–1008, Proceedings of the International Conference on Communications and Mobile Computing. Vancouver, British Columbia, Canada, 2006. doi.acm.org/10.1145/1143549.1143750.

[7] Ming Li, Hua Zhu, Imrich Chlamtac, and Balakrishnan Prabhakaran, *End-to-end QoS framework for Heterogeneous Wired-Cum-Wireless Networks*, Wireless Networks **12**(4):439–450, 2006.

[8] Yong Liao and Lixin Gao, *Practical Schemes for Smooth MAC Layer Handoff in 802.11 Wireless Networks*, pages 181–190, Proceedings of the International Symposium on World of Wireless, Mobile and Multimedia Networks. Niagara Falls, Buffalo, NY, 2006.

[9] Ulrike Meyer, Jared Cordasco, and Susanne Wetzel, *An Approach to Enhance Inter-provider Roaming through Secret Sharing and Its Application to WLANs*, pages 1–13, Proceedings of the 3rd ACM International Workshop on Wireless Mobile Applications and Services on WLAN Hotspots. Cologne, Germany, 2005. doi.acm.org/10.1145/1080730.1080732.

[10] Vivek Mhatre and Konstantina Papagiannaki, *Using Smart Triggers for Improved User Performance in 802.11 Wireless Networks*, pages 246–259, Proceedings of the 4th International Conference on Mobile Systems, Applications and Services. Uppsala, Sweden, 2006. doi.acm.org/10.1145/1134680.1134706.

[11] Anthony J. Nicholson, Yatin Chawathe, Mike Y. Chen, Brian D. Noble, and David Wetherall. *Improved Access Point Selection*, pages 233–245, Proceedings of the 4th International Conference on Mobile Systems, Applications and Services. Uppsala, Sweden, 2006. doi.acm.org/10.1145/1080730.1080732.

[12] Ahmed H. Zahran, Ben Liang, and Aladdin Saleh. *Signal Threshold Adaptation for Vertical Handoff in Heterogeneous Wireless Networks*. Mobile Networks and Appliances **11**(4):625–640, August 2006.

Appendix A
Acronyms and Abbreviations

The acronyms and abbreviations are sorted based on themselves, and not their expansions, except in the case of ties.

A

AAA	Authentication, Authorization, and Accounting
ABET	Accreditation Board for Engineering Technology
ABR	Arbitrary Bit Rate
AC	Access Controller
ACK	Acknowledgment
ACS	Access Control Server
AES	Advanced Encryption Standard
AFSK	Audio Frequency Shift Keying
AIUR	Air Interface User Rate
AMPS	Advanced Mobile Phone Service
ANSI	American National Standards Institute
AOL	America Online
AP	Access Point
AR	Access Router
ARP	Address Resolution Protocol
ARP	Autoradiopuhelin
ARPANET	Advanced Research Projects Agency Network
AT&T	American Telephone & Telegraph
AuC	Authentication Center

B

B-AAA	Broker AAA Server
BSC	Base Station Controller
BSS	Base Station Subsystem
BSS	Basic Service Set
BSSID	Basic Service Set Identifier
BTK	Base Transient Key
BTS	Base Transceiver Station

C

CA	Certificate Authority
CAC	Computing Accreditation Commission
CAPWAP	Control and Provisioning of Wireless Access Points
CBC	Cipher Block Chaining
CBR	Constant Bit Rate
CCA	Clear Channel Assessment
CCCF	Call Control Continuity Function
CCI	Cochannel Interference
CCK	Complimentary Code Keying
CCKM	Cisco's Centralized Key Management
CCLS	Client Configured SSID List
CCMP	Counter Mode CBC-MAC Protocol
CCX	Cisco Compatible Extensions
CDMA	Code Division Multiple Access
CDPD	Cellular Digital Packet Data
CDSL	Current Driver Scan List
CEO	Chief Executive Officer
CEPT	Council of European Postal and Telecommunications Administrations
CF	Contention Free
CFP	Contention Free Period
CGF	Charging Gateway Function
CHAP	Challenge Handshake Authentication Protocol
CIDR	Classless Inter-Domain Routing
CIO	Chief Information Officer
CN	Correspondent Node
COA	Care Of Address
CP	Contention Period
CPU	Central Processing Unit
CS	Circuit Switched
CS	Command Services
CSMA/CA	Carrier Sense Multiple Access with Collision Avoidance
CSMA/CD	Carrier Sense Multiple Access with Collision Detection
CTA	Cisco Trust Agent
CTS	Clear To Send

D

DA	Destination Address
dB	Decibels

dBm	Decibels with a reference level of milli Watts
DCF	Distributed Control Function
DCS	Digital Cellular System
DDoS	Distributed Denial of Service
DECT	Digital Enhanced Cordless Telephone
DFS	Dynamic Frequency Selection
DHCP	Dynamic Host Configuration Protocol
DOI	Digital Object Identifier
DoS	Denial of Service
DS	Direct Sequence
DS	Distribution Service
DSL	Digital Subscriber Line
DSSS	Direct Sequence Spread Spectrum
DYN WEP	Dynamic Wired Equivalency Protocol

E

EAP	Extensible Authentication Protocol
EAPOL	EAP Over LAN
EAP-TLS-KS	Extensible Authentication Protocol, Transport Layer Security, Key Splitting
EDCA	Enhanced Distributed Channel Access
EDCF	Enhanced Distributed Control Function
EDGE	Enhanced Data Rates for Global Evolution
EIR	Equipment Identification Register
EM	Electromagnetic
ES	Event Notification Services
ESS	Extended Service Set
ESSID	Extended Service-Set Identifier
ETSI	European Telecommunication Standards Institute

F

FA	Foreign Agent
FBT	Fast BSS Transition
FCC	Federal Communications Commission
FCS	Frame Check Sequence
FDMA	Frequency Division Multiple Access
FH	Frequency Hopping
FHSS	Frequency Hopped Spectrum
FMC	Fixed Mobile Convergence
4G	Fourth Generation

Appendix A

FRA	Federal Railroad Administration
FT	Frame Type
FTIE	Fast Transition Information Element

G

GERAN	GSM EDGE Radio Access Network
GGSN	Gateway GPRS Support Node
GHz	Giga Hertz
GMSC	Gateway Mobile Switching Center
GMSK	Gaussian Minimum Shift Keying
GPRS	General Packet Radio Service
GPS	Global Positioning System
GSM	Global System for Mobile Communications
GSM	Groupe Spéciale Mobile
GSN	GPRS Support Node
GTK	Group Transient Key
GTP	GPRS Tunneling Protocol

H

HA	Home Agent
H-AAA	Home AAA Server
HC	Hybrid Coordination
HCCA	Hybrid Coordination Function Control Channel Access
HCF	Hybrid Coordination Function
HLR	Home Location Register
HLR/AuC	Home Location Register/Authentication Center
HSCSD	High-Speed Circuit-Switched Data
HTTP	HyperText Transfer Protocol
HTTPS	HyperText Transfer Protocol Secure
Hz	Hertz

I

IBSS	Independent Basic Service Set
ICMP	Internet Control Message Protocol
ICSA	IMS Controlled with Static Anchor
IDS	Intrusion Detection System
IE	Information Element
IEEE	Institute of Electrical and Electronics Engineers
IETF	Internet Engineering Task Force

IM	Instant Message
IMAP	Internet Message Access Protocol
IMC	Integrity Measurement Collector
IMS	IP Multimedia Subsystem
IMTS	Improved Mobile Telephone Service
IMV	Integrity Measurement Verifier
IP	Internet Protocol
IPSec	Internet Protocol Security
IPv4	Internet Protocol Version 4
IPv6	Internet Protocol Version 6
IS	Information Services
ISDN	Integrated Services Digital Network
IS-41	Interim Standard 41
ISM	Industry, Science, and Medicine
ISP	Internet Service Provider
IT	Information Technology
ITU	International Telecommunications Union

J

JDC	Japanese Digital Cellular
J2ME	Java 2 Micro Edition

K

KRK	Key Request Key
KS	Key Splitting

L

LAN	Local Area Network
LCI	Location Configuration Information
LDAP	Lightweight Directory Access Protocol
LEAP	Light Extensible Authentication Protocol
LLC	Logical Link Control
LSAP	Logical Service Access Point
LTE	Long Term Evolution
LWAPP	Lightweight Access Point Protocol

M

MAC	Media Access Control
MAHO	Mobile Assisted Hand-Off

MAN	Metropolitan Area Network
MAP	Mobile Application Part
MCHO	Mobile Controlled Handoff
MCS	Modulation and Coding Schemes
MD	Mobile Domain
MDC	Mobility Domain Controller
MDIE	Mobility Domain Information Element
MDx	Message Digest x
ME	Mobile Equipment
MGCF	Media Gateway Control Function
MGW	Media Gateway
MI	Moving Into
MIB	Management Information Database
MIC	Message Integrity Check
MIH	Media Independent Handover
MILNET	Military Network
MIMO	Multiple-Input Multiple-Output
MITM	Man-In-The-Middle
MLME	MAC Layer Management Entity
MN	Mobile Node
MO	Moving Out
MobileIGNITE	Mobile Integrated Go-to-Market Network IP Telephony Experience
MP3	Motion Picture Experts Group Layer 3
MS	Mobile Station
MSC	Mobile Switching Center
MS-CHAP	Microsoft Challenge Handshake Authentication Protocol
MSDU	MAC Service Data Unit
MSK	Master Session Key
MS-PAP	Microsoft Password Authentication Protocol
MTS	Mobile Telephone Service
MTSO	Mobile Telephone Switching Office
MW	Milliwatts

N

NAA	Network Access Authority
NAC	Network Access Control
NAC	(Cisco's) Network Admission Control
NAP	(Microsoft's) Network Access Protection
NAR	Network Access Requester

NAS	Network Access Server
NAT	Network Access Translation
NAV	Network Allocation Vector
NCHO	Network Controlled Handoff
NDIS	Network Driver Interface Specification
NDS	Network Domain Selection
NIC	Network Interface Card
NIST	National Institute of Standards and Technology
NMT	Nordic Mobile Telephone
NSK	Network Session Key
NSS	Network and Switching Subsystem
NTT	Nippon Telephone and Telegraph

O

ODC	Original Domain Controlled
ODS	Over the Distribution Service
OFDM	Orthogonal Frequency Division Multiplexing
OID	Object Identifier
OKC	Opportunistic Key Caching
OMC	Operation and Maintenance Center
OS	Operating System
OSI	Open Systems Interconnection
OSS	Operation Subsystem
OTA	Over the Air
OTD	Over the Distribution Service

P

PAC	Protected Access Credential
PAN	Personal Area Network
PAP	Password Authentication Protocol
PAR	Project Authorization Request
PBX	Private Branch Exchange
PCF	Point Coordination Function
PCMCIA	Personal Computer Memory Card International Association
PCS	Personal Communication Services
PCS	Personal Communication Systems
PCU	Packet Control Unit
PDA	Personal Digital Assistant
PDC	Pacific Digital Cellular

PDIF	Packet Data Interworking Function
PDP	Packet Data Protocol
PDP	Policy Decision Point
PDU	Packet Data Unit
PEAP	Protected Extensible Authentication Protocol
PEP	Policy Enforcement Point
PHS	Personal Handyphone System
PHY	Physical Layer and Physical Layer Technology
PI	Principal Investigator
PKI	Public Key Infrastructure
PMK	Pairwise Master Key
PMKID	Pairwise Master Key Identity
PN	Pseudorandom Noise
POP	Point of Presence
POP	Post Office Protocol
PPP	Point to Point Protocol
PPT	Power Passing Tap
PRL	Preferred Roaming List
PS	Packet Switched
PSK	Phase Shift Keying
PSK	Preshared Keying
PSTN	Public Switched Telephone Network
PTK	Pairwise Transient Key
PTT	Patents and Technology Transfer
PTT	Post, Telephone, and Telegraph (Government Agency)

Q

QAP	Quality of Service Access Point
QBSS	QoS Basic Service Set
QoS	Quality of Service
QSTA	QoS Station

R

RA	Receiver Address
RADIUS	Remote Access Dial-In User Server
RAI	Routing Area Identity
RAT	Radio Access Technology
RC4	Rivest's Cipher 4

RCPI	Received Channel Power Indicator
RDIE	RIC Data Information Element
RF	Radiofrequency
RFC	Request for Comments
RIC	Resource Information Container
RNC	Radio Network Controller
ROI	Return on Investment
RPI	Receiver Power Indicator
RRB	Remote Request Broker
RRM	Radio Resource Measurement
RSN	Robust Secure Network
RSNA	Robust Security Network Association
RSNIE	Robust Secure Network Information Element
RSS	Radio Signal Strength
RSS	Radio Subsystem
RSS	Received Signal Strength
RSSI	Radio Signal Strength Indication
RSSI	Request Signal Strength Indication
RSVP	Resource Reservation Protocol
RSVP/SBM	Resource Reservation Protocol/Subnet Bandwidth Manager
RSVP/WRESV	Resource Reservation Protocol/Wireless LAN Flow Reservation Protocol
RTP	Real-time Transport Protocol
RTS	Request To Send
RTS/CTS	Request To Send/Clear To Send
RTT	Round Trip Time

S

SA	Source Address
SAP	Service Access Point
SAT	Supervisor Audio Tone
SCCP	Skinny Client Control Protocol
SDMA	Space Division Multiple Access
SGSN	Serving GPRS Support Node
SHA	System Health Agent
SHV	System Health Validator
SIM	Subscriber Identity Module
SIP	Session Initiation Protocol
SME	Station Management Entity
SMS	Short Messaging Service

Appendix A

SMS	System Management Server
SMTP	Simple Mail Transfer Protocol
SNR	Signal to Noise Ratio
SSCAN	Seamless Converged Communication Across Network
SSH	Secure Shell
SSID	Service-Set Identifier
SSL	Secure Sockets Layer
SSN	Safe Secure Network
SS7	Signaling System No. 7
STA	(End-User) Station
SVP	Spectralink Voice Priority

T

TAC	Terminal Access Controller
TACACS	Terminal Access Controller Access Control System
TACS	Total Access Communications System
TAP	Fast Transition Enabled Access Point
TCG	Trusted Computing Group
TCH	Traffic Channels
TCP	Transmission Control Protocol
TDM	Time Division Multiplexing
TDMA	Time Division Multiple Access
TG	Task Group
3G	Third Generation
TGk	Task Group k
3GPP	Third Generation Partnership Program
3GPP2	Third Generation Partnership Program 2
TIA	Telecommunications Industry Association
TIE	Timeout Information Element
TIVO	Name for a Personal Recording Device
TKIP	Temporal Key Integrity Protocol
TLS	Transport Layer Security
TLV	Type Length Value
TNC	(Trusted Computing Group's) Trusted Network Connect
TPC	Transmit Power Control
TSPEC	Traffic Specifications
TSTA	Fast Transition Enabled Station
TTLS	Tunneled Transport Layer Security

U

UDP	User Datagram Protocol
UE	User Equipment
UMA	Unlicensed Mobile Access
UMTS	Universal Mobile Telecommunications System
UNC	UMA Network Controller
URL	Uniform Resource Locator
USB	Uniform Serial Bus
USDC	United States Digital Cellular
USIM	User Service Identity Module
UTRAN	Universal Terrestrial Access Network

V

VCC	Voice Calling Continuity
VLAN	Virtual Local Area Network
VLR	Visitor Location Register
VoFi	Voice over Internet Protocol over Wireless Fidelity
VoIP	Voice over Internet Protocol
VoWi-Fi	Voice over Wireless Fidelity
VoWIP	Voice over Wireless Internet Protocol
VPN	Virtual Private Network

W

W-AAA	WLAN AAA Server
WAP	Wireless Application Protocol
W-CDMA	Wideband Code Division Multiple Access
WECA	Wireless Ethernet Community Alliance
WEP	Wired Equivalent Privacy
WG	Working Group
WG/TG	Working Group/Task Group
Wi-Fi	Wireless Fidelity
WiMAN	Wireless Metropolitan Area Network
WiMAX	World Interoperability for Microwave Access
Wireless MAN	Wireless Metropolitan Area Network
WLAN	Wireless Local Area Network
WME	Wireless Multimedia Extensions
WML	Wireless Markup Language
WMM	Wi-Fi Multimedia Mode
WPA	Wi-Fi Protected Access

Appendix A

WPA2	Wi-Fi Protected Access Version 2
WRESV	Wireless LAN Flow Reservation Protocol
WSM	Wireless Scheduled Multimedia
WTLS	Wireless Transport Layer Security
WZC	Wireless Zero Config

Appendix B
List of Figures

1.1	Three Overlapping Network Cells	2
1.2	Standard Caricature of Roaming, while an SUV Driver Chats on a Cell Phone	6
1.3	A Person Who is Talking on a Cell Phone in a Moving Car	9
1.4	A Person Who is Working on an 802.11-Enabled Laptop and Walking Down a Hallway	9
1.5	A Person Who is Talking on a Cell Phone and Reaching an Area Not Covered by the Home Wireless Provider	10
1.6	A Person Who is Working on an 802.11-Enabled Laptop and Surfing the Internet, While Using the Wireless Internet Access Offered by the Airline	10
1.7	A Person Who is Working on an 802.11-Enabled Laptop and Surfing the Internet, While Using Wireless Internet Access by Making a (Currently Prohibited in Flight) Call Using a Cell Phone with the GPRS Modem in the Laptop	11
1.8	The Relative Sizes of a Cellular Phone Network Cell (Left) and an 802.11 Network Cell (Right)	13
1.9	An AAA Server in an 802.11 Wireless Network	17
1.10	An Example of an Overlapping Cell Coverage Area X, Where a Device could Communicate with Either Antenna 1 or Antenna 2	20
1.11	A Network Consisting of Seven Circular Cells	22
1.12	A Network Consisting of Six Hexagonal Cells	23
1.13	An Example of a Hard Handoff	24
1.14	An Example of a Soft Handoff	24
2.1	An Abstraction and Oversimplification of Antenna Placements in Europe Illustrating the Density of the Coverage Relative to the United States	34
2.2	An Abstraction and Oversimplification of Antenna Placements in United States Illustrating the Density of the Coverage Relative to Europe	35
2.3	A Handoff Scenario in a Cellular Network	40
2.4	A Schematic Representing the Division of 49.6 MHz of Radio Spectrum into 124 Full-Duplex Channels of 200 KHz Each	42
2.5	An Illustration of the Radio Subsystem	43
2.6	An Illustration of the Network and Switching Subsystem	44
2.7	An Illustration of the Operation Subsystem	45
2.8	A Schematic of the GPRS System	55

Appendix B

3.1	The 802.11 Management Frame MAC Header	64
3.2	Basic 802.11 Beacon Frame	65
3.3	Basic 802.11 Probe-Request Frame	66
3.4	Basic 802.11 Probe-Response Frame	66
3.5	Basic Service Set	69
3.6	Illustration of Passive Scanning	70
3.7	Illustration of Active Scanning	71
3.8	Basic 802.11 Association-Request Frame	72
3.9	(Re)Association-Request Frame	73
3.10	(Re)Association-Response Frame	73
3.11	An 802.11 Unicast Data Frame	79
3.12	An Example of an Extended Service Set	80
3.13	The Configuration of a Network with Multiple Extended Service Sets in Operation	81
3.14	Example of Roaming Delay Times. (Courtesy: Azimuth Systems.)	82
3.15	Phases of 802.11 Roaming Delay	83
3.16	A Sample Topology for a Local Roam	86
3.17	An Example of an 802.11 Management-Frame Exchange in a Local Roam	87
3.18	An Example of a Local Roam to Illustrate the Impact on a User Application	89
3.19	An Example of a Complex Roaming Topology	90
3.20	An Example of Global Roaming Showing the Impact on User Data	91
3.21	Some Important Mobile-IP Terminology and Its Relevant Locations in an IP Network	93
3.22	The Basic Steps in Mobile-IP Registration (IPv4)	94
3.23	An Illustration of Mobile-IP Data Flow (IPv4)	94
3.24	An Example of a Global Roam with Mobile IP	96
4.1	The Development Flow Chart for IEEE Standards	104
4.2	Selected 802.11 Task-Group Timelines	106
4.3	Establishing an 802.11 Connection with 802.11i	113
4.4	The 802.11i Personal Mode Key Establishment	115
4.5	The 802.11i Full Enterprise Mode Key Establishment	116
4.6	The EDCF and HCF Functions in 802.11e	119
5.1	Software Implementation in a Typical 802.11 Client Showing the Driver and Interface on a PC	135
5.2	Microsoft's Wireless Client Which is Called Wireless Zero Config	141
5.3	Funk's Wireless Client Which is Called Odyssey	142
5.4	Meetinghouse's Wireless Client Which is Called Aegis	143
5.5	Listing of Background Implicit Scanning Information in Active Mode	145
5.6	Listing of Background Scanning Details for a Specific Probe	146

List of Figures

5.7	Listing of Background Scanning Details for a Wildcard Probe	146
5.8	Listing of Background-Implicit Scanning Details in Active Mode with Beacons Suppressed	147
5.9	Topology for the Basic Global Roaming Tests	154
5.10	Crude Roaming Test Using Ping	155
5.11	Transition from the Background Scan to the Association in the Global Roam for the WZC Case	157
5.12	Acquisition of a New IP Address and the Resumption of the Ping Application for the WZC Global Roam	157
5.13	Microsoft's Wireless Client During a Global Roam	160
5.14	Topology for Basic Local Roaming Tests	161
5.15	Transition to a New AP in the Local Roam with the Odyssey Client	161
5.16	View of the Active Channel as the AP is Powered Down in the Local Roam with the Odyssey Client	162
5.17	View of the Active Channel as the STA is about to Change the Channel in the Local Roam with the Odyssey Client	163
6.1	The 802.1X Paradigm	170
6.2	The Traditional Dial-Up AAA Configuration	173
6.3	The EAP Protocol Hierarchy	178
6.4	Authentication in 802.11 Using Basic 802.1X	184
6.5	802.11 Authentication Using EAP Tunneling	186
6.6	Cisco's Network Admission Control Architecture	190
6.7	The Trusted Network Connect Architecture	191
7.1	The 802.11 Security Staircase	196
7.2	The 802.11i Preauthentication	199
7.3	Topology for WPA-PSK Protected Roaming Tests	201
7.4	Postroam Association and Four-Way Handshake for WPA-PSK	202
7.5	Resumption of User Traffic Following a WPA-PSK Roam	203
7.6	Target AP Authentication Flow for WPA2 Enterprise Mode	204
7.7	Topology for WPA2 Enterprise Protected Roaming Tests	205
7.8	WPA2-Enterprise Mode Monitor M1 Frame Trace	207
7.9	WPA2-Enterprise Mode Monitor M3 Frames 2524–2549	207
7.10	WPA2-Enterprise Mode Monitor M3 Frames 2550–2575	208
7.11	WPA2-Enterprise Mode Monitor M3 Frames 2576–2601	209
7.12	Target AP Authentication Flow with Preauthentication	211
7.13	Topology for Preauthentication Roaming Tests	212
7.14	Preauthentication Monitor M3 Frames 162–187	213
7.15	Preauthentication Monitor M2 Frames 86–111	213

7.16	Preauthentication Monitor M2 Frames 112–138	214
7.17	Preauthentication Monitor M1 Frames 188–213	214
7.18	Preauthentication Monitor M2 Frames 244–269	216
7.19	Preauthentication Monitor M3 Frames 202–227	216
7.20	Preauthentication Monitor M3 Frames 238–253	217
7.21	Preauthentication Monitor M3 Frames 525–550	217
8.1	Generic VoIP Architecture	221
8.2	Spectralink Architecture	225
8.3	Local Roaming via Tunneling	232
8.4	Local Roaming via Tunneling's Convoluted Data Path	233
9.1	Prehandoff Possibilities Contemplated by the 802.11r Standard	238
9.2	The 802.11r Standard's Architectural Entities Introduced. (Courtesy of the IEEE.)	240
9.3	The 802.11r Standard's Entities in a Fat-AP Architecture. (Courtesy of the IEEE.)	241
9.4	The 802.11r Standard's Entities in a Wireless Switch Architecture. (Courtesy of the IEEE.)	242
9.5	The 802.11r Keys, Their Holders, and the Relationships among Them. (Courtesy of the IEEE.)	243
9.6	The 802.11r Standard's Key Hierarchy. (Courtesy of the IEEE.)	244
9.7	Resource Information Container Processing. (Courtesy of the IEEE.)	250
9.8	Fast BSS Transition Over the Air, No QoS, and No Security. (Courtesy of the IEEE.)	252
9.9	Fast BSS Transition Over the DS, No QoS, and No Security. (Courtesy of the IEEE.)	253
9.10	Fast BSS Transition Over the Air, No QoS, and Security. (Courtesy of the IEEE.)	255
9.11	Fast BSS Transition Over the DS, No QoS, and Security. (Courtesy of the IEEE.)	256
9.12	Fast BSS Transition Over the Air, QoS, and No Security. (Courtesy of the IEEE.)	257
9.13	Fast BSS Transition Over the DS, QoS, and No Security. (Courtesy of the IEEE.)	258
9.14	Fast BSS Transition Over the Air, QoS, and Security. (Courtesy of the IEEE.)	259
9.15	Fast BSS Transition Over the DS, QoS, and Security. (Courtesy of the IEEE.)	260
9.16	The 802.11k Standard's Type Definitions for Measurement Reports	261
9.17	Roaming Enhanced by Knowledge Derived from 802.11k	262
9.18	Roaming Enhanced by QoS Information Derived from 802.11k	263
10.1	Horizontal and Vertical Handovers. (Courtesy of the IEEE.)	269
10.2	The Major Organizations Involved in Intertechnology Roaming	275
10.3	IMS-Based Roaming between Cellular and 802.11	278
10.4	CCCF-Anchored Voice Call Continuity Signaling in 3GPP	283
10.5	UMA-Based Roaming between Cellular and 802.11	287
10.6	MIH Function Location and Key Services. (Courtesy of the IEEE.)	291

10.7　MIH Reference Model for Mobile Stations with Multiple Protocol Stacks. (Courtesy of the IEEE.) 292
10.8　MIH Reference Model across Different Networks. (Courtesy of the IEEE.) 293
11.1　Sample Instance of the Mobility Model 311

Appendix C
List of Tables

2.1	Multislot Classes of GPRS Devices	54
2.2	Coding Schemes for GPRS and EDGE	58
2.3	Cellular Generations	59
3.1	The 802.11 Standard's Frames Based on Type and Class: Part 1	78
3.2	The 802.11 Standard's Frames Based on Type and Class: Part 2	78
4.1	The PHY Characteristics of 802.11	109
4.2	The PHY Advantages and Disadvantages of the Various 802.11 Versions	110
4.3	Summary of Other 802.11 Task Groups	127
5.1	Listing of Mandatory NDIS 5.1 OIDs: Part 1	136
5.2	Listing of Mandatory NDIS 5.1 OIDs: Part 2	136
5.3	Listing of WPA NDIS 5.1 OIDs	137
5.4	Listing of WPA2 NDIS 5.1 OIDs	137
5.5	Listing of Optional NDIS 5.1 OIDs: Part 1	137
5.6	Listing of Optional NDIS 5.1 OIDs: Part 2	138
5.7	Implicit Scanning Timing Test Results	151
6.1	Prominent EAP Methods and Their Characteristics	180

Index

Page numbers in italics refer to figures and tables.

2.75G technology 57–59
3Com, 221
8-phase shift keying (8-PSK), 57–59
802 task group, 276–277
802.1X key management 240–243
802.1Xrev standard 172
802.1X standard 170–171
   Flexibility 171–172
802.11 standard 1, 8, 27, 64, 101–102, 297, 312
   AAA issues 17–19
   Basic architectural services 111–112
   Interactions between 802.16 272–273
   Interworking between 3GPP 279–280
   Ongoing work 289–307
   Relationship with 802.15.1 273–274
802.11 task group
   Subgroups 102, 127–128
   Timelines 106
802.11a OFDM PHY channel 68
802.11b DS PHY channel 68
802.11e standard 117–120, 130, 245
   Two-frame elimination 237–238, 264
802.11f standard 237
802.11g PHY channel 68
802.11g standard 109–110
802.11i standard
   Preauthentication 198–200, 218, 239–240
   Security 112–117, 131
802.11k standard 120–124, 235, 257–258, *261*
   Limitations 264
   New information elements 258–261
   Utility 261–264
802.11r standard 124–127, 233–235, 252
   Architectural elements 240–242
   Facets 238–240
   Information elements 247–250
   Key hierarchy 242–245
   Ongoing work 298–299
   Overview 236–238
   Protocol exchanges 250–257
   QoS enhancements 237–238, 264
   Resource reservation issue 245–247
802.15.1 standard 273
   see also Bluetooth technology
802.16 standard (WiMAX) 80, 271–273
   Interworking between 802.11 272–273
802.21 standard 274–275, 289–294
   Interactions between 3GPP2 289
   Scope 290–291

A

Access controller (AC) 231
Access point (AP) 4, 307
   Lightweight 229
   Nonbeaconing (hidden) 144
   Ongoing work 302
   Virtual 79, 143–144
Access point (AP) placement 164–165
   Self-monitoring 802.11 network 166–167
   Site survey 165–166
Access router (AR) 230
Active Directory 172, 176
Active scanning 65–66, 71–72, 142, 144–145, 147–148
Active-scan timeout, 149
Address resolution protocol (ARP), 74, 79, 85, 158, 160
Ad hoc 802.11 network 68
Admission control
   see Authentication
Advanced encryption standard (AES) 114, 198, 304
Advanced mobile phone service (AMPS) 36–39, 41, 46
Aegis client 141, *143*, 145, 147, 150
Agere 110
Airespace 230
Airgo 268
Alcatel 221
America Online (AOL) 51
American Telephone and Telegraph (AT&T) 7, 33, 36, 38
Analog system 38–40
A-Netz 33–34
ANSI-41 39–40, 46
Antenna 2, 4, 11–12, 20–23
   Directional 36
   Placement in Europe vs US 34–35
Antheil, George 48, 67
Application resumption delay *83*, 85, *201*, 204, 210, 215, 236, 265

Index

ARPANET 174
Aruba 230
Association 72–74
Atheros 108, 110–111, *151*
Atheros AR5004G 150–151
Attack
    Denial of service 190
    Dictionary 115, 180–182
    Man-in-the-middle 180
    New 2, 196
    Spoofing 183
        MAC spoofing 19, 187, 223
Audio Frequency Shift Keying (AFSK) 35
Authentication 19, 73–74, 129, 188–189, 192, 268
    Flexible and strong 184–186
    Frame 73–74, 239, 251
    Mac-based 187, 223
    Methodology 187–188
    Model 169–170
    Mutual 180–183, 186
    Ongoing work 306–307
    Open 802.11 125, 158, 160, 251
    Open mode 73, 188
    Port-level 170–172
    Web server 188
    Web-based 187–188
Authentication and key agreement (AKA) 179–180, 183
Authentication, authorization, accounting (AAA) server 173–174, 205
Authentication, authorization, accounting (AAA) technology 17–19
Authentication delay 84–85, 204, 209–212, 215
Autoradiopuhelin (car-radio phone) 33
Avaya 221

B

Background scanning *135*, 144–148, 157–158
    Definition 145
    Enhancement 152–153
    Implicit and explicit 144–145, *147*, 150–152
    Pitfalls 151–152
    Vendor-specific extensions 148–149
Bandwidth 14
Base station 36
    AMPS 37
    Definition 4
Base station controller (BSC) 43–47, *287*
Base transceiver station (BTS) 44
Basic service set (BSS) 68–69
    802.11r 125–127
    Independent 68
    Pre-802.11r 125
    see also Fast BSS transition
    Transition 124
Basic service set identification (BSSID) 64, 68–69, 79
Beacon 64–66, 69–70, 143–144, 151–153, 248, 258
Beacon bloat 248
Beacon frame 64–66, 147
Beacon interval 149–150, 152
Beacon request 121–122
Bearer traffic 23, 25, 39, 279, 286
Bell Laboratories 7, 35–36, 38
Bell Mobile Phone Service 35
Bits per second (bps) 4
Bluetooth technology 31, *110*, 271, 273–274
B-Netz 33
British Telecom 286
Broadcom 108
Broadvoice 222
Broker AAA (B-AAA) server 288
Bunching mechanism 224

C

Caching key 125, 219, 227–229
Call-admission mechanism 224, 225
Call chain 283
Call control continuity function (CCCF) 283–285
Care of address (COA) 93–94, 96
Carrier sense multiple access/collision avoidance (CSMA/CA) scheme 74, 272
Carrier sense multiple access/collision detect (CSMA/CD) scheme 74
CDMA2000 274–275, 292
    Tunnel management in 288–289
cdmaOne 47–48
Cell 5, 12
    Size 12–17, 21–22
Cellular digital packet data (CDPD) service 39
Cellular telephony 12, 60
    AAA issues 18
    First generation 39–40
    Fourth generation 294
    Handoff decisions 37
    Network design and implementation 16
    Second generation 40–51
    Second generation-plus 268
    Third generation system 51–59, 274–276, 303
Centralized wireless switch architecture 219, 227, 229–233, 236–237, 241–242, 244
Centrino 108, 145, 153
Challenge handshake authentication protocol (CHAP) 176–177, 181
Channel
    802.11 49, 66–68
Channel load report 123
Channel overlap
    Limitations to 97–98
Channel utilization 120
Charging gateway function (CGF) 54
Chipping sequence 49
    Numerical 67–68
Chipset market 106–108, 268
Cipher 114–116
Circuit switched (CS) network 51
Circuit switched (CS) voice service 280
Circular network cell model 22–23
Cisco 182–183, 189–190, 197–198, 221–223, 230

Cisco call manager 222
Cisco network admission control 190
Cisco's centralized key management (CCKM) 228–229
Client configured SSID list (CCSL) *135*, 140
Client in 802.11 station 134
    Functions 139–141
Client-server communication protocol 171, 173–174
C-Netz 38
Cochannel interference 68, 97–98, 110
Code division multiple access (CDMA) 47–51, 60, 274–275
    Handoffs 50–51
Code-division multiplexing 49
Cognio 167
Collision 74–75, 77
Co-located foreign agent 93
Computing 29–30
    Accessibility initiatives 30
    Mobile 30–31
Conference of European Postal and Telecommunications Administrations (CEPT) 41
Contention free (CF) access mode 76
Contention free period (CFP) 117–118
Contention-based access 74–76
Control and provisioning of wireless access points (CAPWAP) 227, 231, 237, 265
Control frame 77–79
Control signaling 25–26
Control traffic 25–26
Correspondent node (CN) 93–95
Counter mode CBC-MAC protocol (CCMP) 113–114
Cross-technology roaming 268–271
Current driver scan list (CDSL) *135*, 139, 142, 149, 156

D

Data frame 77–79
    Symmetrical encryption 114
Datagram 56
Decibel (dB) 4
Delay 15
    Application resumption *83*, 85, *201*, 204, 210, 215, 236, 265
    Authentication 84–85, 204, 209–212, 215
    Discovery 83–84, *201*, 204, 215, 264–265
    Infrastructure routing 85–86, 158, 264–265
    Key management *83*, 85, 201–202, *204*, 208–210, 215
    Reassociation 83–84, 201–202, *204–205*, 208, 215–216
    Roaming 81–86, 162–164, 236, 264–265
Dell 108, 145, 153, 205
Denial of service 190
Dense deployment 34–35
Destination address (DA) 64, 79
DIAMETER protocol 171, 173–174, 177
Dictionary attack 115, 180–181, 182
Digital cellular system (DCS1800) 41
Digital divide 30
Digital enhanced cordless telephone (DECT) 48
Direct sequence (DS) 67–68
Direct sequence spread spectrum (DSSS) 47, 49, 109
Directed probe 71, 142–143, 145–147
Discovery delay 83–84, *201*, 204, 215, 264–265
Discovery protocol 74
Distributed control function (DCF) 117, 224–225
Distributed MAC architecture 230
Distribution service (DS) 80, *91*, 96, 124–125

Driver in 802.11 station 134
    Functions 134–139
Dual-mode phone 46, 108, 280
Dual-mode Wi-Fi cellular handset 268
Due process 102
Dynamic host configuration protocol (DHCP) 55, 74, 79, *83*, 85, 88
Dynamic WEP keying 196–198

E

EAP over LAN (EAPOL) 170, 184–185
EAP-TLS-KS 305
Encryption 48, 197
Encryption cipher 129, 198
Enhanced data rates for global evolution (EDGE) 57–59
Enhanced distributed channel access (EDCA) 117–118
Enterprise data user
    Ideal roaming experience 270–271
Enterprise mode 115
Enterprise security mode 113
    PMK derivation 115
Ericsson, Lars Magnus 6
Ethernet 74, 79–80, 109, 153–154, 156, 200, *225*, 272
European Telecommunications Standards Institute (ETSI) 41, 52
Eurospot 18
Explicit scanning *135*, 144–145, 147, 151–152
    Timing 150
Extended service set (ESS) 79–80
    In operation 80–81
Extensible authentication protocol (EAP) 170, 172, 177–178
    Authentication and key agreement 179–180, 183
    Flexible authentication via secure tunneling 179–180, 183–184, 186, *190*, 198

Light extensible
   authentication protocol
   179–180, 182–183
Message digest 5 179,
   180–181
Ongoing work 304–305
Over LAN 170, 184–185
Protected extensible
   authentication 179–180,
   182–183, 197
State machine 178–179
Subscriber identity module
   43–44, 182–183
Transport layer security
   179–181, 186, 197
Tunneled transport layer
   security 179,
   180–182, 197
tunneling 186–187

F

Fading 3, 22, 50
Fast BSS transition 236, 298
   With no QoS and no security
      251–254
   With QoS and security
      254–256, *260*
Fast BSS transition over-the-air
   251–252, *255*, *257*, *259*
Fast BSS transition over-the-DS
   252–254, *256*, *260*
Fast handoff 152–153
Fast roaming 19–21
Fast roaming task group 111–112
Fast session resumption 218
Fast transition information
   element (FTIE) 126,
   247–248, 254–255
Federal Communications
   Commission (FCC) (US)
   17, 33, 35–36, 38, 49
First contact 236
First generation (1G) cellular
   system 39–40
Fixed mobile convergence (FMC)
   268
Flexible authentication via secure
   tunneling (FAST) 179–180,
   183–184, 186, *190*, 198
Forbidden list 288

Foreign AAA (F-AAA) server 93
Foreign agent (FA) 93–94
   Co-located 93
Fourth generation (4G) cellular
   system 294
Frame 64–79, 149
Frame association-request 64,
   69–70, 72–73, *91*, *96*, *113*,
   126, 156, 158, 239, 251
Frame association-response 64,
   72–73, *91*, *96*, *113*, 126
Frame beacon 64–66, 147
Frame check sequence (FCS)
   64, *79*
Frame measurement report 121
Frame probe-request 65–66,
   71–72, *113*, 121–122,
   145–147, 149
Frame probe-response 64–66,
   71–72, 79, *113*, 121–122,
   124, 126, 144–147, 149,
   230, 247–248, 251
Frame request 122
Frame type (FT) action request
   frame 239, 253–254
France Telecom 286
Frequency division multiple
   access (FDMA) 32, 39
   Handoffs 50
Frequency hopped spread
   spectrum (FHSS) 109
Frequency hopping (FH) 67
Frequency reuse pattern 32, 36
Frequency spectrum 32
Frequency-division
   multiplexing 49
Funk 140

G

Gaussian minimum shift keying
   (GMSK) 57–59
General packet radio service
   (GPRS) 52–54
   Coding schemes 57–58
   Components 54–55
   Roaming for data
      applications in 55–57
Generation
   First 39–40
   Fourth 294

Second 40–51
Second-plus 268
Third 51–59, 274–276, 303
Geosynchronous 30
Global positioning system
   (GPS) 49
Global roaming 12, 88–91, 268
   Alternatives to mobile IP 96
   With mobile IP 95–96
Global roaming test 153
   Results 156–160
   Test-bed description 156
   Tools 153–154
Global system for mobile
   communications (GSM)
   41–42, 51–54, 56–57, 279
   Subsystems 43–45
GPRS attach 56
GPRS gateway support node
   (GGSN) 54–56, 278, *283*
GPRS support node (GSN) 54
GPRS tunneling protocol (GTP)
   55–56
Group transient key (GTK)
   114–115, 203, *243*,
   255–256, 259–260
Groupe Spéciale Mobile
   (GSM) 41
GSM EDGE radio access
   network (GERAN) 279, 292

H

H.323 protocol 222
Half-duplex 33
Handoff 16, 23–26, 34, 37–38,
   50–51, 60, 231, *269*
   Fast 152–153
   Hard *24*, 26, 50–51, 198
   Horizontal 268–269, 303
   In 3GPP vernacular 285
   Intersystem 38, 40
   Intra- and inter-BSC 44–45
   Ongoing work 299–300,
      301–303
   Smooth 300–301
   Soft *24*, 26, 49–51
   Versus handover 25, 285
   Versus roam 285
   Vertical 268–269,
      299–300, 303

Index

Handover 25, 290–291
　　In 3GPP vernacular 285
　　Media independent 291–293
　　Paradigms 290
　　see also Handoff
Hard handoff *24*, 26, 198
　　Versus soft handoff 50–51
Head-of-the-queue blocking delay 226
Health policy compliance 192
Health policy validation 192
Hertz, Heinrich 32
Hertz (Hz) 4
Hewlett Packard (HP) 108, 205
Hexagonal network cell-model 22–23
Hidden access-point 143–144
Hidden node problem 74–75
High-speed circuit switched data (HSCSD) 51–55
Home AAA (H-AAA) server 93, 288–289
Home agent (HA) 93–94
Home location register (HLR) 38, 45, 55, 93, 286–287
Horizontal handoff 268–269, 303
Horizontal roaming 269, 277
Hot-spot 18, 89, 92, 188, 305–306
Hybrid coordinator function controlled channel access (HCCA) 118
Hybrid coordination function (HCF) 118
Hypertext transfer protocol (HTTP) 188

I

Implicit scanning 144–145, *147*, 152
　　Timing 150–151
Improved mobile telephone service 33
IMS controlled with static anchoring (ICSA) model 282–286
Incremental redundancy 58
Independent basic service set (IBSS) 68

Information element 65, *113*, 247–249, 256
　　802.21 290
　　New 802.11i 227–228
　　New 802.11k 258–261
　　New 802.11r 236–256
Infrastructure mode network 68, 199
Infrastructure routing delay 85–86, 158, 264–265
Inmarsat's broadband global access network 30
Instant Message (IM) 282
Institute of Electrical and Electronics Engineers (IEEE) 4, 8, 170, 197, 231, 274, 289
　　Ballot group 105
　　Balloting 103–105
　　Consensus 104–105
　　Due process 102
　　Inter-AP protocol standardization 237
　　Openness 102–103
　　Standard *see* Standard
　　Standards development process 102–106
　　Versus IETF 130–131
　　Versus Wi-Fi Alliance 128–130
Integrated services digital network (ISDN) 41
Intel 108, *151*, 271
Intel 2200 miniPCI 150–151
Inter access point protocol (IAPP) *127*
Interference cochannel 68, 97–98, 110
Interim standard
　　IS-41 39–40, 46
　　IS-54 46
　　IS-95 47–48
　　IS-136 46, 57
International Telecommunications Union (ITU) 222
Internet control message protocol (ICMP) 153

Internet Engineering Task Force (IETF) 4, 222, 230–231, 274
　　Liaison 131, 275–276
　　Versus IEEE 802 130–131
Internet key exchange (IKEv2) protocol 95
Internet protocol (IP)
　　IPv4 3, *94*, 174, 281
　　IPv6 3, 174, 281
　　Mobile *83*, 85–86, 92–96, 269, 289, 299–300
　　Multimedia Subsystem 276
　　Tunnel 94, 269, 288–289
Internet protocol security (IPSec) 192, 285–289
Internet protocol version 4 (IPv4) 3, *94*, 174, 281
Internet protocol version 6 (IPv6) 3, 174, 281
Inter-provider roaming 304–305
Intersil 108, 110
Intersystem handoff 38, 40
IP Multimedia Subsystem (IMS) Forum 276
IS-41 39–40, 46
IS-54 46, 48
IS-95 47–48
IS-136 46, 57

J

Japanese digital cellular (JDC) 47–48
Jitter 15
JTACS system 47

K

Kerberos 173–174
Key caching 125, 219, 227–229
Key freshness 115, 129
Key holder 243
Key lifetime interval 249
Key management 129
Key management delay *83*, 85, 201–202, *204*, 208–210, 215
Kilobyte (K) 4

L

Lamarr, Hedy 48, 67
Latency reduction 116–117, 126

Index

Light extensible authentication
protocol (LEAP) 179–180,
182–183
Limitations 197
Light weight access point
protocol (LWAPP) 227,
230–231
Lightweight access point 229
Lightweight directory access
protocol (LDAP) 172, 176
Link measurement report
121–122
Linksys 156
Linux wireless extensions
134, 140
Livingston Enterprises, Inc. 174
Load balancing 133, 263, 305
Local area network (LAN) 4
Virtual 79, 143–144, 160,
171, 184, 192, 223
Local roaming 11, 12, 86–88
Local roaming test 159
Results 159–162
Location configuration
information (LCI)
Report 123
Long term evolution (LTE)
RAT 279
Loss 14–15
Lucent Technologies 174, 221

M

Management frame 64–79,
87, 149
Management information
database (MIB) 120
Man-in-the-middle (MITM)
attack 180
Marconi, Guglielmo 6, 32
Master session key (MSK)
242–245
McAfee 190
Measurement pause 121
Media access control (MAC)
address 4–5, 19, 79,
187, 223
Media access control (MAC)
mechanism
802.16 272

Media access control (MAC)
processing 229–230
Media access control (MAC)
spoofing 19, 187, 223
Media access control
(MAC)-based authentication
187, 223
Media independent handover
(MIH) 291–293
Medium sensing time histogram
request 123
Meetinghouse 141
Meru 230
Message digest 5 (MD5) 179,
180–181
Message integrity check (MIC)
254, 256
Metropolitan area network
(MAN) 4
Microsoft 134, 140–141, 153, *160*
Challenge handshake
authentication protocol
176–181
Network access protection
189, 191–192
Password authentication
protocol 176
MIT 173
Mobile 802.11 297–303
Mobile application part
(MAP) 46
Mobile assisted handoff
(MAHO) 50
Mobile computing 30–31
Mobile device 25–26, 36–37,
172, 183, 290–291, 305,
308–310
Mobile equipment (ME) 43
Mobile integrated go-to-market
network IP telephony
experience (MobileIGNITE)
276
Mobile IP *83*, 85–86, 92,
269, 289
Alternatives to 96
Architecture 92–95
Global roaming with 95–96
Ongoing work 299–300
Mobile network 307–310

Mobile node (MN) 93–96
Mobile station (MS) 43–44
Mobile switching center (MSC)
36–38, 40, 44–45, 50, *283*,
285–287
Mobile telephone service (MTS)
33–34
Security 35
Mobility and Multihoming
Working Group (MOBIKE)
95, 286, 289
Mobility domain identifier 247
Mobility domain information
element (MDIE) 126,
247–248
Mobility domain (MD) 125,
227–228, 236
Mobility model 297–298,
307–310
Instances 310–311
Problems 311–312
Mode
Active 121–122, 145, 147
Beacon table 121–122
Infrastructure 68, *136*,
199, 262
Open-authentication 73,
158
Passive 121–122, 148
Model
Handoff 23–26
Mobility 297–298, 307–312
Network cell 21–23
Modulation and coding scheme
(MCS) 58
Motorola 7, 46, 221
Multiple alternative service set
identification 90–91
Multiplexing 5, 13–14, 49
Mutual authentication 180–181,
183, 186

N

Naked SIP phone 281
NA-TDMA 46, 48
NDIS 5.1 object identifier (OID)
135–139
Negroponte, Nicholas 30
Neighbor report 121, 124,
260–261, 263, 294

Network 4
    Peer to peer 68
    Types 4
Network access control (NAC) 170, 188–192, 198
Network access protection (NAP) 189, 191–192
Network access server (NAS) 170, 173, *190*
Network admission control (NAC) 189–190
Network allocation vector (NAV) 75
Network cell model 21–23
Network domain selection (NeDS) 283–285
Network isolation 192
Nippon Telephone and Telegraph (NTT) 38, 47–48
Noise histogram report 121, 123
Nonbeaconing access-point 143–144
Nordic mobile telephone (NMT) system 38
Nortel 221
nQAP 119
nQBSS 119
nQSTA 119
Numerical chipping sequence 67–68

O

Object identifier (OID) 135–139
Odyssey client 140, *142*, 162–164
OID-802-11-BSSID-LIST 136, 139
OID-802-11-BSSID-LIST-SCAN 136, 138–139, 144, 148
Open authentication 73, *113*, 160
Openness 102–103
Opportunistic key caching (OKC) 219, 227–229
    802.11i 236–237
Original domain controlled (ODC) model 282
Orthogonal frequency division multiplexing (OFDM) 67–68, 109
Overlapping cell coverage area 2, 20
Over-the-air (OTA) 126, 239
    Fast BSS transition 251–252, *255, 257, 259*
Over-the-DS (ODS) 117, 126, 239
    Fast BSS transition 252–254, *256, 260*

P

Pacific digital cellular (PDC) 47–48
Packet control unit (PCU) 54
Packet data protocol (PDP)
    context 56
    Initiation 56–57
Packet delay 222, 226, 300–301, 303
Packet loss 222, 226, 264, 299, 300
Packet switched (PS) voice service 280
Pairwise master key (PMK) 115–116, 199–200, 202–203, 227–228, 242–245
Pairwise transient key (PTK) 114–116, 242–245
    Derivation 125
Passive scanning 65–66, 69–71, 142, 144, 148
Passive-scan timer 148–149
Password authentication protocol (PAP) 176–178, 181
Peer to peer network 68
Personal area network (PAN) standard 273
Personal communication service (PCS) 47
    Low-tie 48
Personal digital assistant (PDA) 31, 189, 223
Personal handyphone system (PHS) 48
Personal security mode 113
    PMK derivation 115
    see also Preshared key mode
Phase shift keying (PSK)
    8-PSK 57–59
Phone
    Car-radio 33
    Dual-mode 46, 108, 280
Dual-mode Wi-Fi cellular handset 268
    VoIP 21, 108, 111, 220, 222
Physical carrier sensing 75
Physical layer (PHY) 108–109
    Variants 109–110
Picocell 97
Ping 153–155, *157*, 159–160, 162–163, *201*, 206, 208
Ping-ponging 82, 265
Plumbing of the key 245
Point coordination function (PCF) 117–118
Point of presence (POP) 175
Point to point protocol (PPP) 175–177, 179, *292*
Policy server 190–191
Port-level authentication 170–172
Preauthentication 116–117, 210, 227
    In 802.11i 198–200, 218, 239–240
Preauthentication test 210
    Results 210–218
Preauthentication-enabled WPA enterprise 200
Predictive association algorithm 305–306
Preferred roaming list (PRL) 289–290
Prereservation 126
Preshared key (PSK) mode 113, 242
Priority queue 120
Private branch exchange (PBX) 220–221
Probe 64–66
    Directed probes 142–143
    Probe requests 145–147
    Undirected probes 71, 142
Problem
    Access-point location 312
    Access-point placement 312
    Hidden node 74–75
    Obstacle removal 312
    Sources reachability 312
    User communication 311–312
Project authorization request (PAR) 103–104, 106

Index

Protected access credential (PAC) 183
Protected channel 179
Protected extensible authentication protocol (PEAP) 179–180, 182–183, 197
Protocol
    Address resolution 74, 79, 85, 158, 160
    Challenge handshake authentication 176–177, 181
    Client-server communication protocol 171, 173–174
    Counter mode CBC-MAC 113–114
    DIAMETER 171, 173–174, 177
    Discovery 74
    Dynamic host configuration 55, 74, 79, *83*, 85, 88
    Dynamic wired equivalency 197
    Extensible authentication 170, 172, 177–187, 304–305
    GPRS tunneling 55–56
    H.323 222
    Hypertext transfer 188
    Inter access point *127*
    Internet 3, 56, 79, 90–91, 94, 269, 276, 288–289
    Internet control message 153
    Internet key exchange 95
    Internet protocol version 4 3, *94*, 174, 281
    Internet protocol version 6 3, 174, 281
    Kerberos 173–174
    Light extensible authentication 179–180, 182–183, 197
    Light weight access point 227, 230–231
    Lightweight directory access 172, 176
    Microsoft password authentication 176

Packet data 56
Password authentication 176–178, 181
Point to point 175–177, 179, *292*
Protected extensible authentication 179–180, 182–183, 197
Real-time transport 222
Remote access dial-in user server 171–177, 184–185
Resource reservation 306
Session initiation 222, 277, 299–300
Simple access point 231
Skinny client control 222–223
TACACS-plus 174
Temporal key integrity 113–114, 129, 198
Terminal access controller access control system 171, 173–174
Transmission control 88, 91, 95–96, 222
User datagram 90, 92, 176, 222
Voice over Internet 5, 220–223, 298
Voice over wireless 219–221, 268
Wireless LAN flow reservation 306
Protocol exchange 250–257
Proxim 110
Pseudonoise sequence 49
Pseudorandom noise (PN) code 67–68
Public key infrastructure (PKI) 181
Public switched telephone system (PSTN) 33, 36

Q

QAP 118
QBSS 118
Qualcomm 47, 268
Quality of service (QoS) 14–16
    802.11e 117–120
    Bandwidth 14
    Delay 15
    Jitter 15
    Loss 14–15
    Ongoing work 305–307
    Overhead reduction during BSS transition 237–238, 264
Quality of service (QoS) metrics report 123
Quality of service (QoS) station 118–120
    Peer QSTA 123

R

Radio common carrier (RCC) 7
RadioCom 2000 38
Radiofrequency technology 3
    Fading 3, 22, 50
    Interference 3, 14, 22, 33, 166, 263
    Signaling 15, 17, 66–67, 166
    Technology 8, 13
    Transmit power 13
Radio network controller (RNC) 279
Radio resource measurement 120–124, 235, 257–258, *261*
    Limitations 264
    New information elements 258–261
    Utility 261–264
Radio signal strength indication (RSSI) 37, *136*, 139
Radio telephone mobile system (RTMS) 38
Random walk 308–309
Randomized arbitrary bit rate (ABR) 272
Rate adaptation 76–77, 97
Real-time transport protocol (RTP) 222
Reassociation deadline interval 249
Reassociation delay *83*, 84, 201–202, *204–205*, 208, 215–216
Received signal strength (RSS) 261–262

Remote access dial-in user server (RADIUS) Protocol 171–177, 184–185
Remote access dial-in user server (RADIUS) proxying 177
Request for comments (RFC) 171, 174, 183
   RFC 2216 120
   RFC 2284 177–178
   RFC 2543 222
   RFC 2865 174, 245
   RFC 2866 174
   RFC 2868 174
   RFC 3220 92
   RFC 3261 222
   RFC 3550 222
   RFC 4137 178–179
Request to send/clear to send (RTS/CTS) 75
Resource information container (RIC) request 246–247
Resource reservation 245–247, 249
RIC data information element (RDIE) 247, 249
Richmond Radiotelephone Company 33
RIM 221
Ring, D. H. 36
Rivest's cipher 4 (RC4) encryption 114, 304
Roaming 2, 5–6, 27, 63, 133–134, 312–313
   802.11r 124–127
   Cross-technology 268–271
   Definition 8–12
   For data applications in GPRS 55–57
   Global (nomadic) 12, 88–91, 95–96, 153–154, 156–160, 268
   Horizontal 269, 277
   IMS-based 277–279
   In 3GPP vernacular 285
   Inter-provider 304–305
   Local 11, 12, 81–88, 159–162
   Seamless 39, 282–286
   Secure 2, 3, 21, 27, 312–313
   Taxonomy 21–27
   Vertical 269, 277–279, 291
   Wireless 2, 6, 11, 16
Roaming delay 81–82, 162–164, 236, 264–265
   Factors 82–86
Robust secure network information element (RSNIE) 247, 249, 254–255
Robust secure network (RSN) 113
Robust security network association (RSNA) key management 240–243
Rogue AP 172, 180
Round trip time (RTT) 302
Routing area identity (RAI) 57
Routing area (RA) 56
Routing area update 56

S

Safe secure network (SSN) 113
Samsung 286
Scanning
   Active 65–66, 71–72, *145, 147*
   Background *135*, 144–148, 152–153, 157–158
   How and when 141–144
   Implicit and explicit 144–145, *147*, 150–152
   Passive 65–66, 69–71, 148
   Pitfalls 151–152
   Vendor-specific extensions 148–149
Scanning timer 149–151
Screen room 226
Seamless Converged Communication Across Network (SSCAN) Forum 276
Seamless mobility 31, 276
Seamless roaming 39, 282–286
Second generation (2G) cellular system 40–51
Second generation-plus (2G-plus) cellular system 268
Secret communication system 48–49
Secret sharing 304–305
Secure roaming 2–3, 21, 27, 312–313
Secure roaming delay Components 83–86
Secure sockets layer (SSL) 191
Security 2, 133, 195
   1G cellular systems 39
   802.11i 112–117
   End-to-end 303–304
   Ongoing work 303–305
Security technology 196–197
   Evolution 197–198
Self-monitoring network 166–167
Semiconductor industry market dynamics 106–108
Sensing
   Physical 75
   Virtual carrier 75, 144, 152
Sequence
   Chipping (pseudonoise) 49, 67–68
Server
   Broker AAA 288
   Home AAA 93, 288–289
   RADIUS *113*, *115–116, 170*, 175–177, 184–187, 199, *201*, 204–206, 209–212, 218, 237, 264
   WLAN AAA 288
Service access point (SAP) 292
Service set 68–69
   Independent 68
Serving GPRS support node (SGSN) 54–57
Session initiation protocol (SIP) 222, 277
   Ongoing work 299–300
Short messaging service (SMS) 42
Signal threshold 303
Signaling system no. 7 (SS7) 39
Signal-strength-based roaming trigger 84, 87, 133, 139
Simonsson, Hilda 6
Simple access point protocol (SLAPP) 231
SIP-capable non-IMS phone 281
Site survey 165–166

Index

Skinny client control protocol (SCCP) 222–223
Smart trigger 301–302
Smartcard 182
Smooth handoff 300–301
Soft handoff 24, 26, 49–50
    Versus hard handoff 50–51
Softphone 223
Sonus 221
Source address (SA) 64
Space division multiple access (SDMA) 32
Space-division multiplexing 49
Spectralink 221–222
Spectralink voice priority (SVP) 224–227
Split MAC architecture 230
Spoofing 183
    MAC 19, 187
Spread spectrum 48–49, 67
Standard
    802.1X 170–172
    802.1Xrev 172
    802.11 1, 8, 17–19, 27, 64, 101–102, 111–112, 272–273, 279–280, 289–312
    802.11e 117–20, 130, 237–238, 245, 264
    802.11f 237
    802.11g 109–110
    802.11i 112–117, 131, 198–200, 218, 239–40
    802.11k 120–124, 235, 257–264
    802.11r 124–127, 233–245, 247–257, 264, 298–299
    802.15.1 273
    802.16 80, 271–273
    802.21 274–275, 289–294
    Advanced encryption 114, 198, 304
    IS-41 39–40, 46
    IS-54 46, 48
    IS-95 47–48
    IS-136 46, 57
    Personal area network 273
Standardization
    Bodies and industry organizations 274–276

    IEEE process 102–106
Static anchor 283
Station statistics request 122–123
Subnet detection 160
Subscriber identity module (SIM) 43–44, 182–183
Supervisor audio tone (SAT) 37
Supplicant 170–172, 180–182, 191, 240
Swedish Telecommunications Administration 33
Symantec 190
SyncScan 152–153
System health agent (SHA) 192

T

TACACS-plus protocol 174
Task group 104–105
    Fast roaming 111–112
Tekelec 221
Telecom Italia 286
Telecommunications Industry Association (TIA) 39
Telephony 4
TeliaSonera 286
Temporal key integrity protocol (TKIP) 113–114, 129, 198
Terminal access controller access control system (TACACS) protocol 171, 173–174
Third generation (3G) cellular system 51–59
    Ongoing work 303
    Standard bodies and industry organizations 274–276
Third generation partnership program (3GPP) 275–277, 289
    Interoperability levels 280–282
    Interworking between 802.11 279–280
    Transitional step 286–287
    Vertical roaming issues 277–279
Third generation partnership program 2 (3GPP2) 275, 287
    Interoperability levels 288–289

Time division multiple access (TDMA) 42, 49
    North-American 46–47
Time-division multiplexing 49
T-Mobile 18
Toshiba 108
Total access communications system (TACS) 38
Traditional 3-color deployment 68
Traffic analysis tool 153
Traffic bearer 23, 25, 39, 279, 286
Traffic class 120, 123
Traffic specification (TSPEC) 120
Transit power control (TPC) 98, 112, *127*
Transmission control protocol (TCP) 88, 91, 95–96, 222
Transport layer security (TLS) 179–181, 186, 197
Trigger
    Signal-strength-based roaming 84, 87, 133, 139
Trusted network connect (TNC) 189–191, 198
TSG CT W63 276
Tunneled transport layer security (TTLS) 179–182, 197
Tunneling 219, 231
    CDMA2000 network 288–289
    In EAP 186–187
    To turn global roam to local roam 231–233

U

Undirected probe 71, 142
United States digital cellular (USDC) 46, 48
Universal mobile telecommunications system (UMTS) 59, 274–275, 277
Universal terrestrial access network (UTRAN) 277, 279
Unlicensed mobile access (UMA) technology 276, 286–287
User authentication
    see Authentication

User datagram protocol (UDP) 90, 92, 176, 222
User service identity module (USIM) 183
User traffic
   From a mobile device to an antenna 25
   In the network infrastructure 23, 25, 39, 279, 286
Utilization
   Channel 120

V

Vertical handoff 268–269, 299–300, 303
Vertical roaming 269, 277–279
   802.21 291
Virtual AP 79, 143–144
Virtual carrier sensing 75
Virtual cell 230
Virtual LAN (VLAN) 79, 143–144, 160, 171, 184, 192, 223
Virtual private network (VPN) 95, 175, 188, 190–192, 271
Visitor location register (VLR) 45
Vocera 222
Voice call continuity (VCC) 39, 282–286
Voice over Internet protocol (VoIP) 5, 220
   Equipment providers 221
   Ongoing work 298
   Protocols 222–223
Voice over wireless fidelity (VoWi-Fi) 201, 223
Voice over wireless IP (VoWIP) 219–221, 268

Voice user
   Ideal roaming experience 271

W

Walled garden 187–188
War driving 65
Web server authentication 188
Web-based authentication 187–188
Wide-band code division multiple access (W-CDMA) 274
Wi-Fi Alliance 102, 128, 187, 197, 268, 273, 276
   Certification 129–130, 200
   CERTIFIED 128–129
   Versus IEEE 802.11 128–129
   WPA simple roaming test 154–156
Wi-Fi cellular handset, dual-mode 268
Wi-Fi multimedia (WMM) mode 117–118
Wi-Fi protected access (WPA) 129, 200
Wi-Fi protected access 2 (WPA2) 129, 200
Wildcard probe 71, 142
WildPackets, Inc. 153
Wired equivalent privacy (WEP) technology 196–197
   Authentication 73
   Limitations 114
Wireless client
   Aegis 141, *143*, 145, 147, 150
   Odyssey 140, *142*, 162–164
   Wireless Zero Config 140–141, 150, 153, *160*
Wireless communication 7, 31–32
   Definition 31
   History 6–8, 35–40
      Precellular era 32–35

Mobility model for studying 297–298, 307–312
Wireless Ethernet community alliance (WECA)
   see Wi-Fi Alliance
Wireless LAN flow reservation protocol (WRESV) 306
Wireless local area network (WLAN) 1–2, 297
   And 3GPP interoperability 277
   Definition 4
   Ongoing work 300, 303
Wireless roaming 2, 6, 11, 16
Wireless switch 167, 219, 227–233, 236–237, 241–242, 244
Wireless Wireline Convergence Group 276
Wireless Zero Config (WZC) client 140–141, 150, 153, *160*
WLAN AAA server 288
WMM power save 130
WMM scheduled access 118
Workgroup 5
Worldwide Interoperability for Microwave Access (WiMAX) Forum 80, 271–273
WPA-PSK roaming 200
WPA-PSK roaming test 201–202
   Results 202–203
WPA simple roaming test 154–156
WPA2 enterprise roaming 200, 203–204
WPA2 enterprise roaming test 204–206
   Results 206–210

X

X.25 54, 56